D0426898

Water Pollution Control in Low Density Areas

The University Press
of New England

Sponsoring Institutions:

Brandeis University
Clark University
Dartmouth College
University of New Hampshire
University of Rhode Island
University of Vermont

Published for the
University of Vermont
by the University Press
of New England
Hanover, New Hampshire
1975

Water Pollution Control in Low Density Areas

Proceedings of a Rural Environmental Engineering Conference

Edited by
William J. Jewell and
Rita Swan

Published for the
University of Vermont
by the University Press of
New England
Hanover, New Hampshire
1975

The papers in this book were delivered at a Rural Environmental Engineering conference on water pollution control technology in low density areas, chaired by William J. Jewell and held September 26–28, 1973, at the Sugarbush Inn, Warren, Vermont. The meeting was sponsored jointly by the Environmental Program and Water Resources Research Center of the University of Vermont and the Environmental Studies Center and Land and Water Resources Institute of the University of Maine.

Co-sponsors:

U.S. Environmental Protection Agency
Agency of Environmental Conservation, State of Vermont
American Society of Civil Engineers
American Society of Agricultural Engineers
American Water Works Association

Library of Congress Catalogue Card Number 74–82975
International Standard Book Number 0-87451-105-4
Printed in the United States of America

Contents

Affiliations

Aulenbach, Donald B.
Professor of Environmental Engineering
Rensselaer Polytechnic Institute
Troy, New York

Bennett, Edwin R.
Associate Professor of Civil and Environmental Engineering
University of Colorado
Boulder, Colorado

Bouma, Johannes
Associate Professor of Soil Science
University of Wisconsin
Madison, Wisconsin

Boyd, Gail B.
Environmental Engineer
Woodward-Envicon, Inc.
San Francisco, California

Boyle, William C.
Professor of Civil and Environmental Engineering
University of Wisconsin
Madison, Wisconsin

Buzzell, Timothy
Research Sanitary Engineer
U.S. Army Cold Regions Research and Engineering Laboratory
Hanover, New Hampshire

Campbell, Michael D.
Research Director (Technical), Commission on Rural Water, and Director of Research
National Water Well Association Research Facility
Department of Geology
Rice University, Houston, Texas

Clayton, John W., Director
Division of Environmental Health
Fairfax County Health Department
Fairfax, Virginia

Clesceri, Nicholas L., Director
Fresh Water Institute
Rensselaer Polytechnic Institute
Troy, New York

Coltharp, George B.
Associate Professor of Range Science
Utah State University
Logan, Utah

Darling, Leslie A.
Graduate Research Assistant, Department of Range Science
Utah State University
Logan, Utah

Felton, John
Graduate student, Department of Civil and Environmental Engineering
University of Colorado
Boulder, Colorado

Ferris, James J.
Research Coordinator
Fresh Water Institute
Rensselaer Polytechnic Institute
Troy, New York

Field, Richard, Chief
Storm and Combined Sewer Technology Branch
U.S. Environmental Protection Agency
Edison, New Jersey

Flanders, P. Howard
Sanitary Engineer
Vermont Department of Water Resources
Montpelier, Vermont

Goldstein, Steven N.
Technical Director, Conset, Inc., and Associate Director, National Demonstration Water Project
Washington, D.C.

Hall, Millard W., Director
Environmental Studies Center
University of Maine
Orono, Maine

Holzer, Thomas L.
Assistant Professor of Geology
University of Connecticut
Storrs, Connecticut

Howley, James B.
Graduate student, Department of Civil Engineering
University of Vermont
Burlington, Vermont

Hutzler, Neil J.
Specialist in Sanitary Engineering
Small Scale Waste Management Project
University of Wisconsin
Madison, Wisconsin

Jewell, William J.
Associate Professor of Agricultural Engineering
Cornell University
Ithaca, New York

Jones, Philip H.
Professor of Civil Engineering
University of Toronto
Toronto, Canada

Lehr, Jay H.
Editor, Journal of Ground Water, and Executive Director, National Water Well Association
Columbus, Ohio

Linstedt, K. Daniel
Associate Professor of Civil and Environmental Engineering
University of Colorado
Boulder, Colorado

Lindström, Carl R.
Head of Section, TJANSTE
Statens Naturvardsverk
Solna, Sweden

Loehr, Raymond C., Director
Environmental Studies Program
Cornell University
Ithaca, New York

Masters, Hugh E.
Staff Engineer
Storm and Combined Sewer Technology Branch
U.S. Environmental Protection Agency
Edison, New Jersey

McGauhey, P. H., Director Emeritus
Sanitary Engineering Research Laboratory
University of California
Berkeley, California

McGowan, Francis M.
Chief, Special Studies Branch, and Director of Northeast Water Supply Study
Army Corps of Engineers
New York, New York

Miller, John C., Hydrogeologist
Delaware Geological Survey
University of Delaware
Newark, Delaware

Mood, Eric W.
Associate Professor of Public Health
Yale University School of Medicine
New Haven, Connecticut

Otis, Richard J.
Project Coordinator in Sanitary Engineering
Small Scale Waste Management Project
University of Wisconsin
Madison, Wisconsin

Perrin, Douglas R.
Graduate student, Department of Civil Engineering
University of Vermont
Burlington, Vermont

Reed, Sherwood C.
Research Sanitary Engineer
U.S. Army Cold Regions Research and Engineering Laboratory
Hanover, New Hampshire

Sartor, James D., Manager
Woodward-Envicon, Inc.
San Francisco, California

Sproul, Otis J.
Professor of Civil Engineering
University of Maine
Orono, Maine

Struzeski, Edmund J.
Staff Assistant
National Field Investigations Center
Denver, Colorado

Tafuri, Anthony N.
Staff Engineer
Storm and Combined Sewer Technology Branch
U.S. Environmental Protection Agency
Edison, New Jersey

Tchobanoglous, George
Associate Professor of Civil Engineering
University of California
Davis, California

Tofflemire, James
Research Engineer
New York State Department of Environmental Conservation
Albany, New York

Whetstone, George A.
Professor of Civil Engineering
Texas Tech University
Lubbock, Texas

Whitsell, Wilbur J.
Groundwater Engineer
Water Supply Division
U.S. Environmental Protection Agency
Washington, D.C.

Wright, John R., Chief
Water Supply Division
Environmental Improvement Agency
Tesuque, New Mexico

Preface

My interest in the problems of lake pollution has emphasized the lack of knowledge about the most widespread method of wastewater disposal in rural America—the septic tank. In discussions with colleagues in New England and elsewhere it became evident that the water quality problems of rural areas were common to many parts of the United States.

In the Spring of 1973 the University of Maine and the University of Vermont, through their respective Environmental Studies and Water Resources programs, agreed to support a proposal by Professor Jewell that a meeting be held to examine the general water quality problems of low population density areas.

Speakers at the ensuing conference very ably provided comprehensive and thought-provoking material on many of the difficult problems of rural pollution. Nearly 300 participants attended, supporting the sometimes stormy discussions with documentation and solutions drawn from their own experience. If the papers presented here assist in stimulating thought and discussion, as their contents did during the lengthy evening meetings, their publication will greatly promote a deeper understanding of the problems of rural pollution.

Millard W. Hall

Orono, Maine
May 1974

Introduction

One of my father's neighboring farmers in Maine asked us in for a holiday drink last Christmas season. Noticing that he used water from the refrigerator for the drinks, I suggested that tap water would be good enough, but he quickly replied that I wouldn't feel that way if I tasted it. Being interested in water supply problems, I pursued this point with him. It seemed that the water was very salty and only fit for washing and flushing the toilet. When I suggested that his problem was likely caused by road salt movement into a shallow well and might be easily corrected, he responded with a rather dejected but tolerant smile. Six years ago, after an exhausting and time-consuming search for a solution, he had gone to the proper state and local authorities. They in turn told him that the state would gladly cover all the costs of a drinking-water source by drilling a new well, provided that it could be shown that road salt was in fact the source of the contamination. After sampling the well for a two-year period, state employees did conclude that road salt was producing an increasing chloride concentration in his shallow well (which had been dug by hand more than fifty years ago). One year later the state contracted to dig a very expensive well 300 feet deep, with much of the length in stone. Unfortunately, the new well also contained undrinkable salty water. At this point, all existing legal obligations of the state to the injured party had been satisfied, but the farmer still had no drinking water on site. Apparently, nothing said that the newly drilled well had to provide acceptable water. The farmer had run out of alternatives and assistance by this time and had resigned himself to a life without easily available drinking water. This story illustrates not only one particular rural environmental pollution problem, which is discussed later in the text, but also the frustration that intimately affects many people in our society.

The rural environment is the last area which most people associate with pollution. But a review of the numbers of people who live in rural America and the problems of providing a clean healthy environment there uncovers one of the most significant challenges for the environmental engineer.

The U.S. Census Bureau is often quoted as indicating that our rural population comprises about 20 percent of the total population. Rural residents are defined as those living in communities of fewer than 2,500 persons. Rural areas and the relatively small geographical areas defined as urban by the Census Bureau are illustrated in Figure 1. It is questionable, however, whether a population of 2,500 constitutes a valid division between urban and rural populations in terms of their capacity to control their environmental quality. It would appear that much higher concentrations of people are required before the social-legal-political-technical interacting forces can be strong enough to maintain and protect a high-quality environment. Places with populations of 25,000 would probably qualify as rural populations in the sense of environmental control capability. More than half the population lives in rural areas if that population level is used as the breakpoint between rural and urban centers.

Wastewater Treatment and Disposal in Rural Areas

Many of the experts involved with domestic wastewater treatment would probably assume that a central municipal system is the most commonly used method for the treatment of domestic wastewater. Statistics suggest, however, that more people depend on septic tanks than on any other system. This is particularly significant when it is noted that there are many factors in septic tank usage which remain relatively unknown. The movement of nitrogen and pathogens through the system and into the receiving waters is almost impossible to predict. Also, desirable methods for disposal of the sludge which is periodically pumped from septic tanks are almost nonexistent. Nearly 400 million gallons of this material must be annually disposed of in the northeastern United States alone, and the septic tank sludge generated in the country as a whole is greater than all sludge generated in domestic secondary treatment facilities. About a third of the papers in the text describe the present status of rural wastewater treatment and provide insight into some alternatives which are available.

Rural Drinking Water Supplies

There are 35,000 communities in the United States without public water-supply facilities, and all but 244 have populations of less than 1,000. Of course, many private wells may provide satisfactory and economical drinking-water supplies. A recent survey reported, however, that 85 percent of the public water-supply systems served popu-

1. Rural areas of the continental U.S. as designated by the U.S. Census Bureau definition. Screened areas are rural, and black areas indicate population centers greater than 50,000.

lations of fewer than 5,000 and that 40 percent of such systems provided water that did not meet the U.S. Public Health Service Drinking Water Standards (DWS), while larger systems had a much better record. In addition, only 10 percent of the smaller systems met the recommended bacterial surveillance standard. Thus it was once again emphasized that the drinking-water problems in low density areas affect large numbers of people throughout the country.

Rural Solid Waste Management Problems

Although the solid waste generated by urban centers creates impressive problems, the total quantities involved are a small fraction of the amounts generated in low density areas. If the inorganic wastes (mostly mining wastes) that are classified as mineral types are excluded, agricultural and animal wastes compose 85 percent of the remaining organic wastes. Although much of the solid wastes generated in the rural sector are dispersed over a wide area, it is necessary to provide proper management to avoid aesthetic and other environmental problems. This text presents very little information on this aspect, but it would serve as a good topic for a future conference.

Point Source Versus Non-Point Source Pollution

Recently, interesting data have been gathered to compare the pollution generated by man with that produced as an indirect result of man's activity, i.e. from urbanization, forest management, animal and crop production, etc. For example, in many communities, it would appear that the total quantity of pollutants that enter a receiving river or lake by natural runoff (non-point source pollution) exceeds that contributed by human wastes. These data raise some confusing questions regarding environmental management and control concepts. Loehr and others offer new understanding of the effects of land usage on environmental quality in this text.

In summary, the water and land environmental problems associated with rural America are surprisingly large scale. These problems are further compounded by the lack of laws and highly organized political units or special-interest groups with strong powers to protect this sector of the population. And the problems associated with low density areas can be expected to accelerate in the future as the population continues to spread from the large cities. Finally, the use of land as a waste-disposal mechanism is rapidly taking on new significance. With the no-discharge regulations of the new pollution control laws, land disposal must be considered as a major waste-treatment and disposal alternative. This alternative is discussed by a number of authors in the following chapters.

The papers presented here by distinguished engineers and scientists represent one of the first attempts to bring experience and knowledge to bear on one of the most neglected areas in terms of environmental quality control—the rural environment. It is hoped that this text represents only a small beginning for what will be a gigantic effort to protect and preserve the quality of our rural environment.

Besides the contributions from the authors who participated in this meeting, several individuals should be acknowledged. Dr. Carl C. Reidel, Director of the Environmental Program of the University of Vermont, provided outstanding leadership, support, and encouragement throughout the development of the conference and in preparation of this manuscript. The conference and the book would not have come about without his assistance. The support of Dr. Millard W. Hall, Director of the Environmental Studies Center, the University of Maine, and his assistance in developing the program and conducting the conference were also essential.

William J. Jewell

Ithaca, New York
May 1974

I
Land Treatment of Wastewater

1. Water and Land Oriented Wastewater Treatment Systems*

Francis M. McGowan

This paper is an overview of the land-treatment technique for wastewater management and of advanced water-oriented techniques as well. Judgments of the best or most appropriate technique are left to the designer and planner. The paper has three sections: (1) some introductory comments on the philosophy of pollution abatement; (2) a description of technologies; and (3) some observations on opportunities for multiple use of land-treatment systems.

Introduction

Pollution abatement will be expensive, particularly when aimed at meeting the standards and conditions of the 1972 Water Quality Amendments Act: high water quality, reuse of treated wastewaters, coordinated basin and metropolitan area wastewater management, use of advanced technologies and alternative systems. The conditions and objectives of the Act are a consequence of the severity of the pollution problem. And the solution to this severe problem, if indeed one can be implemented, lies not only in the technology we will discuss but also in a redirection of our thinking. Let me briefly mention three of these changes. First, our philosophy must change from one of viewing waste products as useless things to get rid of to seeing them as resources out of place. Nature does not allow us to destroy but only to modify and relocate waste products. Choosing the modification and relocation most beneficial to industry, the public, and the environment is the objective of proper planning.

Second, a commitment must be established to improve waste products continually with advanced technologies in the same way that industry strives constantly to improve its products. This commitment will be measured to a large degree by the institutional mechanism

*The views presented herein are those of the author and not necessarily those of the Army Corps of Engineers.

proposed for financing and operating to full scale the achievements of our research laboratories. Laboratory achievement alone does not ensure success.

We must work in the world of full scale. And this means taking some risks, having some failures, and struggling with new systems and techniques at that full scale until the waste management system is working according to plan. In other words, we must implement while we plan. Reliance on a single technique, such as the activated sludge process (which has not been effective in achieving total or even near total pollution control), will never get us to the level of water quality directed by the 1972 Water Quality Amendments Act.

There is a third change necessary in our thinking if successful wastewater systems are to be implemented. These systems must be formulated so that they achieve more than the single technical function of pollution abatement. In other words, there is more to be designed into a pollution-abatement system than pollution abatement. Planners of projects and systems must recognize the interactions possible with waterfront redevelopment, flood control, urban-rural growth management, and the amalgamation of funds which would normally support these purposes separately. They must consider methods which solve environmental problems mutually, such as the ultimate processing of sewage sludge with garbage and other solid waste and the use of thoroughly renovated wastewater to dilute polluted streams and hasten their recovery rather than constructing reservoirs for such dilution releases. We must be sensitive to the impact on the local job market of the rather large expenditures made for pollution abatement systems and the recycling or multiplier effects of such expenditures on the local economy.

Technologies

This section of the paper will describe briefly the technical processes of advanced water and land treatment systems, present some specific technical design criteria, note the anticipated performance of treatment systems in terms of effluent qualities, and indicate the costs associated with some specific designs for system capacities ranging from two to fifty million gallons per day. The information comes directly from the recent Corps of Engineers report on wastewater management alternatives for the Codorus Creek basin in Pennsylvania.

Advanced Water-Oriented Treatment Processes

Both the land-treatment system and combinations of advanced water-process treatment units can achieve high levels of biochemical oxygen demand (BOD), chemical oxygen demand (COD), suspended solids, ammonia, total nitrogen, and phosphorus removal. However, certain performance aspects of each technology differ from the other.

Advanced water-process technology is comprised of a group of unit processes which are capable of removing or converting specific constituents of wastewater not normally removed by conventional secondary treatment. Certain of these processes are largely single-constituent oriented, e.g. chemical precipitation of phosphorus, while others are capable of removing a number of pollutants, e.g. filtration for BOD, suspended solids and insoluble phosphorus removal.

Ammonia Removal. Ammonia concentrations in municipal effluents are typically in the concentration range of 5 to 15 mg/1, and ultimate long-range objectives call for reduction of ammonia discharge concentrations to 0.5 mg/1 or less. Presently, three methods are available for ammonia removal: ammonia stripping, selective ion exchange, and microbial nitrification (conversion to nitrate). For large-scale systems, only ammonia stripping and microbial nitrification appear competitive in cost.

Recent experience with plant-scale ammonia-stripping systems have documented an inability to maintain high-level process performance during the colder months of the year. This is attributed to the increased solubility of ammonia in water at low temperatures and to operational difficulties with stripping towers in cold weather. The aerobic nitrification-sludge system appears to offer a more consistent year-round performance, although it also suffers some reduction in performance at colder temperatures. While the operational experience for the process exists only on the small-scale demonstration level, it appears to be the best available process for achieving ammonia conversion.

Total Nitrogen Removal. The process for removal of total nitrogen must be selected in conjunction with either ammonia removal or conversion. The nitrogen removal process that is most complimentary to the ammonia-nitrogen conversion process is the denitrification-sludge process, incorporating the use of methanol for biological reduction (denitrification) of the nitrate compound. Reductions in total nitrogen discharges to 2 mg/1 or less are achievable with this process. Like the ammonia conversion process, it is not yet in use at large installations which could provide reliability evaluations.

98-Percent Phosphorus Removal. Consistent 98 percent removal of phosphorus to reduce effluent phosphorus concentrations to 0.2 mg/1 or less after final filtration requires a special tertiary chemical treatment step utilizing high chemical dosages with optimum mixing and flocculation actions followed by sedimentation and filtration. Alternative chemicals include lime and alum. The heavy dose of chemical required and the significant disposal problem posed by the light alum waste sludge lead to preference of lime treatment coupled with lime recovery through recalcination.

Residual BOD & Suspended Solids Removal. Filtration in multimedia systems, such as sand and coal, provides a consistent residual BOD and suspended solids reduction for secondary effluents. BOD reduction is achieved through the removal of the portion of the BOD in the non-soluble suspended particle form, which represents up to 50 percent of the total BOD in a good secondary effluent. Effluent BOD concentrations of 3 mg/1 are achievable with filtration of chemically flocculated secondary treated wastes. Filtration after chemical treatment for phosphorus removal increases the total removal efficiency by filtering out the finely suspended chemical flocs that overflow the chemical sedimentation basins.

COD Removal. Carbon adsorption and regeneration processes reduce the concentration of dissolved refractory organics remaining in the effluent, as reflected in the COD concentration, by approximately 50 percent. Typical effluent COD would therefore be reduced from the 20 to 30 mg/1 level achievable with mixed media filtration to 10 to 15 mg/1 with the inclusion of the carbon adsorption step. Other benefits of the carbon step would include a slight improvement in BOD removal and also the removal of a major portion of the residual color.

Advanced Land-Oriented Treatment Processes

The use of land treatment as a technology is another approach to solving the pollution problem and accomplishing other things besides. In a land solution, all wastewater is routed either to an activated sludge unit or to a series of mechanically aerated lagoons where biological treatment occurs similar to the treatment in an activated sludge sewage plant. The treated waste from the lagoons or the activated sludge units is then settled and irrigated onto the land, where the nutrient-rich effluent improves the yield of the land, whether it be forage, timber, or vegetables that require cooking. The land and the crop impose nature's treatment devices to put pollutants that once came from the land back to the land. Underdrainage or well-point systems recover much of the irrigated water while also maintaining an interface or separation between the groundwater table and the irrigated wastewater. Storage lagoons hold the treated wastewater during nonirrigation seasons.

There is a flexibility in the operation of the aeration lagoons which should be noted. Since these lagoons are large, compared to the aeration chambers in an activated sludge plant, the biological reaction takes place over a longer period of time (days vs. hours) and at a less intense rate. This long residence time enables the spill of a toxic waste into the system to be detected and isolated in one lagoon while biological treatment of the continuing incoming wastewater proceeds in the remaining series of lagoons. The shorter residence time and lack of large storage volume in an activated sludge plant make it more difficult to detect and isolate toxic waste spills in time to prevent biological upsets and, consequently, sporadic plant breakdowns.

The land-treatment system, operated as a low-rate spray irrigation process on crop or forested land, utilizes the soil mantle as a multiprocess "living filter." Properly designed, a high level of tertiary treatment appears achievable. In the soil mantle, or "living filter," multiple processes operate simultaneously: physical filtration, ion exchange, crop uptake, volatization, adsorption, chemical precipitation, and chemical complexing.

The performance of the soil process can be most concisely summarized by examining the fate of each constituent that might come into contact with the soil via wastewater irrigation.

Suspended solids are removed by the array of mechanisms one ascribes to a dispersed media filter, i.e., screening, entrapment, gravity forces, coagulation, and flocculation. Organic suspended solids thus captured by the soil mantle slowly break down and solubilize and are converted through microorganism metabolism to new organic cell matter and gaseous carbon dioxide. The new organic matter and the inert residues, together with the inert suspended solids that are also captured, accumulate slowly in the soil mantle. The solid but organic content of soils solubilizes with or without irrigation and is lost to the soil mantle at a net rate of 3 to 4 percent per year. The critical design consideration for an application rate of suspended solids via irrigation is that the soil system does not become clogged and that the rate of organic deposition does not exceed the eventual assimilation capabilities of the soil microorganisms. The literature is replete with documentation indicating that pretreated wastewater, such as that characterized by municipal secondary effluent, does not begin to strain the soil's capacity to assimilate suspended solids.

Dissolved organics, including nitrogen constituents and those constituents characterized as BOD, COD, total organic carbon (TOC), organic-derived color, oils, and grease, are removed in the soil mantle by means of an adsorption mechanism. Two widely different components in the soil are capable of this adsorption mechanism. One of the components, microorganisms, must adsorb the dissolved organics into their exterior enzyme system in order to preprocess them for subsequent metabolic uptake in which new cell matter and carbon dioxide are the final products. The latter part of this process requires aerobic soil conditions. The second soil component, clay, is also capable of adsorbing dissolved organics much as activated carbon does. The organics sorbed on clays are in a sense stockpiled for subsequent processing by microorganisms. In this case, however, the microorganisms must be mobile, since the organic molecules remain fixed to the clay adsorbent until completely assimilated by microorganisms. In the case of either sorbent, the uptake of organics can be rapid. A substantial literature testifies to the adequacy of the soil process for easily assimilating dissolved organics in the range of concentrations encountered in municipal secondary effluents.

Nitrogen, in the form of ammonia and nitrates and nitrites, is captured by a variety of mechanisms and is the constituent in wastewater that usually limits the overall irrigation application rate of typical municipal secondary-effluent wastewater. Ammonia nitrogen is captured by an ion exchange mechanism, commonly referred to as the cation exchange capacity of the soil, which is particularly manifested by organic and clay components in the soil mantle. This mechanism is capable of capturing other cations as well as the ammonium ion, but maintains a rechargeable selectivity for the ammonium ion large enough to give the soil mantle a nitrogen-banking capacity against the future time-distributed nitrogen demands of the growing crop. This is the soil property that permits the farmer to apply his fertilizer in discrete amounts and still supply the time-distributed demand of his crops. The ammonia nitrogen is subsequently attacked by nitrifying microorganisms during the spring-to-fall period of the year and converted to nitrate nitrogen which is no longer held by the soil mantle but is instead free and mobile and capable of migrating to the crop roots where it is assimilated by the growing crop.

Not all of the nitrate nitrogen is absorbed by the crop roots however, and this balance, together with any nitrite nitrogen, is free to migrate down toward the groundwater sump below the soil mantle. Along the way, the nitrate and nitrite nitrogen will encounter denitrifying bacteria that will, as long as a source of dissolved organic carbon is available, partially reduce the oxidized nitrogen forms to nitrogen gas and manufacture some new organic cell material. The remainder of unreduced nitrate and nitrite nitrogen will eventually migrate to the groundwater table, where further change will cease.

Some of the ammonium ion stored in the soil mantle will also be the nitrogen source for the new microbial cells formed by aerobic synthesis in the upper soil mantle. If more dissolved organic carbon substrate were available in the irrigated municipal secondary-effluent wastewater, a much more significant portion of the total nitrogen applied in irrigation would undergo microbial synthesis. Present well-managed fertilizer strategies, as applied to agricultural crops, regardless of whether the fertilizer source is commercial chemical or irrigated municipal secondary effluent, appear to permit approximately 10 percent of the applied nitrogen to percolate through to the ground water.*

It is possible that regional wastewater management, through its bringing together of municipal and industrial waste residues, will produce a combined wastewater with a higher overall carbon-to-nitrogen ratio, which will improve the nitrogen dynamics within the soil mantle and therefore the rate of total irrigating water that can be applied. Most industries producing an organic waste have a wastewater that is nitrogen deficient as contrasted with municipal secondary effluent.

The evidence in the agricultural literature demonstrates that nitrogen applications, in practical balance with crop uptakes, yield agricultural drainage water with up to 2 mg/1 nitrogen, representing 10 percent of the applied nitrogen. Though much higher rates of application can produce much higher nitrogen concentrations in the drainage water, this practice does not reflect good management, regardless of whether the source of nitrogen is commercial fertilizer or pretreated wastewater.

Phosphorus, in the soluble form as orthophosphate, is removed in the soil mantle by adsorption/ion exchange on soil clay constituents. In acid soils, the phosphorus adsorbing constituents are primarily aluminum and iron. In basic soils, the calcium and magnesium content of the clays can contribute strong adsorption sites for phosphorus.

*Editor's Note: It should be pointed out that, in general, fertilizer management in agriculture does not consider a loss to the groundwater, but instead applies fertilizer to maximize yields. In documented cases more than 100 lbs/acre of nitrogen, up to 50 percent of the application, have been lost to the groundwater. Up to 20 mg/1 of nitrogen have been observed in the drainage water in agricultural crop areas.

There is always an equilibrium amount of soluble phosphorus present in the soil solution from which the crop is able to derive its requirements through the root structure. Phosphorus is applied at a rate which exceeds the crop uptake, and thus the soil is called upon to act as an ultimate sink for phosphorus.

The active adsorbing components in soil clays at any one time are on the order of 10 percent of the total components within the soil potentially capable of adsorbing and holding phosphorus. Once the immediate phosphorus adsorbing components of the soil have been saturated, a resting period, such as the winter nonirrigation season, is required to permit the chemical equilibria within the soil mantle to readjust and produce new active phosphorus adsorption sites. Complete adsorption activity can be recovered within three to six months.

From a short-term equilibrium adsorption consideration, the varieties of sandy and clay soils and their respective depths that have been encountered, at least in the Codorus Basin, have exhibited phosphorus removal lives of ten to one hundred years. From the standpoint of the long-term equilibrium adsorption capabilities allowing for appropriate rest and recovery periods, the phosphorus removal lives of these soils is between one hundred and one thousand years. The capability of the soil to adsorb and hold phosphorus is evident both from the literature and from the residual concentrations of phosphorus in groundwater in agricultural areas. In well-designed irrigation systems using municipal secondary effluents, it is possible to produce an agricultural drainage of reclaimed water with background phosphorus concentrations of 0.01 mg/1.

Bacteria, viruses, and pathogens are removed by the same mechanisms as those cited for suspended solids, since they are, indeed, microscopic suspended solids. Various investigations have determined that once these constituents have been captured in the soil mantle, they do not long persist. Apparently, the soil environment is not conducive to their survival, perhaps because the indigenous soil microorganisms are too acclimated and competitive to permit a less than indigenous species to survive. A properly designed soil process irrigation system does an effective job of disinfection.

Heavy metals are ion exchange-adsorbed by the clay constituents of soils and are chelated by the organic constituents of soils. Once captured in the soil, they require varying degrees of acidic leaching to effect their release. Within certain limitations prescribed by agricultural experience, small residual concentrations of most metals are compatible with soils and can be substantially removed by soils. As the organic concentration of soils decomposes, the formation of new solid organic matter, along with the deposition of more clays, is taking place, so that in a "living filter" type of soil system there appears to exist an unlimited life sink for controlled amounts of heavy metals.

Chlorinated-hydrocarbons, pesticides, and phenol-like substances are captured in the soil by adsorption mechanisms, much as other dissolved organics are, and they are subsequently converted to new cell material and gaseous carbon dioxide by aerobic microorganisms. However, the acceptable concentration of these constituents in the soil and wastewater system, like the concentration of heavy metals, must be substantially controlled and regulated by pretreatment. These organic species are largely inimical to the soil microorganisms, and to abuse the soil system with an overload would eliminate the very microorganisms that accomplish the adsorption and ultimate disposal. Pretreatment from the biological systems in lagoons or activated sludge units prevent excessive concentrations of these species.

Total dissolved solids, exclusive of the species heretofore discussed, pass through the soil process unaltered. Typical of constituents in this category are sodium, sulfate, and chloride. Potassium is largely extracted by the crop root system for crop growth.

Design Criteria

The efficiency of advanced water-oriented treatment processes is highly sensitive to process flow rates. As a result, these processes must be designed to provide the desired treatment performance at peak rates of flow frequently encountered. For the land treatment system, the low rates of application in relation to the capability of the soil treatment system to remove organic material greatly mitigates the effects of variations in the secondary treatment performance prior to land application.

A specific set of design criteria for treatment processes could be as follows:

Advanced Water-Oriented Treatment Systems

Regulation of treatment plant flows for advanced treatment processes could be designed to a maximum of 1.7 $\times$ average daily flow through provision of storage prior to or after secondary treatment.

Specific performance and design aspects of advanced water-oriented processes could be:

Process	*Treatment Performance*	*Design Factors*
Nitrification	$NH_3 \leqslant 0.5$ mg/1	Aeration sludge system—3 hours detention at peak flow
Denitrification	Total N $\leqslant$ 2.0 mg/1	Anaerobic sludge system—3 hours detention at maximum flow: methanol added for process stability
98% Phosphorus Removal	Total Phosphorus $\leqslant$ 0.2 mg/1	Chemical flocculation and settling with lime dosages up to 400 mg/1
Filtration	$BOD_5 \leqslant 4$ mg/1 Suspended solids $\leqslant$ 3 mg/1	Mixed-media filtration loading rates should not exceed 6 gpm/ft^2 at maximum flow
Physical chemical treatment with lime clarification, carbon adsorption, and filtration	$BOD_5 \leqslant 3$ mg/1 Suspended solids $\leqslant$ 3 mg/1 Total Phosphorus $\leqslant$ 0.2 mg/1	The carbon columns could be a counter-current flow system using granular carbon, 6 gpm/ft^2 process flow rate

Advanced Land-Oriented Treatment Systems:

The land treatment system for the Codorus Basin was based on the following design criteria:

Pretreatment of wastes: conventional or aerated lagoon secondary treatment

Application rate: 2 inches per week, 8 months per year

Land requirements: 194 net irrigation acres per MGD average annual flow

Winter storage of flows: 4 months storage of total flow

Water application system: rotating rig or a network of distribution pipes with fixed head spray nozzles

Drainage: subsurface drainage of irrigated areas using wells or drain tile as dictated by economic and performance requirements

Storage lagoons: supplemental aeration when total depth exceeds 15 feet.

In summary, the anticipated effluent quality from the advanced water-oriented processes and the land-treatment processes would be:

Treatment Type	*COD mg/1*	*BOD mg/1*[5]	*Susp. Solids mg/1*	*Dissolved Solids mg/1*	*Phos. % rem./ mg/1*
Water Processes	10	3	3	350	98/0.2
Land	5	3	0	400	99/0.05

Treatment Type	*Color Pt-Co Units*	*NH_3-N mg/1*	*NO_3-N NO_2 mg/1*	*Organic N mg/1*
Water Processes	<5	0.5	2	0
Land	<5	0.1	2	0

Costs

The following is a summary of costs estimated for various sized treatment systems considered in the Codorus Creek report.

Average Flows	*Costs/MGD* *Capital Costs/MGD*	*Annual Operation & Maintenance Costs/MGD*
	Advanced Water-Oriented Treatment	
35 MGD	$1,000,000	$ 65,000
6.5 MGD	1,500,000	75,000
2.4 MGD	2,500,000	125,000
	Advanced Land-Oriented Treatment	
51 MGD	1,400,000	34,000
6.5 MGD	1,500,000	22,000
2.4 MGD	2,200,000	22,000

Some Impacts of Land Treatment Systems

Land systems with some of the elements mentioned are in use in a number of locations in this country, but on a small scale. The system initiated for Muskegon, Michigan, is the only one containing all the elements discussed. Almost total unfamiliarity with land-disposal systems on the part of officials and the public, coupled with the use of rather large amounts of land, have generated widespread intuitive skepticism about it. But properly understood, the approach allows considerable flexibility and potential for beneficial impact. For example, the system in most cases will not require changes in existing land use. It will perpetuate needed agricultural and timber production. In one of the New England river basins, the Merrimack, thousands of farms have gone idle in the last ten years. If technology or some other factor should make obsolete all or part of a land system in the future, the worth of the land would leave it with a considerable salvage value. The large number of treatment sites and transmission lines associated with land disposal allows considerable latitude in guiding and planning growth. The land sites, while used only during the irrigation season and then at the relatively low rate of 2 to 2.5 inches per week, actually provide green space, perhaps around urban areas, and could be used to promote recreational activities as well as agricultural production. The storage lagoons, which hold large quantities of treated wastewater prior to irrigation, make excellent sites for nuclear or conventional power plants since the cooling capacity of the lagoon waters could be used in lieu of thermal discharges to lakes or rivers. And the surrounding agricultural or forested land is a desirable buffer between the power plant and the areas of population. When an irrigation system is located in a flood plain, as could be the case in many areas, it not only contributes to flood plain management, but also preserves access to the river and its tributaries. The large geographical area covered by a land system further allows widespread augmentation of the dry weather flow of the many tributaries passing through or near the various irrigation areas. Finally, it makes ecological sense to use wastewater nutrients productively.

2.
Spray Disposal of Treated Domestic Wastewater on Upland Fragipan Soils in a Severe Winter Climate

P. Howard Flanders

On May 27, 1971, the Vermont Water Resources Board adopted "Regulations Governing Water Classification and Control of Quality," which established as "pristine streams" those with flow above an elevation of 1500 feet msl or a rate of flow of less than 1.5 cubic feet per second at any elevation and prohibited discharges to them that would degrade their quality in any respect.

The adoption of these regulations initiated a search for an alternate disposal method in the upland areas of severe soil conditions. The spray irrigation project at Pennsylvania State University was visited and studied. That system operated over unglaciated soil with a water table at 100 to 350 feet below the surface. It had not operated under sustained periods of extreme subzero temperatures as only three periods with temperatures below zero and a low temperature of -11° F had been experienced. A standard authority for design, addendum no. 2, "Ground Disposal of Wastewater," of the "Recommended Standards for Sewage Works" by the Great Lakes–Upper Mississippi River Board of State Sanitary Engineers recommends a minimum of 5 feet of aerated zone below the ground surface and a maximum of 8 percent land slope for year-round operation.

Much of Vermont, however, is subject to high perched water tables, shallow soils to bedrock, and land slopes in excess of 8 percent. Winter temperatures in Vermont are often -25° to -30° F with high temperatures for some days not exceeding 0° F. Because these soils and climatic conditions, common to Vermont, neither comply with established standards for ground disposal nor are similar to conditions of sites intensively studied, it was decided that development of specific guidelines and design criteria for spray disposal in Vermont was necessary.

Soil scientists and geologists familiar with soils and geological conditions in Vermont reported the geological profile of Figure 1 to be typical of Vermont upland areas. The mountain tops are generally exposed bedrock with very shallow soils. Studies indicate that these areas, by their very nature, serve as prime recharge areas for the bedrock aquifers. Further down the slope are the fragipan soils. These soils are underlain at depths of 15 to 40 inches by a dense, nearly im-

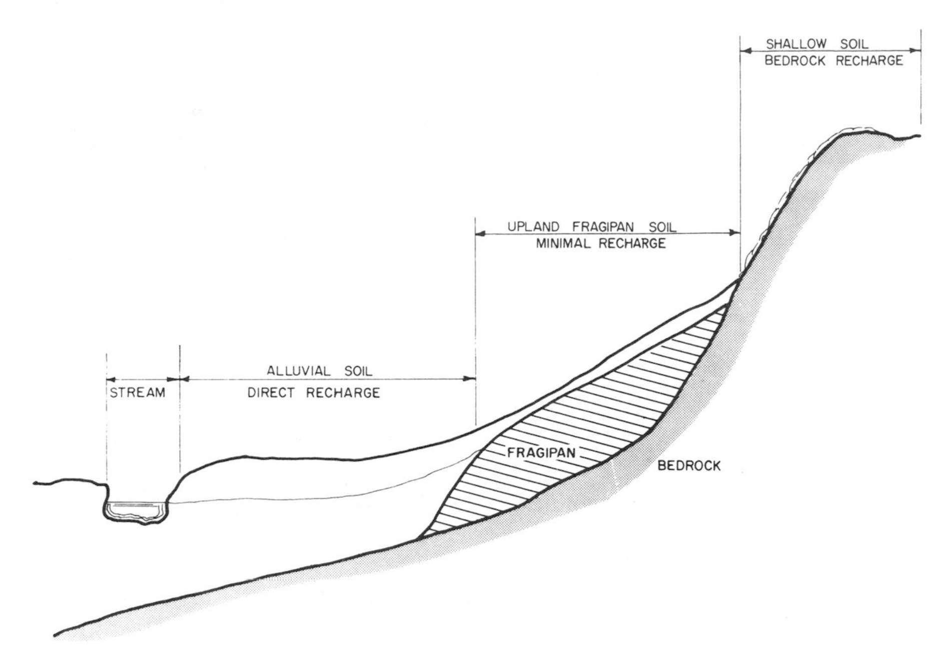

1. Typical Vermont geological profile.

permeable layer of fine sand and silt called a fragipan. Spring typically finds these soils with a perched water table at the ground surface, and excavations in these soils during late spring disclose a saturated soil at the surface of the fragipan. Direct recharge to the bedrock below these soils is minimal because water movement is restricted by the low permeability of the fragipan. Alluvial soils serve both as aquifers and as direct recharge areas. These soils–sands and gravels–generally have high permeabilities with rapid water movement.

A desire to protect the groundwater resources focused investigations on the fragipan soils because of their natural protection against direct recharge. Additionally, the fragipan is believed to facilitate maximum effluent renovation by maintaining the effluent in the root zone above it during the entire time of travel to a groundwater discharge point.

It was next necessary to establish a method for determining the hydraulic capacity of a fragipan disposal site. This capacity has been set as the volume of water that can be conducted down slope over the fragipan and maintain a minimum of 12 inches between the ground surface and the water table. The lateral hydraulic conductivity is utilized along with slope and area values in Darcy's Law, $Q = KA\ hL/L$, to determine the hydraulic capacity. The application of Darcy's Law to a fragipan spray disposal site is shown in Figure 2. The hydraulic conductivity value, K, is adjusted for lower water temperatures during winter operation. Vermont has adopted a weekly application frequency on the basis of the Penn State University studies. Figures 2 and 3 demonstrate that the dimension of the disposal site parallel to the ground surface contours has a direct effect on the hydraulic capacity of the site and resulting weekly application. For equal rectangular disposal areas, the largest hydraulic capacity is obtained when the maximum dimension of the disposal area is parallel to the land surface.

After this manner of evaluating an acceptable spray disposal site was determined, the question of whether spraying during winter months would produce the desired results was still to be answered. A small experimental spray system was operated for two winters to evaluate the effects of winter spraying. One of the first discoveries from observation and ice sampling was that selective freezing was occurring within temperature ranges from 10° F to 32° F, which affected the quality of ice formed and of the effluent entering the soil. Table 1 has a comparison of initial effluent quality and the quality of ice during selective freezing. At temperatures below 10° F the ice quality approached the quality of the effluent.

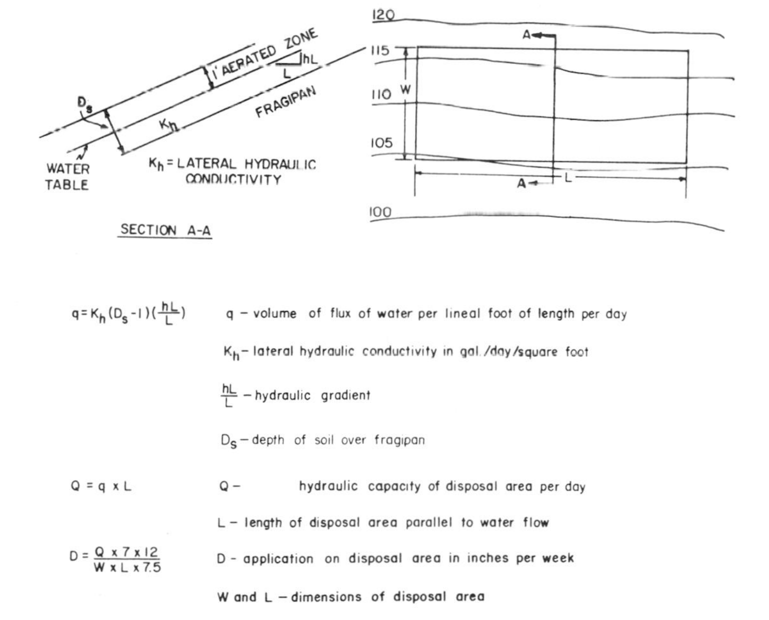

2. Calculation of hydraulic capacity.

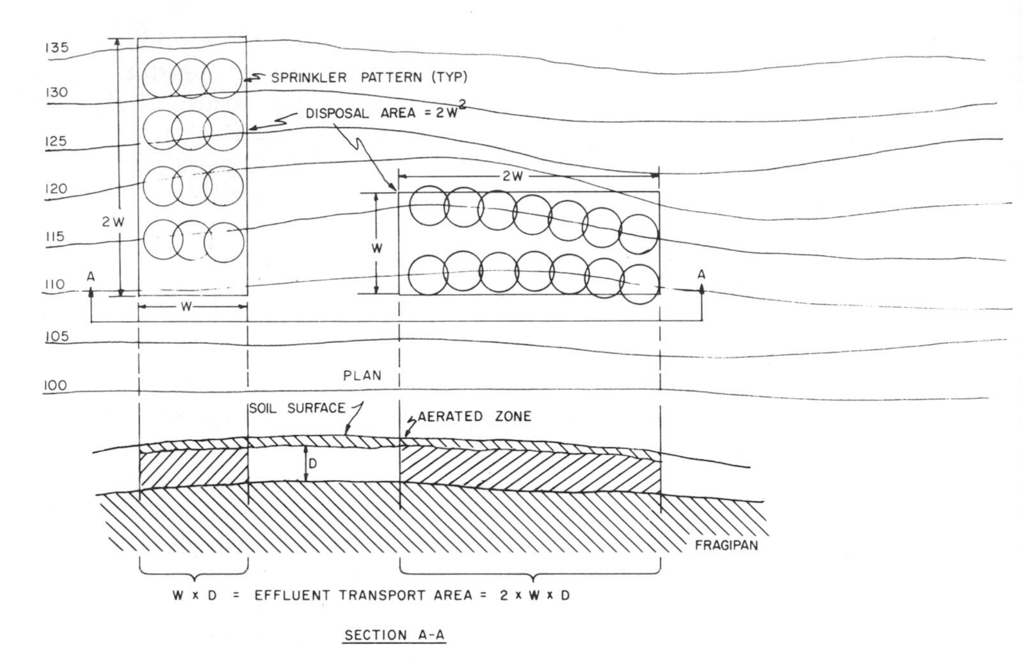

3. Spray disposal layout.

During operation of the system the spray falls on the ice formed the previous week and flows laterally over it to the lower edge where it percolates through the snow and enters the soil. The path is indicated in Figure 4. Selective freezing takes place while the effluent is flowing over the ice. The water lowest in dissolved solids and associated nutrients freezes first while the water highest in dissolved solids reaches the lower edge of the ice and percolates through the snow and into the soil where phosphorus is removed and water is conducted out of the site prior to spring runoff.

To permit effluent movement into the soil, the sprinkler pattern should be as shown in Figure 3 so that the ice patterns formed under each sprinkler do not overlap in the direction of water flow. This maintains a discontinuous ice surface and provides area where the effluent may percolate through the snow and into the soil. The ground must also be unfrozen, which is generally the case in these forested upland areas where the snow cover provides insulation and prevents frost.

When effluent is stored and not sprayed during the spring snow melt and runoff period, the only contribution to the overland runoff is the melt from the ice formed during selective freezing. The average quality of this ice is indicated in Table 1.

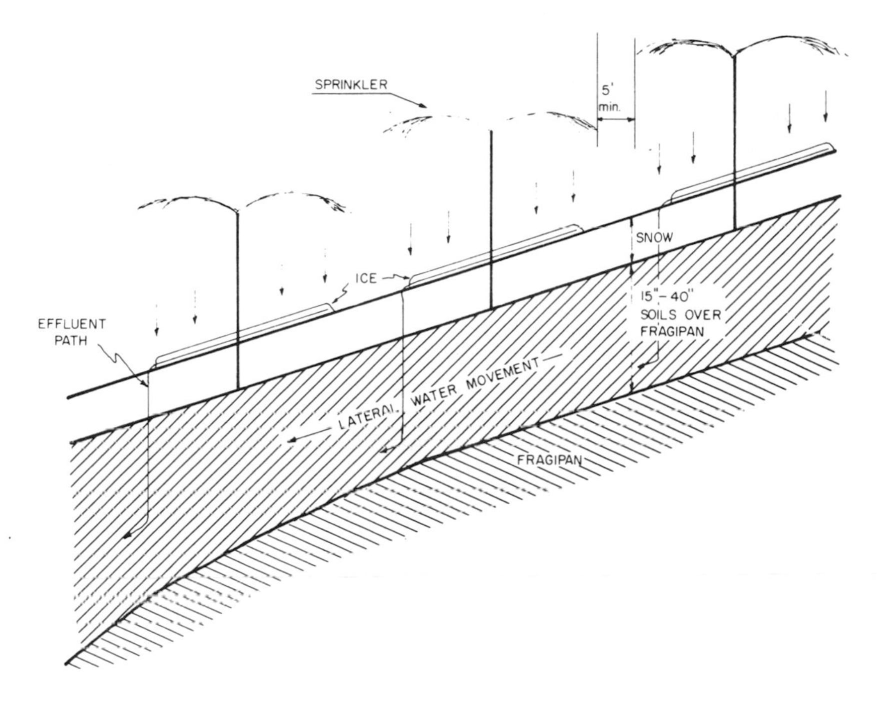

4. Effluent movement path.

It was decided that, with this ice quality, proper design, and selective operation, winter spray disposal on fragipan soils in Vermont would not be likely to violate the "Regulations Governing Water Classification and Control of Quality." These operation and design features, which include an effluent storage facility with capacity to store two months flow for complete holding during April and May, are summarized below:

1. Sprinkler layout with nonoverlapping patterns in the direction of surface water movement.
2. Storage pond with capacity to store two months of wastewater volumes.
3. System pumps and pipes designed to deliver the average daily flow to the disposal field in an eight-hour period. This allows spraying only during the warmest and daylight hours of the day.
4. No spraying when temperatures are below 10° F and during these periods the effluent is diverted to storage.
5. Dual pumps and valving such that two separate areas could be sprayed simultaneously in an eight-hour period to allow disposal of effluent stored during the cold periods.

Table 1
Comparison of Effluent and Ice
(all values in mg/1 except pH)

Constituent	*Effluent Medium Values*	*Ice 2/2/72*	*Ice 2/2/72*	*Ice 4/12/72*
pH	6.9	7.4	7.4	7.2
T.D.S.[1]	425	44	75	24
PO_4-P	8.0	0.55	NA	1.80
T-P	NA[2]	NA	NA	2.70
NO_3-N	17.0	NA	4.0	10.0
Cl^-	180	45	24	3.5
SO_4	24	NA	NA	3

[1] Total Dissolved Solids.
[2] Not available.
[3] Percentage of reduction in constituent from effluent to ice.

Ice 4/26/72	*Ice 5/5/72*	*Ice 5/10/72*	*Average Ice Values*	*Average % Reduction*[3]
6.4	5.7	6.5	—	—
12	15	16	31	93
0.87	0.02	0.03	.65	92
1.10	0.09	0.14	1.0	NA
0.10	0.11	0.06	2.8	84
3.5	3.5	9.5	14.8	92
3	2	12	5	79

References

Chow, V. T. 1964. *Handbook of applied hydrology*. New York: McGraw Hill.

Dufresne-Henry Engineering Corporation. 1972. *Winter spray irrigation demonstration Bromley ski area, Vermont*. Springfield, Vermont: Dufresne-Henry.

Gerdel, R. W. 1954. The transmission of water through snow. *Transactions of the American Geophysical Union* 35: 475–85.

Hill, David E. 1970. *Evaluation of waste water renovation potential in Litchfield County soils.* Special Bulletin—Soils 32. New Haven: Connecticut Agricultural Experiment Station.

Parizek, R. R. 1973. Site selection criteria for wastewater disposal—soils and hydrogeologic consideration. In *Recycling treated municipal wastewater and sludge through forest and cropland*, ed. W. E. Sopper and L. T. Kardos, pp. 95–147. University Park: Pennsylvania State University Press.

Parizek, R. R., Kardos, L. T., Sopper, W. E., et al. 1967. *Wastewater renovation and conservation*. Penn State Studies no. 23. University Park: Pennsylvania State University Press.

United States Soil Survey Staff. 1951. *Soil survey manual*. Agricultural handbook no. 18. Washington, D.C.: United States Department of Agriculture.

Wagner, W. P., and Dean, S. 1972. *Vermont groundwater resources.* Montpelier: Vermont State Planning Office.

3.
Land Treatment of Wastewaters for Rural Communities

Sherwood C. Reed and Timothy Buzzell

Introduction

A gap is apparent if one compares the resources available in a typical rural community to the resources required to implement very stringent water quality and pollution control standards. Such communities may at present have some form of primary treatment or a marginally functional secondary system. Achievement of consistently high-level secondary treatment will be difficult and expensive; achievement of tertiary level quality with advanced waste treatment concepts may impose an intolerable economic burden just in meeting operation and maintenance costs.

A fresh approach with an ancient technique may offer a method for meeting high quality standards at an acceptable cost. Utilization of the land for the treatment of wastewaters is a familiar but, until recently, an ignored possibility.

Concepts and Constraints

Three basic concepts are available for the land treatment of wastewaters: rapid infiltration, spray irrigation, and overland flow. In the physical sense, they differ as to the volume and the pathway of the liquid applied. They are defined as follows (Reed et al., 1972):

Spray Irrigation: the controlled spraying of liquid onto the land at a rate measured in inches of liquid per week with a flow path of infiltration and percolation within the boundaries of the receiving site (Figure 1).

Overland Flow: the controlled discharge by spraying or other means of liquid onto the land at a rate measured in inches per week with the flow path being downslope sheet flow (Figure 2).

Rapid Infiltration: the controlled discharge by spreading or other means of liquid onto the land at a rate measured in feet per week with the flow path being high-rate infiltration and percolation (Figure 3).

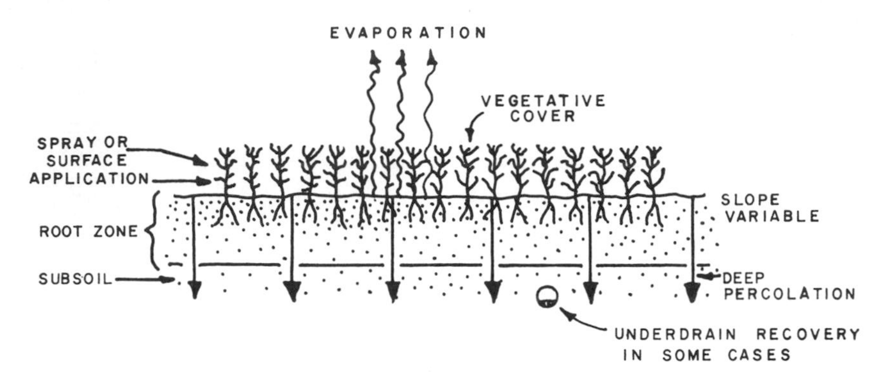

1. Spray irrigation.

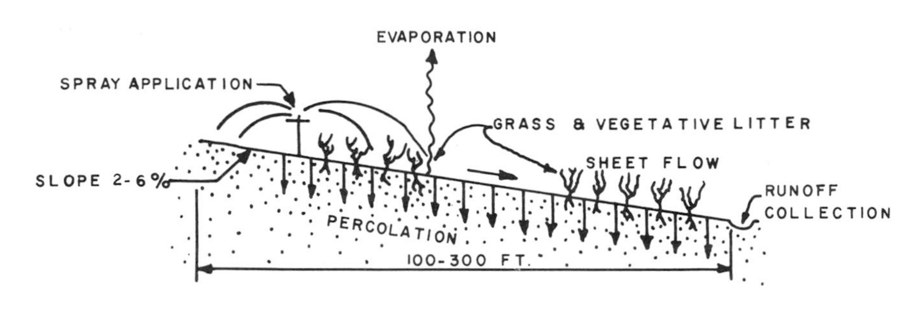

2. Overland flow.

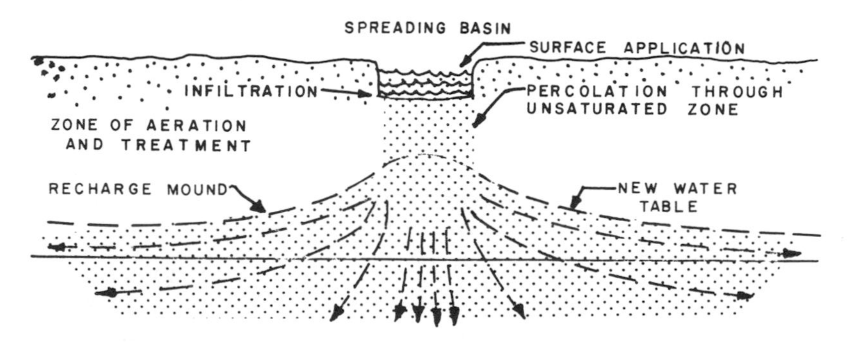

3. Rapid infiltration.

Hybrid systems and combinations of these techniques also exist. An example is the hillside spray systems in use in California where a certain amount of downslope surface flow is acceptable, but where all applied liquid still infiltrates within the site boundaries (Sepp, 1973).

Design criteria, land requirements, and costs for these concepts are discussed in detail in later sections of this paper. It must be remembered that these concepts are proposed as renovative systems and not as convenience disposal or convenience irrigation. Their design, construction, and operational management retain wastewater renovation as the functional intent of all system components and optimization of renovation as the functional goal.

Probably the single most important constraint on the use of these concepts for the present and near future is public acceptance. The American public has become conditioned to expect an almost magical "black box" solution to all problems. If technology can send men to the moon, it can certainly develop a neat container to fit in a closet which would take care of pollution. In addition, there is what might be termed the landfill syndrome. Everyone may agree that a sanitary landfill for solid wastes is the best approach, but then no one wants it in his backyard. Both attitudes can be changed with the proper approach and presentation. Although the black box may be technically possible, the commitment in energy and other resources may be prohibitive. On the second point, operation of a land-treatment site is entirely different from that of a sanitary landfill. A landfill is never totally restored until the end of the operational life of the design, while a land-treatment site does not necessarily impose any significant aesthetic change on the receiving location.

To the engineer, the major concerns are the long and short term reactions of the environment to wastewater constituents. These constituents may be categorized as simple organics, complex organics, metals, nutrients, salts, and the water. At the application level discussed in this paper, there can be no question about the assimilation of simple organic compounds by a land-treatment site. If there were any problem, most of our existing farm pastures and fields would be more than knee deep in animal manures. Thomas and Bendixen (1969) estimate that a well-drained sandy soil could assimilate up to 30 tons of organic carbon per acre per year. This would mean more than 200 million gallons of secondary effluent per acre per year, which far exceeds any loading rate proposed for any of the land-treatment concepts.

The concentration of metals and complex organics should be less in wastewaters from rural communities than from highly industrialized urban centers. A local industry may represent a significant fraction of the total wastewater flow in a rural community. In these cases, however, the current trend is toward in-plant treatment of such wastes to reduce the load on municipal systems. In the concentrations normal to effluents, it is believed that metals or complex organics will not impose any hazard on the land-treatment site, since they are absorbed in the surficial topsoil layer. Further studies are under way to provide additional documentation on long-term effects and build-up. Evidence to date does not expose any critical problems. Long-term operations, such as those in Melbourne, Australia, show no deleterious effects from metals after eighty years of activity. Sludge utilization activities recently initiated in the United States show no harmful effects, and the concentrations of metals and other materials in these sludges can easily be much higher than that expected for wastewater effluents.

Salts may be a problem, particularly in arid climates where the evaporation rates are high. Excessive salts may inhibit plant growth and, by changing the soil structure, may alter the hydraulic capacity of the receiving site. In humid climates with adequate rainfall, however, such effects are not expected. The salt concentration may increase in proportion to the volume of liquid lost by evapotranspiration. But the significant problem is the reuse potential of the product water from a land treatment system. Each cycle of domestic use may be expected to increase the dissolved solids concentration by several hundred ppm. In a closed system with direct undiluted reuse, the salt concentration would rapidly approach unacceptable levels at least for drinking water standards. However, such direct, multicycle reuse is not a consideration for most rural land-treatment operations, and salt effects from single-pass systems should not be a problem except in very arid climates.

The nutrient materials, nitrogen and phosphorus, are the major limiting factor on the operation and useful life of a land-treatment site (Hinesly, 1973). The vegetative cover on the site is the major recognized pathway for the nitrogen applied. Certain forms of nitrogen are adsorbed by soil components, and denitrification is known to occur. Work is under way to define and optimize these mechanisms further, but until management techniques are developed, the plants remain the most significant component for nitrogen removal. Application rates and schedules must therefore recognize the variable plant requirements for nitrogen during the growing season. Use of a single annual crop may significantly limit the application season. Use of perennial grasses can extend the season, since the root systems are in place and are active earlier in the spring. Experiments are under way with multiple crops to extend the application season further and to improve nitrogen removal.

On the long term, the phosphorus contained in the wastewater will be the limiting factor on the useful life of a land-treatment site (Hinesly, 1973). The plant cover does extract some phosphorus, but the major pathway is immobilization by the receiving soils. There is a finite upper limit to this capacity, and when it is reached, the applied phosphorus will move through the soil column. It is possible to obtain a preliminary estimate of this limit by determining the phosphorus adsorption capacity and the actual flow path of the applied liquid through the soil. It is believed that such a determination will underestimate the phosphorus adsorption capacity of a given site. In addition to the short-term adsorption as determined in the laboratory, mechanisms prevail in the natural soil which in time convert the phosphorus into sparingly soluble forms. A combination of long-term laboratory experiments and of analysis of soils from older land-treatment operations will be required for a better definition of this actual phosphorus capacity.

There is a very significant body of data and expertise within the agricultural sciences on optimizing crop production by irrigation. This is valid information and serves as a useful starting point for design of land-treatment concepts. However, a critical point must be recognized. Much of this work was aimed at identifying the minimum amount of water required for irrigation, while land treatment is based upon the maximum amount of water which can be applied without impairing crop growth. The crop is a component in a system whose principal goal is renovation of the wastewater. Such a goal requires vigorous and healthy crop growth, but optimum renovation is still the target. It is valid to use crop needs or "irrigation deficit" as a starting point for land-treatment design, but a design which is limited to this value for wastewater application cannot be termed land treatment and cannot be justified.

Design Criteria

Design of any waste-treatment system is a complex undertaking far removed from the relatively simple hydraulic calculations and tank sizing of former years. Design of an advanced waste-treatment facility requires an understanding of microbiology, physical and organic chemistry, precipitation and filtration mechanisms, adsorption, and ion exchange.

Design of a land-treatment system requires the same level of understanding of exactly the same factors. Land treatment seeks to take advantage, through controlled management, of these phenomena occurring naturally in the receiving ecosystem, while AWT is designed to produce similar reactions in an artificial environment at a higher unit rate.

Design of a land-treatment system is more than just the engineering of a lagoon, a holding pond, and some irrigation equipment. Many such systems could be more properly classified as convenience disposal rather than treatment. The proper design of a land-treatment concept will include a plan for management of all components in the system from the raw sewage pumps to the vegetative cover and the subsoils on the receiving site. The procedure must start with a simultaneous and converging assessment of product quality goals, intended pathways, and the capacity of potential receiving sites.

More than 500 operational systems in the United States incorporate a land area for the reception of wastewaters (Seabrook, 1973). It is not always possible to extrapolate from these systems design criteria for land treatment, for many of them are convenience-disposal activities or convenience irrigation of crops. Optimizing crop production is not the same thing as optimizing wastewater renovation with a growing crop. The rates and periods of wastewater application can be entirely different for the two cases.

Extrapolation from actual land-treatment concepts is valid. However, as with AWT technology, experience with full-scale systems or even pilot plants is somewhat limited. A further caution is necessary, since land treatment deals with a living system. Extrapolation from one geographical area to another or to a different ecosystem must be done with care to avoid the risk of either overdesign or failure. An example is the recommendation of "two inches per week at a quarter inch per hour," which is approaching the status of a theorem for spray irrigation in much of the current literature. It worked at Penn State and was reported accordingly with the appropriate qualifications. This does not mean that it will work everywhere, nor does it mean that higher rates or different schedules might not be more effective elsewhere.

With such qualifications in mind, the following summaries of criteria for land-treatment concepts are offered:

Rapid Infiltration

a. Site characteristics. Deep, continuous deposits of coarse-textured soils without impermeable layers. Primary components should be sand and gravels with small percentage of silts or clays acceptable and even desirable for their adsorption contribution. Soil must have high hydraulic conductivity to permit rapid movement of applied liquid, e.g. the Flushing Meadows subsoil with 282 ft/day horizontal and 17.6 ft/day vertical (Bouwer et al., 1971). Soil deposit should be of sufficient extent vertically and/or laterally to permit 200–300 ft. of travel for the applied wastewater if recovery wells are used. Maintenance of grass cover in basins provides benefits.

b. Pretreatment. Primary or secondary treatment has been used. Secondary treatment probably allows higher rates and longer inundation periods. Though disinfection is not technically necessary, since there are no aerosol losses, state agencies may still require it.

c. Application. Varies with climate and effluent characteristics. At Flushing Meadows with secondary effluent there are 14 days of inundation and 14 days of drying in the summer with 14 days wet and 20 days dry in winter. There are approximately 13 application cycles per year with total application in 1972 being 250 ft. of water, or 19–20 ft. per cycle, or an average of 1.4 ft. of wastewater per day during inundation.

At Ft. Devens, Massachusetts, with primary effluent there are 2 days of inundation and 14 days drying on a year-round basis with approximately 23 cycles per year for a total of 72 ft. of water, or 3 ft. per cycle, or an average of 1.5 ft. of wastewater per day during inundation.

At 1 MGD and 1.5 ft. application rate per day, a 2-acre basin would be required. To operate on the Ft. Devens schedule with 14 days for drying, eight such basins would be required for a total of 16 acres.

Overland Flow

a. Site Characteristics. Relatively impermeable subsoils on gentle slopes with grade from 2 percent to 6 percent. Slope should be sufficiently long to permit 150 to 200 feet of travel distance for wastewater. Longer slopes could be terraced with intermediate collection ditches. A grass cover is the most desirable crop with Reed Canary the most commonly used species to date.

b. Pretreatment. To whatever level required by state agencies. Secondary pretreatment is probably best, although lesser levels should be adequate. Disinfection is required by most state agencies because spray nozzles are the most common technique for distribution.

c. Application. Generally similar in unit rates and operational season to spray irrigation. The overland flow method may require slightly more land than does spray irrigation due to the difficulty of ensuring uniform coverage on slopes. A low-trajectory, low-pressure, large-droplet spray nozzle is desirable to avoid aerosol risks. Buffer zone requirements are dependent on the state agency but are probably similar to those for spray irrigation. The overland flow method should need only a minimal zone on the downslope side of the operation. Site preparation should include grading and drainage features to control runoff both onto and from the site.

Spray Irrigation

a. Site characteristics. Fine to medium textured soils. In general, any "good" agricultural soil is potentially suitable. Slopes are not critical except on agricultural sites where normal farm equipment must operate and erosion during bare soil periods must be considered. Forested sites in California operate successfully with slopes up to 40 percent. It is essential to maintain a 4 to 5 foot depth of soil at the receiving site in an unsaturated aerobic condition. In many soils the use of tile underdrains or draw-down wells may be required. Such elements provide control over the system. A detailed site investigation must characterize the soils, their lateral and vertical extent, their hydraulic capacity, and the location and pathway of the natural ground water.

b. Pretreatment. Most state agencies probably require secondary level treatment, although primary treatment can be used and raw sewage has been applied in some foreign countries. A distinction should be made between requiring secondary type pretreatment and demanding stringent secondary effluent characteristics prior to spraying. When such effluents are discharged into streams or lakes, stringent requirements on BOD and suspended solids are understandable since the receiving waters are not part of the treatment system. To require the same levels of removal prior to land application is illogical and unnecessary, since the land is a component in the treatment system. Its capacity for assimilation of the simple organics generally comprising BOD and suspended solids is beyond question. The land site could renovate raw sewage if safe and economical techniques for delivery and application could be worked out. Disinfection is generally required by all state agencies for systems which depend on spray devices for application. Investigations are under way on alternatives to the conventionally used chlorine, and further studies should be undertaken on the natural die-away where long-term holding ponds are used.

c. Application. Rates and schedules are dependent on site characteristics and climatic conditions. Average rates on projects in use or under design and construction vary from less than 1 inch per week to over 6 inches per week. This is usually applied during one or two days per week at unit rates ranging from less than 0.1 inches per hour to 0.3 inches per hour. Translated into gallons and acres, a 2 inch per week application at 0.25 inches per hour means a one day a week for eight hours application and 54,700 gallons for every acre sprayed. With proper controls it is possible and desirable to vary the application schedule in response to crop needs and climatic conditions.

The total volume that can be applied in a given year is a function of climate, soil characteristics, vegetation responses, and wastewater composition. This total includes the irrigation deficit (the amount required for plant growth) plus the amount intended for deep percolation and/or lateral flow to the subdrain system. In cool humid regions the irrigation deficit might be less than 15 inches per year while most "good" agricultural soils are believed to be able to receive and renovate five to seven times this amount of wastewater.

The existence of such hydraulic capacities for the general case can be established by considering the design of septic tank–leachfield systems which also depend on soil absorption for water movement. Assuming a 100 inch per season land-treatment design for a 30-week season and subtracting 15 inches as the irrigation deficit, the system must then move by deep percolation 85 inches per season. At a one-day-per-week application during the 30-week season, this is equal to 1.8 gallons applied per square foot of surface per day, which is not an unreasonable value for soil absorption capacity. Leach fields are designed for such capacities in many soils, and it is assumed they will receive this quantity seven days per week, fifty-two weeks per year for a fifteen to twenty year design life. By contrast, the land-treatment site receives one application per week for thirty weeks per year in this example. If failure mechanisms were identical, which is unlikely, the land-treatment site would therefore have a design life of between 180 and 243 years. The situation should be even more favorable at land-treatment sites, since there will be careful management to avoid physical clogging at the surface and the soils will be maintained in an aerobic condition.

Ultraconservative estimates of only 20 to 40 inches per year for land-treatment applications cannot be supported on the basis of hydraulic capacity for the general case. If in fact the soil had such poor characteristics, a state agency could never approve the construction and use of household septic tanks and leach fields. In fact, such systems are probably the dominant form of liquid disposal in the same rural areas being considered for land-treatment concepts.

Determination of application period and storage requirements are directly related. The former may be set by the state agency but can be derived from an analysis of climatic records and responses of the intended vegetative cover. If the treatment system is based on the use of a single crop and that crop is depended on for nutrient removal, then wastewater applications can be considered only during the growing season. The use of multiple crops or perennial grasses can extend the season significantly. Climatic constraints to be considered include: periods of intense rainfall, frozen ground, presence of snow or ice on the surface, and below-freezing air temperatures. These affect most significantly the infiltration capacity of agricultural sites. At the famous Penn State experiments, spraying continues throughout the winter in the forested areas.

Buffer zone requirements are generally established by the responsible state agency. These zones can be minimized by proper site selection and the choice of minimum aerosol application equipment.

Land Requirements

The total land required is directly dependent on both the application mode selected and the operational period. The engineering and scientific analysis described in previous sections will establish these criteria for a particular area. An estimate of land requirements may then be made using the following equations:

(1) $$\text{Wetted Area (acres)} = \frac{(13{,}444)\ (\text{MGD})}{(\text{A})\ (\text{O})}$$

where: MGD = daily wastewater flow (million gallons per day)
A = Application rate in inches/week
O = Operational period in weeks

(2) Application Area (acres) (A) = Wetted Area/X
where: x = % efficiency for coverage of site with liquid. For rapid infiltration, x = 1.0; for spray irrigation, x = .75 to 1.0; for overland flow, x = .50 to .85.

(3) Total Site $= (0.02)\ (\text{A})^{1/2}(\text{w}) + \text{A}$
where: A = application area (acres)
w = width of required buffer zone in ft.

This assumes a square plot, with a buffer zone on each side. For a final estimate the actual area should be plotted and the buffer should be measured directly.

Example: 1 mgd flow, 2″/week, 30 week season, 75% efficiency, 250 ft. wide buffer zone.

Wetted Area $= \frac{(13{,}444)\,(1)}{(2)\,(30)} = 224$ acres

Application Area = (224)/(.75) = 299 acres

Total Area $= (0.02)(299)^{1/2}(250) + 280 = 366$ acres (with 250 ft. buffer)

Under these assumed conditions almost 40 percent of the total area is either noneffective or required for buffer zones. It is essential in planning the application system to design for maximum possible coverage efficiency within the constraints imposed by site topography. In this example, 100 percent site utilization would reduce total requirements to 300 acres, including a 250-foot buffer strip. The size of the buffer zone will generally be set by the responsible state agency. Its function is to prevent off-site transmission of bacteria, viruses, and other materials affecting health and hygiene. The need for such protection is the subject of current research concerned with aerosol transmission and viral survival. Such data should resolve the existing controversy on the need for buffer zones and their size. Qualitative data obtained during the recent APWA survey of existing systems provided no evidence of any health problems associated with land-treatment concepts (Seabrook, 1973).

Rapid infiltration basins can function on a year-round basis even in northern climates, and, because no aerosols are present, require no buffer zones. At an assumed 10 inch-per-week application, 52-week season, and 100 percent utilization, calculations similar to those above would require a total area of approximately 26 acres. This represents an order of magnitude savings in land area for rapid infiltration.

In addition to the calculated estimates presented above, space for the pretreatment steps and for storage ponds if required, will be necessary.

Identification of suitable land areas of the magnitude estimated above should not be difficult in rural areas. The estimated values would serve the needs of a community of 10,000. Since 85 percent of the public water supplies in the nation serve populations of less than 5000, it seems that land treatment should be a viable alternative for almost all rural communities. The APWA survey (Seabrook, 1973) indicated that 73 percent of the existing land treatment systems studied have capacities under 5 MGD. Presumably, most of these small communities selected land application in preference to more conventional treatment technology. This choice was generally made before the establishment of current water quality standards and may have been based on economics.

The unit cost relationships for conventional and advanced waste treatment concepts favor consolidation into larger single units serving a number of communities. As shown in a later section of this paper, land costs are a major factor for land-treatment concepts. Since these costs are relatively stable regardless of system size, consolidation of land treatment systems is not so advantageous as with AWT systems. Rural communities can therefore construct and control land systems within their own boundaries. In some cases, a combination of communities or transmission to a remote site may be necessary or economically desirable. In the latter case, public acceptance at the receiving site will be improved if the local community is included as a user of the system.

There appear to be no technical barriers to the use of land treatment by any community regardless of its size. Very large cities may require significant transmission distances to suitable land. The constraints in this case are still social, institutional, and economic rather than technical. The need to cross state or county boundaries is eliminated for the rural community. The riparian rights of others must be considered, however. If operation of a new land-treatment system will result in the depletion of water resources at the former disposal or discharge point, it may be necessary to consider recovery and return of the water.

The area selected as a land-treatment site must be committed for this purpose for the life of the system. These lands are a component of the system and must be properly managed. Such a commitment does not necessitate land ownership by the wastewater authority or exclude beneficial uses besides wastewater treatment. Management for agricultural crops or for forest products will provide an economic return while retaining the rural integrity of the site. Parks, greenbelts, and other recreational uses may be planned for spray irrigation or overland-flow systems. Either type of system will impose minimal changes on existing landscape features and will tend to preserve rather than destroy the aesthetic character of rural land.

Obviously, the amount of land required will increase in direct proportion to the size of the community. It is not possible, however, to identify a general rule or relationship between the availability of such land and community size. In general, it is believed that suitable land will be available and that the constraints on its utilization will be economy and public acceptance rather than technical aspects or environmental concerns.

Cost Comparisons

A cost analysis is an essential element in any planning or design operation. It will become even more critical in the future since under the most recent water quality legislation it will be necessary to identify the most cost-effective treatment approach.

Of concern in this paper is a comparison of costs for the various land-treatment concepts and for advanced waste treatment (AWT) required to produce comparable effluent quality. Such comparisons should be approached and accepted with caution. A large body of cost data on either type of concept does not yet exist. Extrapolations are usually based on small scale pilot plants or on the very few existing systems such as the South Tahoe AWT.

Direct comparison of such published data is also not usually possible. One author may include certain factors which others ignore. The amortization period and the interest rate chosen to calculate annual costs are quite often different. The cost of land and of site preparation are very critical factors in land-treatment concepts. Values in the literature vary from \$140 to \$2000 per acre for land costs and from less than \$1000 to over \$4000 for site development. A variation in land costs is to be expected, reflecting regional real estate values and current land use at a specific location. The difference in development costs is partially due to different application techniques adopted.

The original sources for all the cost data used in this paper are cited. However, in an attempt to provide a common base for comparison the authors have adjusted the original data in some cases. In general the adjustment is limited to a substitution of a common value for land costs or a proportional increase in the original capital costs. These total capital costs were then divided by design capacity in MGD × 365 to generate a unit cost per million gallons of annual capacity. The resulting values do not take into account the amortization period or interest rate but should be valid for the comparisons intended.

Selection of a particular land-treatment concept will not usually be based on costs alone. Site topography, soil conditions, hydrogeology, and quality goals for product water will influence and may determine process selection. When more than one option is technically feasible, costs may govern. The comparisons presented below are based on the

Paris, Texas, operation (Thomas, 1973) for overland flow; Flushing Meadows (Bouwer, 1971) for rapid infiltration; and Muskegon, Michigan (Bauer and Matsche, 1973) and Penn State (Nesbitt, 1973) for spray irrigation. The published data from Paris, Texas, and Flushing Meadows do not include a factor for pretreatment or storage. These figures have been adjusted to allow for aerated lagoon pretreatment for both concepts and storage for overland flow. An estimated land cost of $2000 per acre was used for all concepts.

Concept	*Unit Capital Cost $/MG*	*O&M Costs $/MG*
Rapid Infiltration	400	80
Overland Flow	2,800	110
Spray Irrigation		
Original data source:		
Bauer	11,000	90
Nesbitt	10,000	30
Powell	14,473	251

These figures demonstrate the very strong capital-cost advantage for rapid infiltration, because of the much smaller land area required. It is reasonable to expect that overland flow might be slightly more economical than spray irrigation, since subsurface drainage is not required, but the significant difference in costs shown on this table may not be realistic. Conditions at the Paris, Texas, site may have been unusually favorable, permitting low construction costs. The reported cost was approximately $1000 per acre for site development, including installation of pipes and irrigation equipment. In their recently published analysis, Powell and Culp (1973) estimate a cost of $4446 per acre for spray-irrigation site development in Montgomery County, Maryland. Of this total, $2200 was assigned to distribution piping and irrigation equipment and $750 to subdrainage.

The very strong influence of land costs was apparent in the adjustments made to the original spray irrigation cost data. Bauer assumed a cost of $320 per acre, Nesbitt $140, and Powell $2000. Adjusting the total capital costs reported by the first two by a simple proportion ($2000/reported land cost) produced the unit costs tabulated above. The relative closeness of the three values tends to verify that land costs are more significant than other variables in the original estimates.

Powell and Culp (1973) have compared AWT to spray irrigation land treatment for Montgomery County, Maryland, at a 60 MGD design capacity. Their unit capital costs were $6324 per MG for AWT and $14,473 for land treatment. Very high land and site preparation costs were assumed for land treatment. In addition, engineering costs were assumed at 15 percent for land treatment and only 10 percent for AWT.

The AWT process chosen for their analysis is essentially similar to that in use at Lake Tahoe with the exception of break-point chlorination for nitrogen removal instead of the stripping tower used at Tahoe. The published cost data for the Tahoe system (Culp and Culp, 1971) do not include land or engineering costs. Assuming a 15 percent factor for engineering, the total capital cost would be $6,356,050, excluding land, for construction and equipment for the 7.5 MGD system. The unit cost, as calculated for this paper, would be $2321 per MG. This does not include the cost of the effluent transmission line or construction of the Indian Creek Reservoir for final effluent storage.

These estimates of AWT costs are based on relatively large-scale systems and are not directly applicable to rural communities with populations of 10,000 or less. The cost of land would be relatively stable regardless of system size, but the unit costs for equipment and AWT construction will go up as the size of the unit goes down. Applying the six-tenths power rule which is commonly used for process design (Byrd, 1969), the unit costs for the Tahoe system would be approximately $5000 per MG for a 1 MGD system.

Since land costs would be a stable factor and since these are the critical element in land treatment costs, a linear extrapolation from these data is considered valid. On this basis, extrapolation from the previous estimates of Bauer and Nesbitt indicates that land treatment would be favorably competitive with AWT at land costs up to approximately $1000 per acre for a 1 MGD system.

Williams (1973) reports costs on several land-treatment systems actually constructed in Michigan. At 1 MGD capacity his data show a unit cost of $2100 per MG based on approximately $500 per acre land costs. His data also indicate that activated sludge, including chemical precipitation of 80 percent of phosphorus, would at 1 MGD have a unit cost of $1500 per MG. For AWT to produce an effluent comparable to that of land treatment, phosphorus-removal efficiency would have to be increased and some process for nitrogen removal included as well as filtration and carbon adsorption. The addition of these extra steps would then produce a unit cost approaching the $5000 per MG cited for the AWT model at Lake Tahoe.

Data from all sources confirm that operation and maintenance costs are significantly lower for land treatment than for AWT alternatives. Powell and Culp project 1980 O & M costs of $251 per MG for land treatment and $434 per MG for AWT. Williams projects an O & M cost for 1 MGD at $100 per MG for land treatment and $273 per MG for activated sludge with only 80 percent phosphorus removal. Bauer estimates $90 per MG for O & M costs at Muskegon. The projections based on experience at the Tahoe plant are $246 per MG for a 7.5 MGD capacity (Culp and Culp, 1971). These data represent different-sized systems in different parts of the country. Without any attempt to make adjustments, it seems conservative to assume that AWT will be at least twice as expensive to operate and maintain as land-treatment systems. This is a critical factor for small rural communities, since such costs are a direct local responsibility.

Selection of the most cost-effective approach must consider both capital costs and O & M costs. Capital costs for land treatment can actually be higher than those for AWT. But in those areas where land costs and site preparation costs are reasonable, the capital costs for land treatment should be less than AWT.

All of the previous discussion assumes that the required land area will be purchased and owned by the local waste-treatment authority. Purchase of such lands, which are a component part of the treatment sequence, can now be subsidized by federal grants under current legislation. Such ownership provides total control over operations and may be desirable, but other options seem possible. The land could remain under private ownership with a contractual arrangement for the reception of treated effluent. The authority would be responsible for installation and maintenance of irrigation and subdrainage equipment. Such a plan might realize a significant savings in capital costs.

Research Needs

Research needs have been cited throughout the various sections of this paper. They include: identification of actual long-term phosphorus capacity, quantification of aerosol risks, improving nitrogen removal through crop management or other techniques such as denitrification, definition of responses and long-term capacities in cool humid climates, optimizing pretreatment, setting storage and disinfection requirements, and developing cost effective criteria for design, construction, and operation of the various concepts in different regions with different ecosystems. Principal emphasis in the latter area should be in the northern U.S., where humid climates and freezing conditions prevail. Such work is essential, since the bulk of current experience is based on disposal operations in warm arid regions.

In recognition of these needs the Corps of Engineers, in cooperation with EPA, USDA, USGS, the Army Surgeon General, and a number of state agencies and academic institutions, has initiated a comprehensive research program. This includes study of the three major application techniques, research in different regions with different climatic conditions and different ecosystems, and evaluation of performance of existing operational systems. The U.S. Army Cold Regions Research and Engineering Laboratory (CRREL) located in Hanover, New Hampshire, is the leading laboratory for this program. Visits and inquiries are welcome.

Conclusions

Land-treatment concepts for rural wastewater management are not the universal panacea for all problems.

Under proper conditions they can offer higher level treatment at lower costs than existing AWT technologies.

Land treatment should be given consideration as a viable alternative during the early planning and process selection stages for wastewater management in all rural communities.

References

Bauer, W. J., and Matsche, D. E. 1973. Large wastewater irrigation systems. In *Proceedings of the Penn State Symposium on recycling municipal wastewater and sludge through forest and cropland*, ed. W.E. Sopper, pp. 345-363. University Park: Pennsylvania State University Press.

Bouwer, J. C., et al. 1971. *Renovating sewage effluent by ground water recharge*. Washington, D.C.: USDA Water Conservation Laboratory report.

Byrd, J. F. 1969. All parties can benefit from joint municipal industry treatment. *Water and sewage works* 116: 42-43.

Culp, R. L., and Culp, G. L. 1971. *Advanced wastewater treatment*. New York: Van Nostrand Reinhold Company.

Hinesly, T. D. 1973. Water renovation for unrestricted reuse. *Water spectrum* 5: 1-8.

Nesbitt, J. B. 1973. Cost of spray irrigation for wastewater renovation. In *Proceedings of the Penn State symposium on recycling municipal wastewater and sludge through forest and cropland*, ed. W. E. Sopper, pp. 334-338. University Park: Penn State University Press.

Powell, G. M., and Culp, G. L. 1973. AWT versus land treatment: Montgomery County, Maryland. *Water and sewage works* 120: 58-67.

Reed, S. C., et al. 1972. *Wastewater management by disposal on the land*. USACRREL Special Report no. 171. Hanover, New Hampshire: USACRREL.

Seabrook, B. L. 1973. Land application of wastewater with a demographic evaluation. In *Proceedings of the joint conference on recycling municipal effluents and sludges on land*, pp. 9-24. Washington, D.C.: National Association of State Universities and Land-Grant Colleges.

Sepp, E. 1973. Disposal of domestic wastewater by hillside sprays. *Journal Environmental Engineering Division, American Society of Civil Engineers* 99: 109-121.

Thomas, R. E. 1973. An overview of land treatment methods. *Journal of Water Pollution Control Federation* 45: 1476-91.

Thomas, R. E., and Bendixen, T. W. 1969. Degradation of wastewater organics in soil. *Journal of Water Pollution Control Federation* 41: 808-813.

Williams, T. C. 1973. Utilization of spray irrigation for wastewater disposal in small residential developments. In *Proceedings of the Penn State symposium on recycling municipal wastewater and sludge through forest and cropland*, ed. W. E. Sopper, pp. 385-395. University Park: Pennsylvania State University Press.

II
Septic Tanks and Their Effects on the Environment

4. Septic Tanks and Their Effects on the Environment

P. H. McGauhey

Introduction

During the late 1950's and early 1960's there was a major eruption of national interest in the individual-household septic tank, particularly with respect to the causes and prevention of failure of the associated percolation system. At that time researchers who, during the preceding decade, had developed a significant body of knowledge of the phenomena which govern the travel of chemical and biological pollutants with percolating water turned their attention to the specific situation of the domestic septic system. Then there followed a period of nearly ten years of relative quiescence at the national level, although, as with an active volcano, there continued a great deal of turbulence within the crater. Now as the 1970's get under way, interest in septic tank systems has once again erupted nationally, this time in the context of groundwater quality.

To explain the volcanic nature of the subject of septic tank systems it might be useful to recall something of their history. The septic tank was first developed to make it possible for the rural isolated home to take advantage of the relatively inexpensive pressurized water system. Because the rural dweller generally obtained his water supply from his own shallow well and because typhoid fever and other gastrointestinal diseases were common in early America, public health authorities were understandably concerned. Normally, public health interest in the septic system developed as a matter for local (usually county) health departments. These sought to protect the householder by specifying some minimum distance (usually 50 to 100 feet) between the owner's well and the septic tank percolation system as a prerequisite to approval of the installation. The system itself soon became standardized, albeit on the basis of erroneous assumptions, and codified at the local health department level, with state and federal health agencies providing leaflets or manuals describing how the system should be sized and installed.

Failure of septic tank systems occurred from the beginning, due to clogging of the soil by continuous application of liquid or to neglect of the tank until sludge and scum accumulation carried over into the percolation system (the tile field or cesspool) and clogged it.

In the isolated rural situation, failure of the percolation system was not a serious matter. Generally, the householder himself had the means and the land area necessary to increase the size of the system. Or if he simply chose to ignore the presence of sewage effluent outcropping on the land surface, no major threat to the public health occurred.

The percentage of septic systems which failed in any given period was unknown because failures were either unreported or the information was filed in hundreds of county health department offices. As a result of such scattered and unevaluated information, the close of World War II arrived with the general notion that septic tank systems were good for fifteen to twenty years of service and without public laws or policies effectively limiting their use in urban subdivisions.

In the absence of constraints upon its use, the septic tank system was particularly attractive to land developers because their objective was to convert land profitably from single ownership of large parcels to multiple ownership of small plots without retaining any residual responsibility for the whole, and to do so before the rains came.

The era of postwar expansion of subdivisions was made possible by a federal policy of insuring loans through the Federal Housing Administration. Thus, for the first time, the statistics on septic system failures accrued to a single agency. The unfortunate fact was that in many situations 30 percent or more of the installations failed within two or three years and on sites too small to permit enlargement of the system. It was this fact that triggered extensive studies of the failure of septic tank systems under the sponsorship of the United States Public Health Service and the Federal Housing Administration in the 1950's and 1960's.

What followed was an increasing reluctance to approve septic tank installations in subdivisions unless five to ten year limitations were stipulated in their permits or sewers were provided in the streets for future use. Such an approach was feasible in the large urban subdivision; hence, the federal agencies found it relatively easy to manage the situation that triggered the eruption of interest previously cited. However, it did not solve the problem of the rural dweller or of the dispersed community which characterizes much of New England. Moreover, because restrictions such as the foregoing are enforced at the county level, it is difficult to maintain the planning and regulatory perspective necessary to exempt the rural dweller from what may be catastrophic limitations in his individual situation. One difficulty is that large, politically influential land developers often propose "estates" and "ranches" ranging in size from a quarter acre to one or two acres and local officials starving for tax monies find the line of demarcation between the rural and urban situation a bit fuzzy. As a result, either urban subdivisions with septic systems are unwisely approved or restrictions are so rigidly interpreted as to work a hardship on the rural dweller for no conceivable good.

The foregoing predicament characterized the interlude between the initial and the present eruption of interest in septic systems on a national scale. It carries over the same tendency of regulatory agencies to react to categories rather than to situations. In fact, it pervades the federal approach even more than the local. But it is another aspect of the federal approach that has awakened the new national interest in septic tank systems. Specifically, Public Law 92-500, Federal Water Pollution Control Act Amendments of 1972, states that it is federal policy to eliminate the discharge of all liquid wastes into navigable waters by the year 1985. This has generated a vast interest in land disposal of wastewater. At first flush, one might think all augurs well for the rural septic system. However, the same PL 92-500 requires that EPA issue guidelines for identifying and evaluating the disposal of pollutants in many circumstances, including "disposal of pollutants in wells or in subsurface excavations." Septic systems are among those being studied. And if the ultimate determination is that sewage effluent should be excluded from ground water, who knows whether a distinction will be made between urban and rural systems? And on what historic precedent might optimism rest?

Closely and dangerously allied with federal interest in groundwater protection, perhaps by eliminating the septic tank system, is the ever-present notion in today's cultural climate that "innovative" systems will replace conventional ones if the latter are prohibited. Unfortunately, innovation is not a synonym for ingenuity or invention. More often it is simply a buzzword denoting some formless absurdity in somebody's dream world.

The foregoing historical and other factors either are in themselves a part of the state-of-the-art or are a background to an understanding of its more specific details. Of particular significance are such matters as the following:

1. The behavior of septic tanks per se
2. The types of percolation systems in use
3. Present knowledge of systems design, operation, and performance
4. The environmental effects which will accompany the best application of what is known.

Septic Tanks

A septic tank is a satisfactory treatment system which produces a somewhat offensive anaerobic effluent well suited to aerobic treatment by soil bacteria. Its physical structure is generally a water-tight tank constructed in conformity with local, state, and federal codes or specifications and buried in the earth. The size of the tank is usually a minimum of 600 gallons with a capacity to retain floating solids and settled solids for two to five years. Occasionally, however, septic tanks are built without a bottom on the erroneous assumption that the underlying earth will continually accept liquid and so do some of the work of the percolation system.

A septic tank is customarily and inappropriately "pumped out" completely when solids removal is necessary. This thoroughness is a physical and economic inconvenience to the scavenger and a detriment to the system which, without seed sludge, may then require six months or more to re-establish the biological balance on which good pretreatment of the effluent depends. In addition, the septic tank is hampered by a great wealth of folklore on what is necessary to maintain a biological system. For example, excluding detergents, adding yeast cakes, and feeding it vitamins.

The problems of septic systems, as well as their effects on the environment, are associated with the percolation system rather than with the septic tank. Annual inspection and removal of scum and a foot of sludge when the sludge is 1.5 or 2 feet thick will generally insure satisfactory operation of the tank. The scavenger who services the tank must, of course, dispose of the pumpage in the manner prescribed by the authorized regulatory agency. A septic tank per se, if constructed and operated in accord with present knowledge of the state of the art, should be satisfactory to both the owner and the environment.

Percolation Systems

Several types of percolation systems are in use, some approved and some unapproved, although approval is no insurance that they are workable or that the future will not declare them environmentally detrimental. Most common are the following:

1. *The Cesspool.* A covered large diameter (5–6 ft.) sump, often filled with stones to give support to the sidewalls. Such installations gener-

ally serve as both septic tank and percolation system, receiving raw sewage directly from the plumbing fixtures. They are used in many parts of the United States where soil conditions make such systems workable, e.g., in New York (where there are probably 100,000 such installations), New England, the Southwest, the Northwest, and Hawaii. They are, however, patently unacceptable by today's environmental standards and are no longer approved for new installations.

2. *The Narrow Trench.* A trench system 12 to 18 inches wide, with four-inch open-joint distribution tile located in an eight-inch crushed-stone fill at the trench bottom. This system may be installed either in the normally aerobic biologically active zone of the soil, that is, the top two to three feet, or below the frost line without regard to the biologically active soil mantle. The former method is widely used throughout the United States where freezing does not make the system inoperative during the winter. The latter is used in northern and northeastern areas where the frost line is several feet deep.

When the shallower trench system is properly operated, soil bacteria provide good secondary treatment of the septic tank effluent. The deep trench system may simply discharge biologically unstable dissolved solids to the groundwater after some removal of suspended particles. Improperly operated, installed in tight soil, or inadequately sized, either of the two systems will fail, with sewage outcropping on the surface.

Failure of the narrow trench system in otherwise suitable soils may be traceable to the following:

a. anaerobic clogging of the infiltrative surface due to continuous inundation
b. overloading of the system because of inadequate surface area (most installations are only 50 to 75 percent of the needed size as a result of the erroneous assumptions which abound in the codified witchcraft by which they are conceived)
c. consolidation of trench bottom by human feet or by trenching machines while constructing trenches without regard to soil moisture conditions
d. consolidation of trench sidewalls by trenching machines (which are purposely designed to compact sidewalls in order to minimize the need for shoring walls during pipe-laying operations)
e. dependence upon codes, percolation tests, and political clout, rather than soil science, to identify soils capable of accepting water at a satisfactory rate.

3. *The Seepage Bed (sometimes called a wide trench system).* A type

of installation widely used in urban subdivisions with row houses. Typically, the bed is as wide as a dozer blade, as deep as the narrow trench (2–3 feet), and as long as the lot size will permit (or as short as the health department, in absentia, will allow). It is constructed in any type of soil for which a permit can be obtained by any device. The construction method involves proceeding down a line in back of a row of houses, preparing all systems at one pass by appropriate raising and lowering of the dozer blade. Construction is done at the convenience of the contractor. Smearing of the trench bottom and compaction by dozer weight may in some soils and at some moisture conditions make the bed almost totally impervious to infiltrating water. Failure may be further ensured by passing heavy equipment over the bed during the placement of stone and tile and final surface grading. It is difficult to apply design criteria to the seepage bed. Its basic structure appears the most undesirable of all systems because it has almost no sidewall area and its bottom area has been demonstrated to be a poor infiltrative surface.

4. *The Seepage Pit.* Used in a few particular situations where pervious soil strata are deep and the groundwater table is at a low horizon. The pit is a hole bored 30 inches in diameter and of any depth (30 to 70 feet in some instances in California), fitted with a wooden core box which maintains a fill of crushed stone against the pit wall. It may introduce biologically unstable material underground below the aerobic soil mantle. Its infiltrative surface cannot be loaded fully at one instant without an impractically large holding tank and dosing arrangement.

5. *The Sand Filter.* Used primarily where the groundwater table is at or near the soil surface, for example in Louisiana. It is an earth-covered sand bed, typically three feet in depth, equipped with influent open joint tile on six-foot centers near its top surface and similar collection tile near the bottom surface. It is constructed above ground, and therefore the filtered effluent is discharged upon the land or in the surface waters.

6. *The Evapotranspiration System.* Used primarily in situations where unsuitable soil or impervious strata preclude a trench system and where the septic systems are operated only during the dry summer season of abundant vegetation. Summer homes and forest camps in the western mountains have used it successfully. Its potential is limited only by the size of the percolation system and the transpiration capacity of vegetation, which is 760,000 to 1,500,000 gal./acre/season for grass and 100,000 to 270,000 gal./acre/season for trees. In

some of the clays of the Southeast, septic tank trench systems operate well during the summer and fail completely when ET is no longer the dominant phenomenon. Although such trench systems are not designed for ET, they have at times been installed and functioned long enough for the builder to sell the house and terminate his responsibility.

Present Knowledge of Percolation Systems

A few of the highlights of present knowledge about percolation system design and operation are given below.

1. The Standard Percolation Test can measure the infiltrative capacity of a soil initially and so identify a soil in which a percolation system is feasible, but cannot predict the permeability of a zone of clogging which may overlay that soil in the future. When this test is interpreted for determining size, it allows inadequate bottom area because it was originally used for the construction of narrow trench systems. The originator of the percolation test, Henry Ryon, intended his information to aid in his local situation in New York State. In the absence of other measures, the "perc. test" became gospel all over the world before its practical and theoretical limitations were realized.
2. Because of the limitations of the percolation test and other factors discussed in the references cited, codes cannot take the place of soil scientists and engineers in evaluating soils and designing percolation systems. Persistence of builders in total reliance on codes and the perc. test is turning society against the use of septic tanks altogether.
3. Any soil continuously inundated will lose most of its initial infiltrative capacity. Any percolation system designed on the basis of initial rather than maintainable infiltration rates is therefore liable to fail.
4. A soil continuously inundated, even with clear fresh water, will become clogged. The reason is that the aerobic bacteria with the organic matter already in the soil become deprived of oxygen when pore spaces are filled with water. The system then becomes anaerobic, clogs with bacterial slimes, and precipitates.
5. Water cannot be made to dissolve enough oxygen to maintain an aerobic system in inundated soil. The soil must be allowed to drain, the pore spaces to fill with air, and the aerobic organisms to re-establish themselves. Because of this, the aerated septic tank is no answer to the problem of soil clogging.

6. Alternate loading and resting may permit a loading rate of ¼ to ½ the initial infiltration rate of a soil.

7. Soil clogging is a surface phenomenon (top 0.5–1.0 cm), resulting from anaerobic slimes, precipitation by ferrous sulfide, and sedimentation, which can be overcome by proper operating techniques, including alternate loading and resting of soil

Regardless of the type of percolation system used, several criteria must be satisfied to the maximum practicable degree of compatibility. These may be expressed as follows:

1. Continuous inundation of the infiltrative surface must be avoided.
2. Aerobic conditions should be maintained in the soil.
3. Initially, the infiltrative surface should be typical of an internal plane in the undisturbed soil.
4. The entire infiltrative surface should be loaded uniformly and simultaneously.
5. There should be no abrupt change in particle size between trench fill material, which is rock, and soil at the infiltrative surface.
6. The leaching system should provide a maximum of sidewall surface per unit volume of effluent and a minimum of bottom surface.
7. The amount of suspended solids and nutrients in the septic tank effluent should be minimized.

Environmental Effects

Because septic systems are underground and the earth is an excellent medium to absorb odors, a well-functioning septic tank and percolation system do not affect the environment esthetically. However, the environment in which a septic system can be made to maintain its environmental acceptability must be properly selected. The system must be used only in suitable soils. The percolation system must be designed of such a size and operated in such a manner as to preclude failure.

The environmental consequences of failures in septic tank percolation systems vary with the situation. At its worst, failure may occur in a subdivision, with owners piping or ditching effluent into the streets and with attendant health hazards, foul odors, and nuisance. In an isolated situation, failure may mean only a swampy spot in the orchard or something as serious as surface wash of sewage effluent into waters which serve as potable supplies.

Assuming proper installation and successful operation of a septic system, the environmental effects are largely upon the quality of groundwaters. Effects with various types of percolation systems are considered below.

1. *Cesspool.* Because this primitive system will work only in coarse or highly fissured material, raw sewage may move directly into groundwater supplies carrying tastes and odors from products of anaerobic decomposition of sewage solids, bacteria and viruses, undegraded detergents, partially degraded organic compounds, which will later degrade when water is withdrawn or outcrops on the surface, and nutrients capable of enriching surface waters.

2. *Narrow Trench and Seepage Bed in Biologically Active Zone.* A properly operated aerobic soil system will remove from sewage effluents such factors as sewage bacteria and other particulate matter, which are removed by such processes as straining, sedimentation, entrapment, adsorption, and die-away in an unfavorable environment; viruses, which are probably removed principally by adsorption; phosphates, which are immobilized on soil colloids at normal values of soil pH, and synthetic detergents, which are broken down by biodegradation. However, a considerable fraction of the 300 mg/1 total dissolved solids added by domestic use of water moves with the percolating water. These appear as increases in the anions and cations normally found in ground water, e.g., the chlorides, nitrate, sulfates, and bicarbonates.

3. *Deep Trench and Seepage Pit.* These systems are installed in soils below the aerobic biologically active zone but above the maximum water-table elevation. They change the quality of tank effluent percolating downward by removal of bacteria, viruses, and suspended solids, and by immobilization of phosphates and ammonia on soil particles. The soluble products from the anaerobic decomposition of sewage solids in the septic tank may reach the groundwater, adding to it unpleasant tastes and odors and also compounds that will support biodegradation processes in groundwater after the leachate outcrops in springs or is withdrawn through wells and is once again in an aerobic environment.

4. *Sand Bed.* Effluent from the sand bed is discharged upon the surface. After treatment, it is similar in quality to that in the aerobic soil system. In dissolved solids it is similar to the percolate from a shallow trench or bed. It may produce tastes and odors. Where the sand bed is most used, surface waters in the drainage ditches along the roadway or in natural drainage channels are often highly colored by organic

extractives from luxurious natural vegetation. In these cases, the sand-bed effluent is a better quality than the native surface water, and hence has no detrimental environmental effects.

5. *Evapotranspiration Systems.* The ET system can enhance the growth of vegetation and in this sense have a positive esthetic effect upon the environment. The concentration of dissolved solids is reduced by incorporation into the vegetation. Solids left in the root zone may be picked up by infiltrating rainwater in the wet zone and transported downward. However, the concentration of solids which do travel to the groundwater is far lower than that of solids moving from the narrow shallow trench or seepage bed. Dead vegetation still contains many of the quality degrading factors originally in the septic tank effluent, factors which might theoretically appear underground later. In reality, however, this is not the case because degradation of organic matter in nature returns some of the nitrogen directly to the atmosphere and surface wash carries away nutrients, perhaps to increase the growth of harvestable timber or other crops.

Summary

At best the septic system increases the total dissolved mineral content of local ground waters and at worst may introduce bacteria, viruses, and degradable organic matter as well.

From an environmental viewpoint the "best" septic tank system is not the best of all possible alternatives in an urban situation. Rationally, it would seem undesirable to concentrate 2,000 to 15,000 septic systems on the roof of a single groundwater basin or along the margin of a recreational lake. Nor is it necessary today. On the other hand, the "best" septic system is certainly adequate for the isolated dwelling, where service to man far exceeds any possible detriment to the environment.

At the present state of the art, septic systems can be sited, designed, and operated so that these best conditions pertain. This is, however, not being done because:

1. Engineers have been too busy with larger treatment works to give attention to the septic system or even to learn what attention is needed;
2. Soil scientists have been largely concerned with agriculture and have not been called upon to work on sewerage problems;
3. Historic ignorance of the need for soil scientists and engineers

in the field has led to the installation of septic tanks on the basis of codes with inspections by an overloaded member of the local health department.

In the interest of the isolated rural dweller or isolated vacationer, care must be taken lest in correcting the unwise use of septic systems in urban situations, their use be categorically banned altogether and thus some citizens deprived of their standard of living or means of livelihood without achieving any significant betterment of the environment. Unless alertness is maintained within the profession, convenience in enforcing administrative fiat, rather than environmental protection, may become the goal of regulatory activity.

References

Anon. 1971. Septic tank ban begins in shore critical areas. *New Jersey Environmental Times* 4:3.

Bendixen, T. W., et al. 1962. *Study to develop practical design criteria for seepage pits as a method of disposal of septic tank effluent.* Cincinnati: Robert A. Taft Sanitary Engineering Center, United States Public Health Service.

Cotteral, J. A., and Norris, D. P. 1969. Septic tank systems. *Journal Sanitary Engineering Division, American Society of Civil Engineers* 95: 715–46.

Coulter, J. B., et al. 1960. *Study of seepage beds.* Cincinnati: Robert A. Taft Sanitary Engineering Center USPHS.

Environmental Protection Agency. 1973. *Polluted groundwater: some causes, effects, controls, and monitoring.* Office of Research and Development, Report no. EPA 600/4-73-001b.

McGauhey, P. H., and Krone, R. B. 1967. *Soil mantle as a wastewater treatment system.* Berkeley: Sanitary Engineering Research Laboratory final report, SERL report no. 67-11.

McGauhey, P. H., and Winneberger, J. H. 1964. *Causes and prevention of failure of septic-tank percolation systems.* Federal Housing Administration publication no. 533.

McGauhey, P. H., and Winneberger, J. H. 1967. *A study of methods of preventing failure of septic tank percolation systems.* U.S. Department of Housing and Urban Development.

Perlmutter, N. M., and Guerrera, A. A. 1970. *Detergents and associated contaminants in ground water at three public supply well fields in southwestern Suffolk County, Long Island, New York.* U.S. Geological Survey water supply paper no. 2001-B.

Schmidt, K. D. 1972. Nitrate in groundwater of the Fresno-Clovis metropolitan area, California. *Ground water* 10: 50–64.

5. A Conceptual Model of Nutrient Transport in Subsurface Soil Systems

Millard W. Hall

Introduction

An important question confronting those concerned with preserving the integrity of rural environments is the degree to which pollutants originating with septic tank systems may harm such environments. Traditionally, this question has been posed in terms of the effects of such pollutants on public health. Increasingly important, however, are indications that such systems may be a significant source of the plant nutrients—nitrogen and phosphorus—entering surface waters (Robert A. Taft Sanitary Engineering Center, 1966; Hasler and Ingersoll, 1968; Dudley and Stephenson, 1973).

A number of investigators have reported on the concentration of nitrogen and phosphorus in septic tank effluents (Dudley and Stephenson, 1973; Patterson et al., 1971; Popkin and Bendixen, 1968; Bennett, 1969; Lake George Water Research Center, 1971; Sanborn, 1973). Table 1 presents some typical data, which indicate that nitrogen and phosphorus concentrations in septic tank effluent closely approximate those of domestic sewage. Thus, in isolated rural settings, where other large sources of these nutrients are not present, such effluents may very well contribute materially to the enrichment of surface waters. The fate of these nutrients after they leave the septic tank becomes, therefore, quite significant.

Nutrient Movement in Soil Systems

There are several pathways by which effluents from soil systems might enter surface waters, including:

1. Piping, through fractured rock or through eroded rock channels.
2. Overland flow, resulting from clogging of the absorption field and the subsequent surfacing of the effluent.
3. Flow through porous media, that is, the normal, desirable situation wherein the effluent moves under unsaturated and/or saturated flow condition through the soil.

Table 1
Nitrogen and Phosphorus in Septic Tank Effluent

Parameter	*Average Concentration*	*Reference*
Nitrogen:	(mg/1 as N)	
Organic	10	Dudley[a]
Ammonia	35	"
Nitrate	} 0.5	"
Nitrite		"
Organic	10	Biggar
Ammonia	25	"
Nitrate	0.15	"
Nitrite	0.003	"
Organic	5.6	Popkin
Ammonia	24.6	"
Nitrate	0.2	"
Nitrite	0.01	"
Phosphorus:	(mg/1 as P)	
Total	25	Dudley
"	10.4	Bennett
"	8.2	Lake George
"	26.4	Sanborn
"	38.3	"
Phosphates	20	Biggar
"	20.8	Sanborn
"	35.5	"

[a]See References, p. 63, below.

Obviously, the first two of these examples provide little or no removal of nitrogen and phosphorus from the effluent. However, soils are known to react readily with phosphorus and with certain forms of nitrogen. Moreover, under certain conditions, biological denitrification may occur in soil systems. Finally, it is possible that significant amounts of these materials are withheld from surface waters through storage in aqueous solution within the saturated zone. Thus the third possibility given above offers some hope of removal and/or storage of both nitrogen and phosphorus, and the possible fate of these pollutants when exposed to porous media deserves more thorough review.

Nitrogen

As shown in Table 1, the preponderance of the nitrogen leaving the septic tank is in the organic or ammonia (NH_3 or NH_4+) form. In a properly operating absorption field, these forms, for the most part, will be converted to nitrate (NO_3) within the first few inches of aerobic soil surrounding the absorption trench (Dudley and Stephenson, 1973; Preul and Schroepfer, 1968; Kurtz, 1970; Lance, 1972; Walker et al., 1973). Nitrate is very soluble and chemically inactive in aerobic soil environments. As a result, it is easily mobilized by soil and ground waters. Unless biological denitrification occurs, nitrate will not undergo further transformation. Denitrification within a properly designed and operated absorption field is unlikely because it requires the concurrent existence of anaerobic conditions and a biologically useful source of organic carbon.

In summary, most of the nitrogen in septic tank effluent ultimately is converted to the nitrate form, which moves through soil-water systems with ease. Thus it is likely that most of the nitrogen introduced to soil systems by septic tanks will eventually enter ground or surface waters.

Phosphorus

The literature is by no means as clear on the fate of phosphorus in soil systems. There are numerous reports attesting to the fact that soils are capable of "fixing," i.e., immobilizing, large amounts of phosphorus. (See Bennett, 1969; Dudley and Stephenson, 1973; Kurtz, 1970; Biggar and Corey, 1969; Ellis and Erickson, 1969; Urresta, 1970; Holt et al., 1970; Hill, 1972; Hinesly, 1973.) The mechanisms by which this immobilization occurs are not well understood and appear to be quite complex. It is known, however, that most forms of phosphorus are retained in soil systems by a combination of adsorption, replacement, and precipitation reactions.

Table 2 presents data on the phosphorus retention capacity of several soils. Most of these data were obtained from batch studies where known amounts of soil were exposed to aqueous phosphorus solutions of known initial strength until equilibrium, or near equilibrium, was reached. Such studies demonstrate that phosphorus retention in soils is a function of soil characteristics and the concentration of phosphorus in the solution being exposed to the soil. Most likely, temperature, pH, and other chemical and biological variables also have an effect on the rate, extent, and permanence of this retention (Bennett, 1969; Urresta, 1970; Kurtz et al., 1946). Because of the many variables influencing the retention of phosphorus by soils, it is fair to assume that, in some cases, the system will not be able to immobilize permanently all of the phosphorus to which it is exposed.

Table 2
Selected Observed Values of Phosphorus Retention on Soils

Soil	*Initial Conc. (mg/1 as P)*	*Amount Retained (mg P/100 g soil)*	*Source*
Adams	153	15	Bennett[a]
Plaisted	153	34	"
Paxton	153	> 70	"
Plaisted	48 to 387	110[b]	Urresta
Muscatine	6	13	Kurtz
Paxton	65 to 390	325[b]	Foster
Buxton	"	350[b]	"
Adams	"	353[b]	"
Hartland	"	255[b]	"
Colton	"	85[b]	"
Rubicon Sand	3 to 15	1.07[b]	Ellis
Warsaw Loam	"	49[b]	"

[a]Column studies, all others are batch studies. See References, p. 63, below.
[b]Maximum retention computed from Langmuir equation.

Some investigators (Ellis and Erickson, 1969; Hill, 1972) have used data on phosphorus retention to estimate the volume of a given soil necessary for immobilizing the phosphorus in a given effluent. This approach seems reasonable in static situations if it can be assured that all the soil in the volume is exposed for a sufficient time to the effluent. Such approximations should be useful, for example, in the design of soil-dosing systems for effluent disposal. However, the assumption that septic tank absorption fields approximate such conditions is questionable. The application of aqueous solutions of phosphorus to subsurface soils on a more or less continuous basis is a different case from the intermittent application of phosphorus to surface soils. Undoubtedly, soils can retain large amounts of phosphorus. However, the real question is, how much phosphorus does the soil *not* retain in an absorption field situation?

If the entire system is considered, rather than just the soil-phosphorus reactions, it is possible to imagine that phosphorus may have at least some significant mobility in subsurface soil systems. The system is a dynamic one, and changes in water movement through the system as well as changes in the amount of phosphorus and other chemicals contained in this water can undoubtedly alter the effect of the soil-phosphorus reactions. For example, while soil reactions tend to immobilize phosphorus, other occurrences, such as bulk transport by water movement, diffusion, dispersion, and dissolution of phosphorus, could all tend toward mobilization of phosphorus. In other words, there are offsetting occurrences within the system. Additionally, because the system is dynamic, the relative kinetics of these occurrences must be considered along with their relative magnitudes.

In Table 3, the time to equilibrium between phosphorus in solution and that associated with the solid phase is compared for four soils. These data were obtained with batch systems where the soluble phosphorus concentration was measured over periods of several days until a near equilibrium was reached. With one exception, these experiments are the only ones known to this author which deal, even in a gross way, with the kinetics of the soil-phosphorus reactions. These data exhibit large differences between the various soils in the time required to reach equilibrium, even for soils subjected to the same phosphorus loading. Thus soils differ not only in their phosphorus-retention capacity, but also in the rate at which this capacity may be satisfied. This fact, although largely overlooked to date, would seem to be of importance in the development of a model or models to predict the distribution of phosphorus within the vicinity of an absorption field.

It seems reasonable to conclude that although most soils can retain large amounts of phosphorus, the system, like most such systems, can be overloaded either by applying too much phosphorus or by applying it too rapidly. Although little data on this point is available, Dudley and Stephenson (1973) discuss four systems wherein this situation apparently has occurred. At two of these locations, soluble phosphorus concentrations significantly larger than background were detected in groundwaters at locations of 100 feet and more down gradient from the sites. Undoubtedly, as additional investigations of this type are conducted, similar situations will be encountered.

This subject is further complicated by the possibility of leaching. In an operating absorption field, leaching may result from a disturbance in the equilibrium between phosphorus in solution in the soil water and that associated with the solid phase of the system. This could be brought about by a fluctuating water table or by percolation of phosphorus-poor precipitation. In such cases, phosphorus could leave the solid phase, go into solution, and be transported to another location. Several investigators have reported results in support of this argument. For example, Bennett (1969) concluded that phosphorus retained by soils in batch systems could be redissolved by short exposure to waters of low phosphorus concentration. Hsu (1964) showed that significant amounts of the phosphorus retained on the sand columns used to treat secondary effluent could be removed by elutriation with tap water. Kurtz et al. (1946) reported that phosphorus retained on soils in batch tests could be removed by

Table 3
Time to Equilibrium in Selected Soil—Aqueous Phosphorus Batch Study Systems

Soil	*Amount Grams*	*Initial P Conc. (mg/1 as P)*	*Initial Loading (mg P/ 100g soil)*	*Time to Equilibrium (Days)*	*Source*
Adams	100	13	2.6	4[a]	Bennett[c]
Plaisted	100	13	2.6	1[b]	"
Adams	100	1527	305	9[a]	"
Paxton	100	1527	305	5[a]	"
Muscatine	2	6	15	78[a]	Kurtz

[a]Tests conducted for this number of days; data show system nearing equilibrium.
[b]Tests conducted beyond this time, but apparent equilibrium reached before this time.
[c]See References, p. 63, below.

water washing as much as three years later. Sanborn (1973), studying phosphorus distribution in septic tank absorption fields, observed that the phosphorus associated with the soil increased during dry weather and decreased during wet weather. He concluded that phosphorus was being transported and redistributed by leaching and water movement. These results all suggest that phosphorus retained on soils under one set of conditions may be removed and transported when these conditions are altered.

In summary, although soils clearly can "fix," or immobilize, large amounts of phosphorus, there are several reasons to suspect that phosphorus also can be transported through soil systems. It would appear that much additional information on this subject is needed if ground and surface waters are to be protected against phosphorus pollution by septic tank effluents.

A Conceptual Nutrient Transport Model

The available information regarding the fate of nitrogen and phosphorus in soil systems indicates that these pollutants can be transported to some significant degree from one point to another within this system. Thought should be given, therefore, to development of a model which describes the rate and extent of this transport—that is, a model which, given certain information about the system, will predict, as a function of time, the concentration of nitrogen and phosphorus at any point within the system.

In order to develop such a model, it is necessary to define the system and the forces acting on it which might affect the transport of these materials. A portion of a typical system, which involves an absorption field located near a lake, is shown in Figure 1. Of course, the total system is considerably larger than illustrated here, extending to the regional groundwater divide (Born and Stephenson, 1969). For this discussion a single increment of the absorption trench is considered a point source. The cross-hatched area represents an approximation of the expected effluent path.

It should be possible, using mass transport concepts and appropriate boundary conditions, to describe this expected effluent path to whatever precision desired. That is, mass transfer concepts, properly applied, should permit prediction of the concentration of pollutants with respect to time for any point within the system. For example, consider the rate of change of mass ($\Delta P/\Delta t$), in the aqueous phase for some pollutant within some incremental volume at some point within the system. This can be conceptually illustrated for any dimension as in Figure 2 and expressed as:

$$\frac{\Delta P}{\Delta t} = \frac{P_{in}}{\Delta t} - \frac{P_{out}}{\Delta t} - \frac{P_{on}}{\Delta t} + \frac{P_{off}}{\Delta t} \qquad (1)$$

The quality $[(P_{in} - P_{out})/\Delta t]$ represents the net amount of pollutant being transferred to or from the incremental volume during some time period and is a function of (a) water movement, (b) the concentration of pollutant in the water, (c) dispersion, and (d) diffusion. The quantity $[(P_{off} - P_{on})/\Delta t]$ represents the net change in pollutant in the aqueous phase resulting from reactions with the solid phase. These reactions and their rates are functions of the soil and water chemistry, surface properties of the soil, and geometry of the system. Thus, in a more general sense:

$$\frac{\Delta P}{\Delta t} = f\,(\text{Bulk water velocity, Diffusion, Dispersion, Reactions}) \qquad (2)$$

In order for the model to be utilized, then, data on water movement, diffusion, dispersion, and reactions must be available.

Given certain information about effluent input and knowledge of these parameters, this model could be used to predict the concentration of some pollutant, in the aqueous phase, at some point in the system, as a function of time. This would allow decisions to be made on the length of time a given absorption field could be operated without the concentration of pollutant in the aqueous phase exceeding some specified safe value at some critical point in the system. Un-

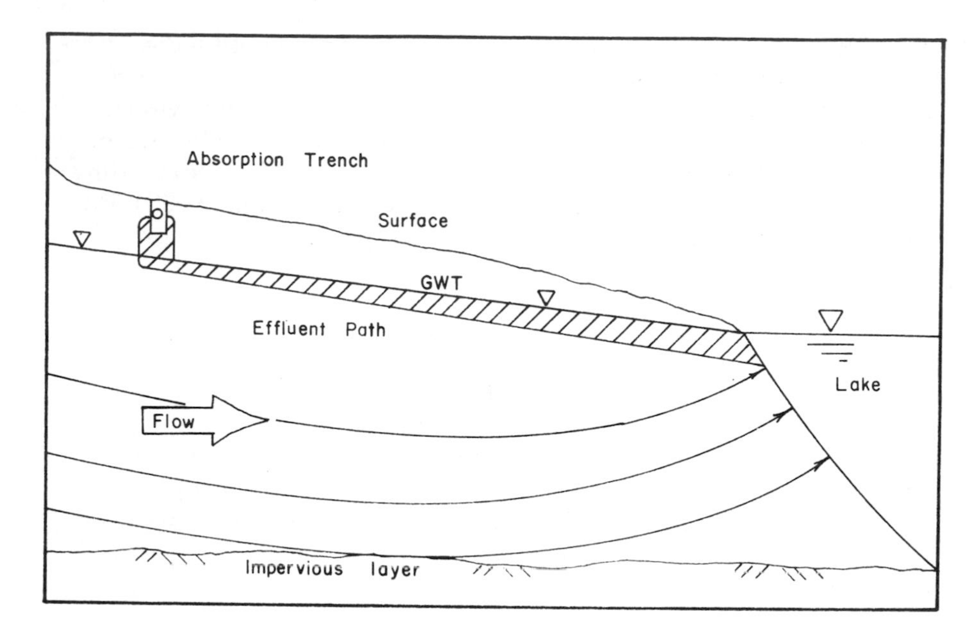

1. Illustration of soil disposal system showing prospective effluent path (not to scale).

fortunately, information on many of these model parameters, especially for field conditions, is extremely sparse at present.

Obviously, the real system is far more complex than the idealized system illustrated and discussed here, and some of these complexities deserve mention. First, in properly installed absorption fields, it is quite likely that the effluent will pass through an unsaturated flow zone prior to its entry into the saturated zone below the surface of the ground-water table (Walker et al., 1973). Water movement, diffusion, dispersion, and even reaction characteristics may well be different for these two flow regimes. Thus in considering the fate of nutrients in the overall system it may be necessary to deal with these two zones separately. Second, fluctuations in effluent and ground-water discharge, as well as the lack of homogeneity within the subsurface system, must be considered and their effects on the model discerned. Third, in most cases, it is doubtful if an absorption field should be considered a point source, as was done earlier for simplification. However, it may be possible to approximate the real system by breaking it down into a series of hypothetical point sources and then summing their effects on the point in question. Finally, success-

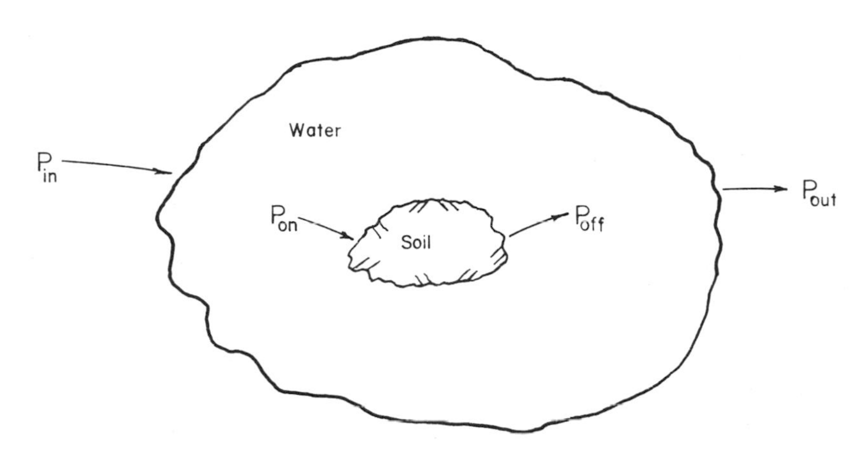

2. Idealized incremental system volume; soil particle surrounded by water.

ful use of this model will depend to a large degree on careful definition of the real system's boundary conditions.

It will likely be some time before researchers can supply enough details about the parameters required by this model to permit its rigorous use. However, even though there is relatively little hard data on these model parameters, examination of the literature, with this model in mind, may reveal enough about these parameters to allow good engineering assumptions about them to be made. This might permit the model to be applied to some limited degree. At any rate, an understanding of the model's principles should be useful in guiding research in this area of inquiry.

References

Bennett, B. D. 1969. Phosphorus retention by soils. M.S. thesis. University of Maine at Orono.

Biggar, J. W., and Corey, R. B. 1969. Agricultural drainage and eutrophication. In *Eutrophication: causes, consequences, correctives,* pp. 475–480. Washington, D.C.: National Academy of Sciences.

Born, S. M., and Stephenson, D. A. 1969. Hydrogeologic considerations in liquid waste disposal. *Journal of soil and water conservation* 24: 52–55.

Dudley, J. G., and Stephenson, D. A. 1973. *Nutrient enrichment of ground water from septic tank disposal systems.* Inland Lake Renewal and Shoreland Management Demonstration Project report. Madison: University of Wisconsin.

Ellis, B. G., and Erickson, A. E. 1969. *Movement and transformation of various phosphorus compounds in soils.* East Lansing: Department of

Soil Science, Michigan State University.

Foster, J. A. Unpublished data. University of Maine at Orono, Department of Civil Engineering, Orono, Maine. 1973.

Hasler, A. D., and Ingersoll, B. 1968. Dwindling lakes. *Natural history* 77: 8-19.

Hill, D. E. 1972. Waste water renovation in Connecticut soils. *Journal of environmental quality* 1: 163-167.

Hinesly, T. D. 1973. Water renovation for unrestricted reuse. *Water spectrum* 5: 1-8.

Holt, R. F., Timmons, D. R., Latterell, J. J. 1970. Accumulation of phosphates in water. *Journal of agricultural and food chemistry* 18: 781-784.

Hsu, C. 1964. Removal of phosphates in secondary sewage treatment effluent by sand filtration. M.S. thesis. University of Illinois.

Kurtz, L. T. 1970. The fate of applied nutrients in soils. *Journal of agricultural and food chemistry* 18: 773-780.

Kurtz, L. T., et al. 1946. Phosphate adsorption by Illinois soils. *Soil science* 61: 111-124.

Lake George Water Research Center, Rensselaer Polytechnic Institute. 1971. Physical-chemical treatment of septic tank effluents. *Newsletter* 1: 2.

Lance, J. C. 1972. Nitrogen removal by soil mechanisms. *Journal of Water Pollution Control Federation* 44: 1352-61.

Patterson, J. W., et al. 1971. *Septic tanks and the environment*, IIEQ 71 2. Chicago: Illinois Institute for Environmental Quality.

Popkin, R. A. and Bendixen, T. W. 1968. Improved subsurface disposal. *Journal of Water Pollution Control Federation* 40: 1499-1514.

Preul, H. C., and Schroepfer, G. J. 1968. Travel of nitrogen in soils. *Journal of Water Pollution Control Federation* 40: 30-48.

Robert A. Taft Sanitary Engineering Center. 1966. *Fertilization and algae in Lake Sebasticook, Maine.* Cincinnati: Technical Services Program, Federal Water Pollution Control Administration.

Sanborn, D. J. 1973. Retention of phosphorus by soils in septic absorption fields. M.S. thesis. University of Maine at Orono.

Urresta, L. F. 1970. Phosphorus retention by plaisted soil of Maine. M.S. thesis. University of Maine at Orono.

Walker, W. G., et al. 1973. Nitrogen transformations during subsurface disposal of septic tank effluent in sands. I. Soil transformations. *Journal of environmental quality* 2: 475-480.

6. Limits to Growth and Septic Tanks

Thomas L. Holzer

Introduction

"Low density area" is a potentially misleading demographic phrase from the standpoint of groundwater pollution by septic tanks. Population densities conventionally are computed for areas defined by political boundaries. The flow of effluent from septic tank systems, once the effluent reaches the water table, is determined by hydrologic boundaries and the hydraulic properties of the material through which the groundwater carrying the effluent is moving. These determinations suggest that the areal unit for computing population densities should be a hydrologic one. Based on hydrologic analysis of the groundwater system, an accurate assessment of the maximum population of a region in which residences must rely on septic tank systems for waste disposal and on groundwater for water supply may be feasible. In the following analysis, eastern Connecticut is used to illustrate the approach. Many parts of New England, as well as other areas of the world underlain by fractured crystalline bedrock at shallow depth, have groundwater systems which are similar in basic outline to those of Connecticut, so the conceptual approach should have widespread applicability.

Appraisal of the potential pollution of groundwater by septic tank systems requires an understanding of the groundwater system into which the effluent is discharged. First, flow paths of groundwater and recharge areas must be delineated. Second, the quantity of groundwater recharge must be estimated because the recharge is a measure of the amount of water available for dilution of effluent. And third, the capability of the natural system to renovate effluent from septic tank systems must be known.

Hydrogeology and Residential Development in Eastern Connecticut

The geology of eastern Connecticut consists of dense crystalline bedrock overlain by a blanket of poorly sorted till ranging from clay to boulders and averaging ten to fifteen feet thick (Figure 1). Thicker deposits of well sorted and more permeable stratified glacial drift commonly overlie the till and bedrock in major valleys. Hills between major valleys have a topographic relief of a few hundred feet relative to the valleys and are usually covered by till. Approximately 80 percent of eastern Connecticut is covered by till.

Groundwater in bedrock flows through fractures or joints. Although the average permeability of bedrock is less than 0.53 ft/day, the porosity of bedrock is less than one percent, so that seepage velocities in fractures can be large. Hence, transport of pathogens in fractured bedrock can be rapid (Allen and Morrison, 1973), and removal of pathogens by the overburden is essential.

Residential development in most of eastern Connecticut depends upon groundwater for water supply and on-site sewage disposal. Domestic water supplies are usually obtained from wells drilled into bedrock. At present, very few dug wells are constructed by contractors. On-site waste disposal is by septic tank systems buried in the surficial deposits.

Groundwater System

Analysis of yields from water wells drilled into bedrock reveals that only insignificant amounts of groundwater can be obtained from depths exceeding a few hundred feet. The quality of this data for eastern Connecticut is poor, and statistical trends can be better demonstrated with data reported by Ryder et al. (1970) for southwestern Connecticut (Figure 2), which has hydrogeological conditions similar to those in eastern Connecticut. Statistical analysis indicates that the water-bearing portion of the bedrock does not extend below a depth of approximately 300 feet, i.e., water-bearing fractures are infrequent below this depth. Because the topographic relief of eastern Connecticut is comparable to the thickness of the water-bearing portion of the bedrock, drainage basins and groundwater basins are identical, and therefore widespread disposal of wastes into the surficial deposits poses the threat of contaminating the deepest aquifer available to residents, the bedrock aquifer. Bedrock beneath areas covered by till is particularly susceptible to this threat, because these areas are predominantly areas in which the bedrock is recharged (Figure 3). This analysis also reveals that the surface water drainage basin to a first approximation may be used as the natural areal unit on which to base residential densities, because this basin coincides approximately with the groundwater basin.

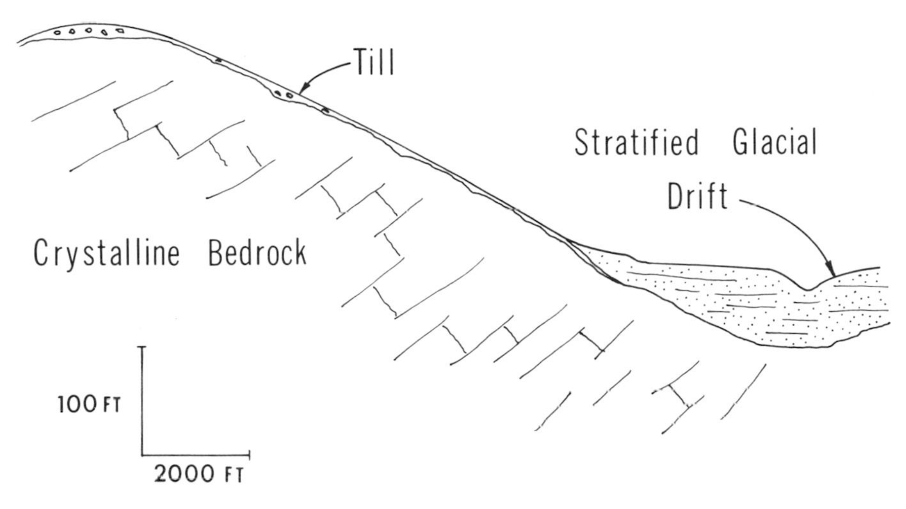

1. Schematic geologic cross section through a hill and major valley in eastern Connecticut.

Groundwater Recharge

The quantity of recharge to groundwater must be estimated in order to evaluate the capability of the groundwater system to dilute effluent from septic tank systems. Two techniques applicable to eastern Connecticut for estimating recharge are: (1) modeling of the ground water system, and (2) identification of the groundwater runoff component of gaged streamflow. The latter technique was used in the present investigation, although the former is now under study. Under suitable conditions, groundwater runoff for a drainage basin can be estimated by correlating streamflow during periods of no precipitation with ground water levels (Walton, 1970). This correlation establishes a groundwater rating curve. The groundwater rating curve makes it possible to separate the groundwater runoff from total streamflow during floods if groundwater levels are monitored in wells sufficiently offset from streams so as not to be influenced by bank storage effects.

A correlation between groundwater runoff and surficial geology is found for drainage basins in eastern Connecticut (Figure 4). Basins completely covered by till have lower amounts of groundwater runoff than do basins partially covered by stratified glacial drift. Groundwater runoff from basins covered by till is equivalent to approximately seven inches of precipitation. The amount of groundwater runoff

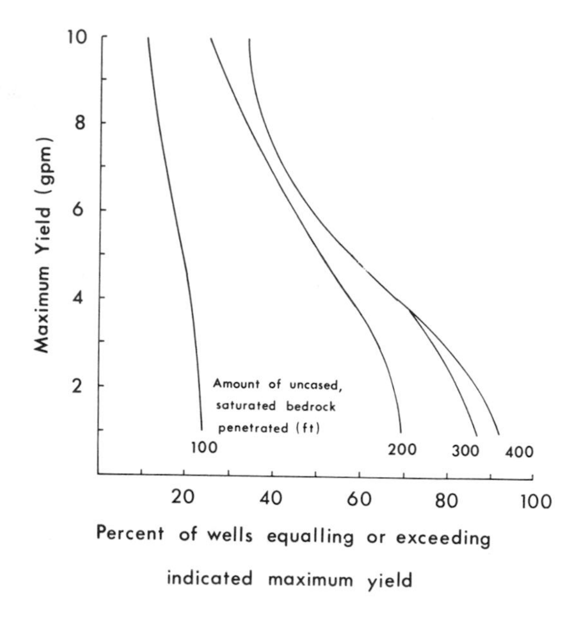

2. Cumulative frequency distributions of maximum yields of wells penetrating different amounts of uncased, saturated crystalline bedrock in southwestern Connecticut (after Ryder et al., 1970).

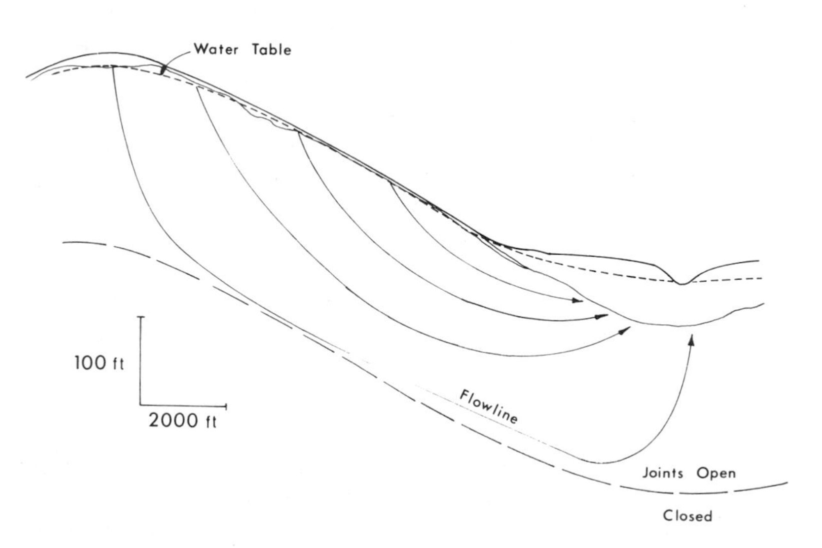

3. Flow paths of groundwater in crystalline bedrock. Geologic units are identified in Figure 1.

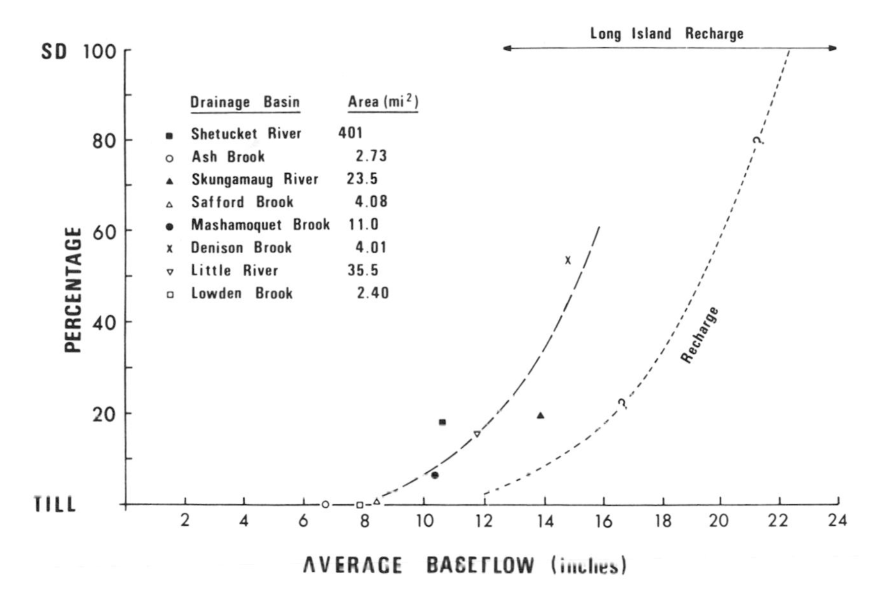

4. Baseflow, groundwater runoff, of streams in eastern Connecticut versus the percentage of area in drainage basin covered by stratified glacial drift, SD (after Thomas et al., 1967).

from areas completely covered by stratified glacial drift cannot be determined in eastern Connecticut, because of the absence of such basins, but investigations of basins underlain by stratified glacial drift in Rhode Island (Allen et al., 1966) and Long Island, New York (Fetter, 1972; Pluhowski and Kantrowitz, 1964) indicate that groundwater runoff may exceed the equivalent of 20 inches of precipitation, although the results of some of these investigations have been questioned.

Groundwater runoff is a minimum estimate of recharge for drainage basins in eastern Connecticut. Groundwater can also discharge directly to the atmosphere by evapotranspiration when the groundwater table is close to the land surface. Hence, the sum of evapotranspiration from groundwater and of groundwater runoff is a more meaningful estimate of groundwater recharge than runoff alone. Estimates of the magnitude of evapotranspiration can be made by comparison of groundwater rating curves for winter and summer months. Average annual evapotranspiration from groundwater in the Shetucket River Basin has been estimated as 4.23 inches (Thomas et al., 1967). The Shetucket River Basin is approximately 18 percent covered by stratified glacial drift. Comparison of monthly groundwater rating curves for Safford Brook Basin, which is almost entirely covered by till, revealed no difference between monthly rating curves and therefore negligible evapotranspiration from groundwater during the period of investigation.

For the purpose of this presentation, groundwater recharge will be assumed to be equal to groundwater runoff. Although groundwater recharge probably is underestimated, the amount of error is most likely less than the equivalent of a few inches of precipitation.

Comparison of Domestic Water Usage with Recharge

Estimating domestic water usage is complicated by social and economic factors. A modern residence with a family of four and modern appliances can easily use 250 gallons of water per day. If this usage is converted to an annual amount and compared with the estimated average annual quantity of recharge infiltrating an acre of land (Table 1), the fate of a till-covered basin completely developed at an average density of one residence per acre is apparent. In till-covered basins, effectively one half of the average annual recharge is used by residences and one half remains unused and available to dilute the water used by the residences. Another way of visualizing this is to consider development on one-half acre lots. In this situation virtually all recharge would be used and no dilution of effluent would occur.

How much Dilution Is Required?

In order to estimate the minimum average lot size sufficient to maintain groundwater quality adequate for human usage, knowledge of the effluent from septic tank systems after it has passed through soil is required. Renovation by the soil of effluent from septic tank systems is complex because both biological and chemical processes, strongly dependent on both soil and hydrologic conditions, are involved. Consequently, the end products entering the groundwater after renovation of the effluent can be expected to have considerable areal variation because of the range of soil and hydrologic conditions usually encountered within a drainage basin. If the site or host conditions of a septic tank are suitably controlled, end products from the chemical processes occurring in the septic tank and soil should be the significant contaminants. The chemicals potentially

Table 1
Comparison of Domestic Water Usage with Ground Water Recharge

Domestic Water Usage:
250 gallons per day or
91,000 gallons per year
Equivalent to 3.36 inches of precipitation falling on one acre.

Ground Water Recharge:

Surficial Geology of Basin	*Equivalent Amount of Precipitation*
Till	7 inches
Stratified Glacial Drift	12 to 23 inches

most hazardous to public health (United States Public Health Service publication no. 956, 1962) can be identified from analyses of domestic sewage (Table 2). When conditions in the soil become sufficiently oxidizing, nitrogen compounds leaving the leaching field may be converted to nitrate and pose a most serious health problem, since most soils do not remove nitrates (Hill, 1972; Thomas, 1972). Hazardous concentrations of nitrates in groundwater entirely attributable to pollution from septic tank systems have been reported but are surprisingly rare (Bouma et al., 1972; Miller, 1972 and this text). Although high concentrations of nitrates in groundwater have been frequently reported, the relative proportions of nitrates from their different sources are usually not fully evaluated (Crabtree, 1972; Kimmel, 1972). Based on the concentrations of nitrate measured by Bouma et al. (1972) and Miller (1972), a dilution of renovated effluent of at least one to one may be required to reduce nitrate concentrations to the public health standard of 45 mg/1 NO_3 in permeable soils with deep water tables. This indicates that residential development in eastern Connecticut should not occur at densities greater than an average of one residence per acre on well-drained sites in areas covered by till.

Site Suitability and Septic Tank Systems

The preceding analysis assumed that septic tank systems are installed at sites which cause chemical contamination of groundwater but not pathogenic contamination or premature surfacing of effluent. If soil, hydrologic, and topographic conditions within an area are unsuitable for development relying on septic tank systems, the limiting population may be less than the population predicted by analysis of the

Table 2
Normal Range of Mineral Pickup in Domestic Sewage (after Feth, 1966)

Mineral	*Mineral Range (ppm)*
Dissolved solids	100–300
Boron (B)	0.1–0.4
Sodium (Na)	40–70
Potassium (K)	7–15
Magnesium (Mg)	3–6
Calcium (Ca)	6–16
Total nitrogen (N)	20–40
Phosphate (PO_4)	20–40
Sulfate (SO_4)	15–30
Chloride (Cl)	20–50
Alkalinity (as $CaCO_3$)	100–150

groundwater system. Evaluation of these three conditions in eastern Connecticut suggests this possibility. Figure 5 is a map of the town of Mansfield, Connecticut, in which areas with limiting host conditions for septic tank systems are delineated. Such areas were determined from the soil survey of the town (Ilgen et al., 1966) and are the areas with either seasonal high-water tables within two feet of the surface, crystalline bedrock within two feet of the surface, or slopes exceeding 15 percent. These conditions are not necessarily the most restrictive but are the ones which can be delineated with the soil survey, the most comprehensive data available. One could argue that sites which are unsuitable based on these criteria could be engineered to overcome their shortcomings, but this alternative probably is unfeasible because approximately 37 percent (or 17 square miles) of the town has limiting host conditions. Thus, special conditions of soil, hydrology, and topography may restrict development far more stringently than standards for maintaining groundwater quality alone.

Conclusions

In regions where groundwater circulates in a relatively thin zone near the surface, an estimation of the quantity of annual recharge to groundwater makes it feasible to compute the maximum population which the region can support if waste disposal is via septic tank systems and water supplies are extracted on site. Since final treatment by the groundwater system primarily is dilution with uncontaminated

5. Areas of Mansfield, Connecticut, indicated in black, which are unsuitable for septic tank systems under natural conditions. See the text for limiting conditions.

recharge, the ability of soils to renovate effluent must be known. If septic tank systems are situated in areas with well-drained soils and relatively deep water tables, nitrate is the final product of primary concern after renovation by the soil. In eastern Connecticut, the dilution required to reduce concentrations of nitrate to 45 mg/1 indicates that average residential densities within groundwater basins covered by till should not be greater than one residence per acre.

Such an estimate of maximum density is based on the requirement that septic tank systems cause only chemical contamination of groundwater by effluent. In some regions, soil, hydrologic, and topographic conditions unfavorable to septic tank systems may place more stringent limitations on development than the limitations imposed by safeguarding groundwater quality. These conditions may prevail in much of eastern Connecticut.

Regardless of which analysis places the greater limitation on development, regional planning of the distribution of residential development is necessary in order to minimize the public health problems caused by community dependence on septic tank systems for waste disposal.

References

Allen, M. J., and Morrison, S. M. 1973. Bacterial movement through fractured bedrock. *Ground water* 11: 6–9.

Allen, W. B., Hahn, G. W., Brackley, R. A. 1966. *Availability of ground water, upper Pawcatuck River basin, Rhode Island*. United States Geological Survey, Water supply paper no. 1821.

Bouma, J., Ziebold, W. A., Walker, W. G., Olcott, P. G., McCoy, E., Hole, F. D. 1972. *Soil absorption of septic tank effluent.* Wisconsin Geological and Natural History Survey, Information circular no. 20.

Crabtree, K. T. 1972. *Nitrate and nitrite variation in ground water.* Wisconsin Department of Natural Resources, Technical bulletin no. 58.

Feth, J. H. 1966. Nitrogen compounds in natural water—a review. *Water resources research* 2: 41–58.

Fetter, C. W., Jr. 1972. Position of the saline water interface beneath oceanic islands. *Water resources research* 8: 1307–15.

Hill, D. E. 1972. Waste water renovation in Connecticut soils. *Journal of environmental quality* 1: 163–167.

Ilgen, L. W., Benton, A. W., Stevens, K. C., Shearin, A. E., Hill, D. E. 1966. *Soil survey of Tolland County, Connecticut.* United States Department of Agriculture, Soil Conservation Service, Series 1961 no. 35.

Kimmel, G. E. 1972. *Nitrogen content of ground water in Kings County, Long Island, New York*. United States Geological Survey, Professional paper no. 800–D.

Miller, J. C. 1972. *Nitrate contamination of the water-table aquifer in Delaware*. Delaware Geological Survey, Report of investigations no. 20.

Pluhowski, E. J., and Kantrowitz, I. H. 1964. *Hydrology of the Babylon-Islip area, Suffolk County, Long Island, New York*. United States Geological Survey, Water supply paper no. 1768.

Ryder, R. B., Cervione, M. A., Thomas, C.E., Thomas, M.P. 1970. *Water resources inventory of Connecticut, Part 4, southwestern coastal river basins*. Connecticut water resources bulletin no. 17.

Thomas, G. W. 1972. *The relations between soil characteristics, water movement and nitrate contamination of ground water.* Lexington: University of Kentucky Water Resources Institute, Research report no. 52.

Thomas, M. P., Bednar, G. A., Thomas, C. E., Wilson, W. E. 1967. *Water resources inventory of Connecticut, Part 2, Shetucket River basin.* Connecticut water resources bulletin no. 11.

United States Public Health Service. 1962. *Public health services drinking water standards*. United States Department of Health, Education, and Welfare, Publication no. 956.

Walton, W. C. 1970. *Groundwater resource evaluation*. New York: McGraw-Hill Book Company.

7. An Analysis of Septic Tank Survival Data from 1952 to 1972 in Fairfax County Virginia

John W. Clayton

History of Septic Tanks

Septic tank systems have been used for many years in Fairfax County, Virginia, and appeared first in the United States in the late 1800's. A septic tank system essentially is made up of two basic parts, a watertight enclosed vault for retention of the solids and a means for discharge of the remaining water-carried wastes to the subsoil.

Originally the septic tank system was used only in country or rural areas where the sanitary sewer system was unavailable. Today, it has expanded to the point where large subdivisions with hundreds of lots side by side are utilizing this method of sewage disposal. Unfortunately, many failures have occurred in these systems, with sewage coming to the surface of the ground, flowing into streams, or, in many instances, backing up into the plumbing fixtures, thus creating health hazards to all who may be exposed to these conditions.

Septic Tanks in Fairfax County

In Fairfax County, Virginia, officials have records of septic tank systems installed in the 1930's and are aware that some were installed even earlier.

In 1949, when the author first joined the Fairfax County Health Department, septic tank systems were installed with a minimum of supervision and concern for the types of soil conditions. Soils testing was not yet established as a prerequisite to designing a septic tank system. At that time "percolation tests" were run only when an FHA loan was involved. This test is a method for determining the absorption area needed to dispose of sewage from a given structure. It is still essential for proper design of a subsurface disposal system, but is not the only criteria to be considered. Of more concern is the groundwater table. Groundwater tables may be present in many soils during the wet season of the year and not be normally recognizable in the dry season. This became clear to the Fairfax County Health Department in the early 1950's when a water table rose into the drainfield area and caused many failures in septic tank systems. In one area of the county there were hundreds of malfunctioning systems with no means for correction. The problem necessitated a $20 million dollar bond issue for sanitary sewers in 1953, the first real expansion of the county sewer construction and extension program.

At this time the department decided that it simply must find a way to identify these questionable soils. A "Seminar on Household Sewage Disposal Systems", held September 23 through September 25, 1953, in Cincinnati, Ohio, by the U.S. Public Health Service, brought out the fact that there are basic characteristic differences in soils, which encouraged the department to support the proposed soils survey of the county by soil scientists from Virginia Polytechnic Institute, Blacksburg, Virginia. During 1954 and 1955 while the soils survey was being conducted and mapped, environmental health officials worked closely with the soil scientists to make correlation studies comparing percolation rates with certain soils, which showed that water table soils could be identified by type and eliminated from future consideration for individual sewage disposal systems with subsurface absorption. This was a great step forward.

In the initial survey, 100 square miles of the county were left out because of the existence of sewers or the proposed extension of sewers for this area in the immediate future. Over 300 square miles of the county were mapped. The soil scientists bored into the soil with earth augers, reading slopes and digging pits occasionally to study soil profiles in detail. The soil maps were actually made in the field (scale: 4 inches per mile) by sketching directly on aerial photographs the location of the different soils and other features of the landscape. Rock outcroppings, streams, roads, quarries, and ponds were indicated. Each acre was studied thoroughly.

The advantages of the soil survey in septic tank development are:

1. The developers have a means for determining soil uses before they buy land.
2. In problem lots it makes possible a quick decision as to what action should be taken.
3. It provides a method to determine readily whether a proposed subdivision can be expected to pass the soils evaluation tests and approximately what percentage of lots can be expected to be utilized satisfactorily with septic tank systems.

Mr. T.W. Bendixen, a soil scientist formerly with the Robert A. Taft Environmental Engineering Center, Cincinnati, Ohio, stated that Fairfax County was the first in the nation to make such use of the soils survey information and to have correlated percolation tests with the soil maps in such a way as to make them useful in determining the proper soils for septic tank absorption systems.

It also became apparent that the county's septic tank ordinance at that time (a 1951 version) was quite inadequate and a committee was appointed by the Board of Supervisors to prepare a more modern ordinance incorporating the newer concepts in the design and installation of septic tank systems. This was done, and the Board adopted the present ordinance on June 6, 1956. It has been amended several times since (June 1, 1960; November 4, 1964; June 29, 1966; February 7, 1972; May 14, 1973; and May 21, 1973). In addition, the State Board of Health adopted Regulations Governing the Disposal of Sewage effective July 1, 1971. A key paragraph in these regulations is as follows:

> *Soil Evaluation*—Soil evaluation for a drainfield shall follow a systematic approach including consideration of physiographic province, position of landscape, degree of slope and soil profile (thickness of horizon,color, texture). Such evaluation shall indicate whether or not the soil has problems relative to the position in the landscape, seasonal water table, shallow depths, rate of absorption, or a combination of any of the above. If absorption rate problems are suspected and there is no indication of a water table, percolation tests should be made but their results shall not be presumptive, prima facia, or conclusive evidence as to the suitability for effluent absorption. Such percolation tests may be considered and analyzed as one of many criteria in determining soil suitability for absorption of effluent.

Some of the major factors included in Fairfax County's Sanitary Inspection Ordinance are:

1. The use of percolation tables for sizing drainfields and seepage pit systems.
2. Specific construction requirements for tanks, drainfields, and pit systems.
3. Bonding and licensing contractors before permitting them to engage in septic tank system construction.
4. Prohibiting construction of sewage disposal systems in impervious soils, low swampy areas, areas with high water tables, or areas subject to flooding.
5. Requiring a thorough inspection of all septic tank systems before they are covered and after final cover and grading.

Another important feature of Fairfax County's program is a full-time county soil scientist who serves all county departments, including the Department of Assessments. He is used whenever the departments must deal with questionable soils (borderline-transitional) or areas that were never mapped.

By 1971–1973, building in Fairfax County had reached a peak and the sanitary sewers and sewage treatment facilities were crowded to maximum use. The political leaders of the county were considering ways to control growth and limit construction on the sanitary sewer systems as well as on individual sewage disposal systems. In response to requests from the Fairfax County Board of Supervisors about data on septic tank failure rates in the county and recommendations for improving methods of design, installation, and maintenance of these systems, the Health Department researched its files and analyzed records dating back to 1952. This was a tremendous task and one that had never been requested before. It was time consuming and the prospect of gaining new information was uncertain. But a surprising fact emerged: septic tank systems, properly designed and installed under fairly good controlled conditions, were surviving not an average life of 15 to 20 years as the department expected on the basis of literature and field experience, but more like 20 to 30 years. It is true that of those which did fail, the average life span was about 10 to 15 years, a fact leading officials to underestimate the life span of all septic tank systems.

The department made a chart of survival data (Table 1) which shows the number of systems installed each year from 1952 through 1971 and the percentage of systems surviving (continuing to function in a satisfactory manner) each year for twenty years or to date. It can be noted that of the 230 systems installed in 1952, 94 percent were still functioning at the end of 20 years, 92 percent of the 455 systems installed in 1955 were surviving after 17 years, 96 percent of the 350 systems installed in 1959 were surviving after 13 years, and 100 percent of those installed in 1968 were surviving after 4 years. Table 2 shows septic tank system failures with the number analyzed, the number and percent failure by year.

The two tables represent data obtained from files of all systems installed in Fairfax County subdivisions within the last twenty years. Since then officials have analyzed the data from all systems installed in areas other than subdivisions during the last twenty years, and the data are very comparable—so much so that separate charts were not made.

Table 3 is based on data obtained from all homeowners who made connections to the sanitary sewer during fiscal year 1972. The center column shows the life span in years. Of the 75 systems reported by these homeowners, 21 lasted at least 15 years, 20 lasted 20 years, and one survived 32 years.

Table 4 indicates the number of systems reported still functioning in 1972, when they were discontinued and sewer connections made. It should be pointed out that many of these were made because the county code at that time required connection to the sewer within five years of the availability of such sewer. This has since been changed and connection is no longer compulsory unless the system has begun to malfunction. Table 4 indicates that 9 systems were still in use after 15 years, 9 after 20 years, and one still functioning after 39 years.

All these data have changed officials' thinking tremendously. They believe their success in high survival rates has been due to careful planning and careful evaluation of the soils where the absorption system is to be installed. They have not relied on the percolation test per se as a factor determining whether or not a permit should be issued for a septic tank and absorption system. Rather, they first determine soil conduciveness to absorption and infiltration and the potential for a high water table in the wet season of the year. After that, a percolation test is run on every building site at the precise location where the system is to be installed. The percolation rate is used only as a guide or design criteria for the size of the absorption system. Although the percolation test is not perfect, it is the best method now available.

On the basis of data obtained from Mitre Corporation, FHA, and particularly McGauhey and Winneberger's *Study of methods of preventing failure of septic tank percolation systems* (U.S. Dept. of Housing and Urban Development, 1967), the Health Department has proposed to the Board of County Supervisors that sanitary ordinances be amended to:

(1) Permit installation of drainfield laterals on six-foot centers rather than eight-foot centers to conserve valuable absorption area.

(2) Require installation of all future absorption systems, whether they be drainfield laterals or seepage pits, in two separate sections with separate distribution boxes and a flow diversion valve or some satisfactory method for utilizing half of the total system for a period of time, then switching to the other for a similar period. This procedure will permit the first one to rest and provide a high rate of aerobic decomposition, biochemical degradation of clogging materials to liquids and gases, and drying and cracking of the sidewall mat and soil, which will improve the soil texture and infiltrative capacity.

(3) Require an area to be reserved for possible future expansion

Table 1
Septic Tank Systems Survival Data[a]

No. Systems	*230*	*276*	*250*	*445*	*390*	*341*	*281*	*358*	*327*
Yr. Installed	*1952*	*1953*	*1954*	*1955*	*1956*	*1957*	*1958*	*1959*	*1960*
Years Survival (percentages)									
1	99	99	99	99	99	99	99	99	99
2	98	99	98	98	99	98	99	98	99
3	98	99	98	98	98	98	99	98	98
4	97	99	98	98	96	98	98	98	98
5	97	99	98	96	96	97	98	98	97
6	97	99	97	96	95	97	98	98	97
7	97	99	97	95	95	97	98	97	96
8	97	99	97	95	95	97	98	97	96
9	97	98	97	94	94	97	98	97	96
10	95	98	97	94	94	96	98	97	95
11	95	97	97	94	94	95	98	97	95
12	95	97	95	94	93	95	98	97	95
13	95	96	94	94	93	95	98	96	
14	94	96	94	93	93	94	98		
15	94	96	94	93	93	94			
16	94	96	94	92	93				
17	94	96	93	92					
18	94	96	93						
19	94	96							
20	94								

*Data compiled from records of systems installed in subdivision developments.

of the absorption system equal to fifty percent of the area required for the initial system.

On May 14, 1973, the Board of Supervisors adopted these amendments. With them and with continued control and care over the installation of all new systems along with mandatory maintenance of all systems, including regular cleanings and alternation of diversion valves, systems should function for 30 to 50 years or even indefinitely. Excellent performance of septic systems is critical because high costs and higher water quality standards for streams make sewers infeasible for many areas.

In summarizing the work of the Fairfax County Health Department, the following points should be recapitulated:

(1) Officials were surprised to find in 1972 that the average life of an individual sewage disposal system with an absorption field or

355	*297*	*254*	*329*	*333*	*325*	*189*	*204*	*235*	*211*	*218*
1961	*1962*	*1963*	*1964*	*1965*	*1966*	*1967*	*1968*	*1969*	*1970*	*1971*
100	99	100	99	100	100	100	100	100	99	100
99	99	100	99	100	100	100	100	100	99	
99	99	100	99	99	100	100	100	100		
98	98	100	99	99	99	100	100			
98	98	100	99	99	99	100				
97	98	100	99	98	99					
97	98	99	99	98						
97	98	99	98							
96	97	99								
96	97									
95										

seepage pits was then averaging about 20 to 30 years instead of the 15 to 20 years they had been quoting.

(2) The analysis of almost 6,000 systems on an annual basis yielded survival rates of at least 94 percent (1952 and 1953), with at least 92 to 96 percent survival from 1954 through 1961 and 97 to 100 percent survival since 1962.

(3) Of 75 systems whose failures were resolved by connection to public sewers (rather than by repair), the median life span was about 18 years, and 88 percent had life spans of at least 10 years.

(4) These survival data are based on permits for installation and repair for systems in subdivisions. It is possible that some systems may have been malfunctioning for years without having been repaired, and some may have been repaired without the department's knowledge (to avoid the necessity of connecting to public sewers). Although the true survival of these systems may be more or less than indicated

Table 2
Septic Tank System Failures[a]

Year of Installation	*No. Systems Analyzed*	*No. Reported Failures*	*Percentage of Failures*	*Permits Issued for Original Systems*
1952	230	14	6.1	N/A
1953	276	10	3.6	N/A
1954	250	18	7.2	N/A
1955	445	36	8.1	N/A
1956	390	26	6.6	N/A
1957	341	19	5.6	N/A
1958	281	6	2.1	N/A
1959	358	15	4.2	N/A
1960	327	16	4.9	N/A
1961	355	18	5.1	N/A
1962	297	8	2.7	N/A
1963	254	2	0.8	452
1964	329	2	0.6	424
1965	333	3	0.9	530
1966	325	0	0.0	343
1967	189	0	0.0	302
1968	204	0	0.0	359
1969	235	0	0.0	335
1970	211	0	0.0	355
1971	218	0	0.0	N/A
1972[b]	107	0	0.0	N/A

[a]These data represent systems installed in subdivisions only.
[b]6 months data

by the records, the department feels that the apparent survival rates are surprisingly high in light of news media reports of mass failures in many areas of the country, and the experience would appear to support its program of conservative standards and strict inspection.

Finally, new amendments to the code permit what the department believes to be an improvement in the design of individual sewage absorption systems which can only improve the survival rate of these systems in the future.

Several figures illustrate features of the Fairfax County program. Figure 1 shows how a typical septic tank system is designed and constructed. A typical building site with topography lines, the proposed building (dwelling), and the location of actual percolation test holes is shown in Figure 2, and the same building site showing the design of septic tank system with tank, flow diversion valve, and split drainfield is given in Figure 3. Figure 4 shows an effluent diversion valve manu-

Table 3
Survival Data of Systems That Failed Prior to Dwellings Being Connected to Public Sewer[a]

No. of Systems	*Life Span Years*	*Percentage of System Survival*
2	2	97
1	6	96
6	10	88
2	11	85
2	12	82
2	13	80
7	14	70
8	15	62
3	16	58
1	17	57
4	18	51
5	19	44
20	20	17
4	21	11
1	22	10
2	25	7
1	27	6
2	28	3
1	30	1
1	32	0

[a]Based on data obtained from homeowners who made sewer connection during FY-72.

factured by Franklin Research in Oakland, California and used in Fairfax County.

Table 4
Septic Tank Systems—Years of Service Prior to Connecting to Public Sewer (No Reports of Failures)[a]

No. Systems	*Years Used*	*Percentage of Years Used*
1	5	99
3	7	95
1	9	94
2	10	91
1	11	90
4	12	85
5	13	79
1	14	78
9	15	67
4	16	62
3	17	58
7	18	50
2	19	48
9	20	37
5	21	30
6	22	23
4	23	18
5	25	12
2	26	10
1	28	9
1	29	7
2	30	5
1	31	4
1	32	2
1	37	1
1	39	0

[a]Based on data from homeowners who connected to sewer during FY-72.

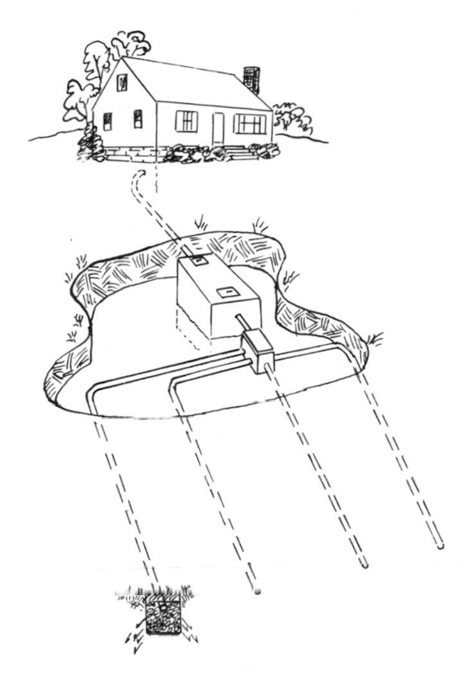

1. A conventional septic tank system.

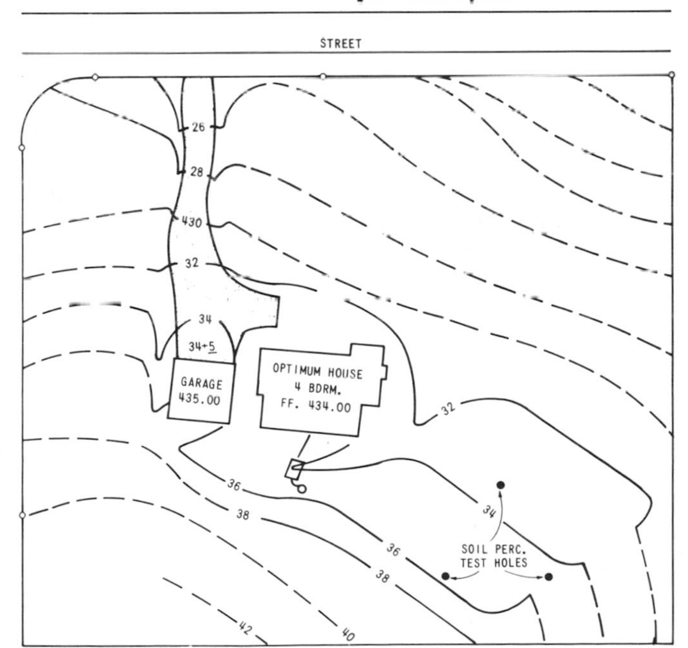

2. Typical building site with septic tank, flow diversion valve, and split drainfield.

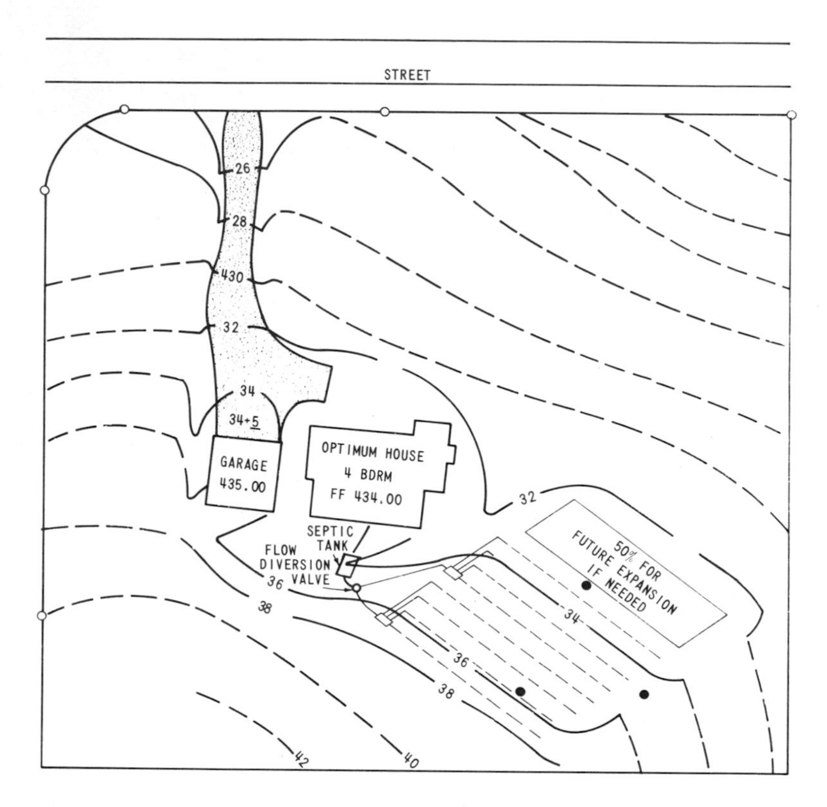

3. The effluent diversion valve switches effluent flow from one septic tank leaching field to another separate field so as to permit alternate periods of loading and resting. When this switching is done annually, natural aeration during the rest periods restores the system's leaching capacity. Courtesy of Franklin Research Septic System Controls, 4009 Linden Street, Oakland, California 94608.

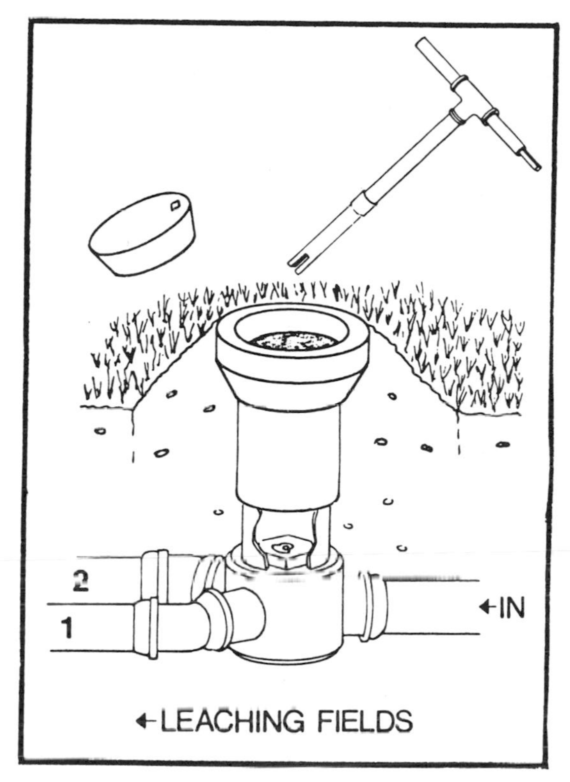

4. Typical building site with 2-foot interval topo lines and test hole locations.

8.
Use of Soil for Disposal and Treatment of Septic Tank Effluent

Johannes Bouma

Introduction

According to state law, the soil must be used for disposal of septic tank effluent in Wisconsin. Many soils, however, do not accept or purify liquid effluent well. Among these are slowly permeable soil (defined as having a percolation rate slower than 60 min/inch), thin permeable soil over very permeable creviced bedrock, and soil with periodic or permanent high ground water. In addition, slope requirements have been defined, and construction in flood plains is not allowed (United States Public Health Service, 1967; Wisconsin State Board of Health, 1969). The general requirement for on-site waste disposal is the availability of at least three feet of sufficiently permeable unsaturated soil below the bottom of the seepage bed. An estimated 50 percent of the soils of Wisconsin do not meet this requirement and are thus unsuitable for on-site disposal.

Figure 1 shows a small-scale soil map of Wisconsin with some of the problem areas delineated. Large-scale soil maps, as produced by the Soil Conservation Service of the United States Department of Agriculture in a nationwide program, are very useful for planning purposes because the suitability of different soils, as defined by the Health Code, and their areal extent can be determined from these maps, and development can proceed following the suitability analysis (Beatty and Bouma, 1973). However, the assumptions behind the Health Code are not always based on scientific data and are often essentially empirical in nature.

For example, effluent disposal in soils that are defined as "non-problem" soils by the Health Code may pose many problems. Pollution of groundwater with nitrates in sands may be serious (Walker et al., 1973a, 1973b), and crusting and clogging of seepage beds in such soils as loams and silt loams, which are highly or moderately permeable initially, may result in insufficient absorption followed by unacceptable surface discharge of contaminated raw effluent (Bouma et al., 1972).

The general aim of on-site soil disposal of liquid wastes is to achieve

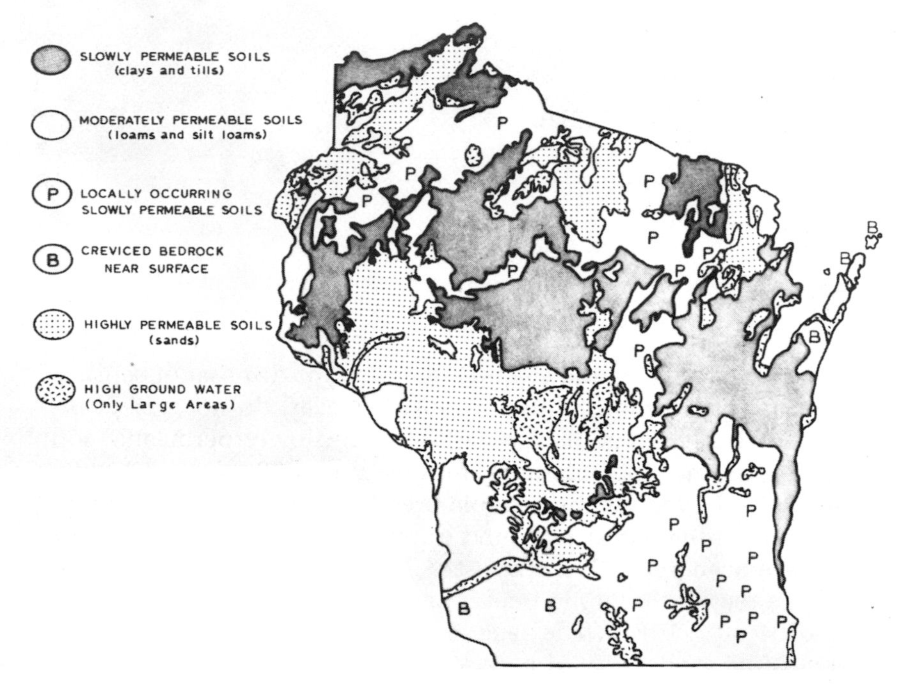

1. Soil condition in Wisconsin (after Hole et al. 1966, Wisconsin Geological Society, Natural History Survey, Univ. Ext.).

adequate purification of effluent while maintaining adequate infiltration into and percolation through the soil. Purification of septic tank effluent or any other type of effluent must include removal of pathogenic bacteria and viruses, BOD, COD, and suspended solids. Significant reductions in N and P contents are necessary as well. The capacity of a soil to accept and transmit effluent is insufficiently expressed by the mandatory percolation test, which is useful only as a parameter ranking different soils (Bouma, 1971). More sophisticated physical techniques are available now to measure exactly the hydraulic properties of soil *in situ* under unsaturated conditions as well as saturated ones (Bouma, Improved field techniques, this text).

The Small Scale Waste Management Project

The interdisciplinary Small Scale Waste Management Project at the University of Wisconsin is developing treatment and disposal systems that are an alternative to the conventional septic-tank soil-absorption system for all soils. A summary of current field investigations is presented in Figure 2. Dosing, the application of effluent once a day rather than as a continuous trickle, is done through a newly designed pressure distribution system (Converse, 1973) which ensures good distribution of effluent over the entire seepage area. Soil purification is thus optimized while soil crusting becomes less probable (Bouma et al., 1972, 1973b). A mound system consisting of a soil-covered seepage bed on top of at least 60 cm of sandy fill over the original soil surface was designed for areas with creviced bedrock near the surface (Bouma et al., 1973a; Figure 2, no. 1). Field monitoring and associated laboratory column experiments have shown that 60 cm of sandy fill over 30 cm of loamy topsoil can remove fecal bacteria and pathogenic viruses from percolating septic tank effluent (Magdoff et al., 1973a, 1973b). However, removal of nitrogen, occurring as nitrate because of the aerobic soil environment, may be inadequate. The microbial process of denitrification, which removes nitrates in anaerobic environments, offers an attractive means to reduce the potential

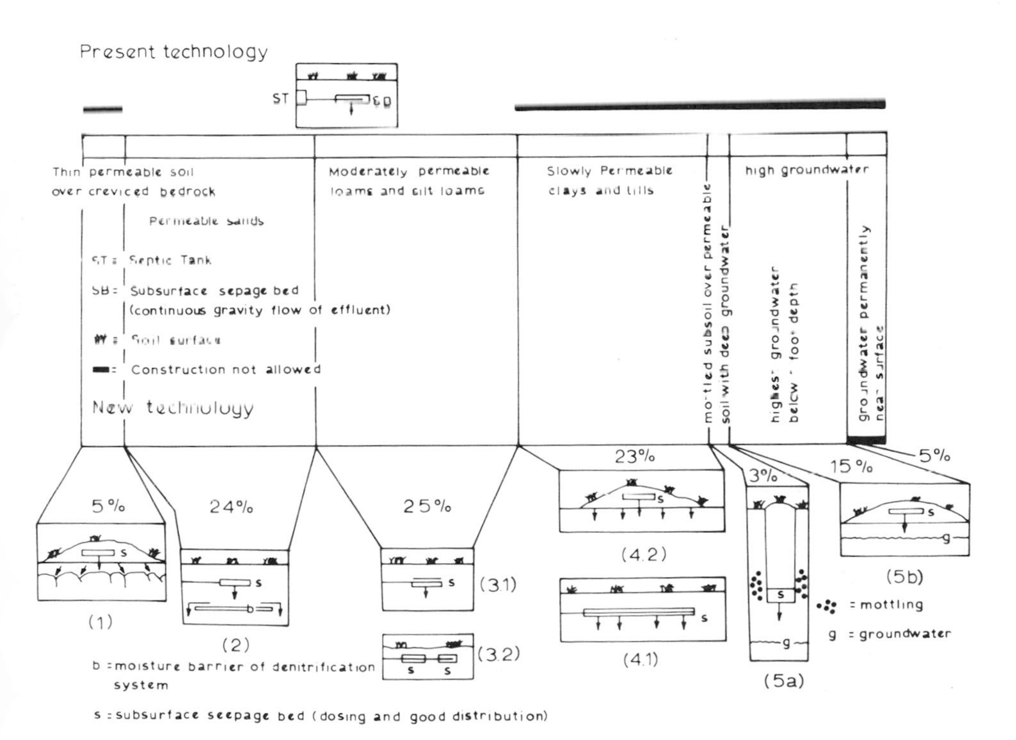

2. Diagram with summary of current field investigations.

problem of nitrate pollution of groundwater. A subsurface denitrification system, which is planned to incorporate improved phosphorus removal as well, is now being developed in the project following the work of Erickson et al. (1971). Such a system would be used with mounds over shallow soils on creviced bedrock and with seepage beds in sands (Figure 2, nos. 1, 2).

Dosing of effluent and the use of dual-bed systems are being explored for moderately permeable loams and silt loams to reduce clogging problems (Bouma et al., 1973b). The dual bed system (Figure 2, no. 3.2) is composed of two seepage beds, each one large enough to accept the effluent produced. Loading is periodically shifted from one seepage bed to the other to allow decomposition by oxidation of clogging compounds that accumulated in the "resting" seepage bed during the previous loading cycle. This decomposition also occurs in resting periods between dosages in a single seepage bed (Figure 2, no. 3.1), but these periods are relatively short since dosing usually will occur once a day.

Two alternatives are considered for slowly permeable soils: (1) a large subsurface seepage bed sized on the basis of soil physical data (Bouma, Improved field techniques, this text) and dosed with high-quality mechanically pretreated effluent (Figure 2, no. 4.1) and (2) a mound system which partially purifies septic tank effluent in the fill and distributes it over a wide area so as to make absorption possible (Figure 2, no. 4.2). A typical mound for a single household may have a bottom area of 4000 square feet or more. The original soil surface beneath the mound is plowed and stabilized before the sand is applied, to ensure adequate infiltration into the soil *in situ*.

Two alternatives are being explored for soils with high groundwater. The seasonal wetness due to ponding of percolating rain in slowly permeable silty subsoil layers can be eliminated by excavation of the subsoil if these layers are underlain by more permeable soil and if the groundwater is deep. The seepage bed can then be built in the more permeable soil (Figure 2, no. 5a). Seasonal wetness in these subsoils can be either directly observed or derived from the occurrence of soil mottling. Mottling consists of grayish and reddish spots in soil, the former formed by reduction processes during periods of water saturation and the latter formed by oxidation of iron compounds upon reentry of air into the soil. Mottling is an important criterion in the Soil Survey Program to estimate the soil hydraulic regime. A mound system can be built on soils with continuous relatively high groundwater (Bouma et al., 1972, p. 190) (Figure 2, no. 5b). Soils with groundwater at or near the surface all the time should preferably not be developed (Figure 2, no. 5c).

Once alternatives are developed for different soils, soil maps can again be used to estimate the potential impact of new technology on land use patterns (Beatty and Bouma, 1973).

A system is acceptable for on-site treatment and disposal of liquid waste only if it is safe, economical, and reliable to construct and operate. Systems incorporating soil are usually attractive, because soil is very effective as a "living filter" and normal maintenance costs are relatively low.

References

Beatty, M. T., and Bouma, J. 1973. Application of soil surveys to selection of sites for on-site disposal of liquid household wastes. *Geoderma* 10: 113–122.

Bouma, J. 1971. Evaluation of the field percolation test and an alternative procedure to test soil potential for disposal of septic tank effluent. *Soil Science Society of America proceedings* 35: 871–875.

Bouma, J., Ziebell, W. A., Walker, W. G. Olcott, P. G., McCoy, E., Hole, F. D. 1972. *Soil absorption of septic tank effluent*. University of Wisconsin Extension Service, Geological and Natural History Survey, Information circular no. 20.

Bouma, J., Converse, J. C., Magdoff, F. R. 1973a. A mound system for disposal of septic tank effluent in shallow soils over creviced bedrock. Paper read at International Conference on Land for Waste Management, October 1973, Ottawa, Canada. To be published in proceedings.

Bouma, J., Converse, J. C., Magdoff, F. R. 1973b. Dosing and resting to improve soil absorption beds. *Transactions of the American Society of Agricultural Engineers* 17. (In press)

Converse, J. C. 1973. Distribution of domestic waste effluent in soil absorption beds. *Transactions of the American Society of Agricultural Engineers* 17. (in press)

Erickson, A. E., Tiedje, J. M., Ellis, B. G., Hansen, C. M. 1971. A barriered landscape water renovation system for removing phosphate and nitrogen from liquid feedlot waste. In *Proceedings of the International Symposium on Livestock Wastes*, pp. 232–234. St. Joseph, Michigan: American Society of Agricultural Engineers.

Magdoff, F. R., Bouma, J., Keeney, D. R. 1974. Columns representing mound-type disposal systems for septic tank effluent. I. Soil-water and gas relations. *Journal of environmental quality*, in press.

Magdoff, F. R., Keeney, D. R., Bouma, J. Ziebell, W. A. 1974. Columns representing mound-type disposal systems for septic tank effluent. II. Nutrient transformations and bacterial populations. *Journal of environmental quality*, in press.

United States Public Health Service. 1967. *Manual of septic tank practice*. Publication no. 526.

Walker, W. G., Bouma, J., Keeney, D. R., Magdoff, F. R. 1973a. Nitrogen transformations during subsurface disposal of septic tank effluent in sands. I. Soil transformations. *Journal of environmental quality* 2: 475–480.

Walker, W. G., Bouma, J., Keeney, D. R., Olcott, P. G. 1973b. Nitrogen transformations during subsurface disposal of septic tank effluent in sands. II. Ground water quality. *Journal of environmental quality* 2: 521–525.

Wisconsin State Board of Health. 1969. Private domestic sewage treatment and disposal systems, section H62.20.

9. Comparison of Septic Tank and Aerobic Treatment Units: The Impact of Wastewater Variations on These Systems

Edwin R. Bennett,
K. Daniel Linstedt,
and John Felton

Nearly one third of the homes in the United States are located in unsewered areas and must rely on some form of individual treatment and disposal system for dealing with household wastewater. In many such areas, for example in the mountainous resort areas of Colorado, the treatment conditions are not conducive to the optimum performance of these units, and numerous problems have been encountered with wastewater disposal. These problems have been compounded by the boom in suburban residential expansion into mountainous areas and the increased popularity of second homes in ski and resort areas.

Currently, three types of home disposal systems are most commonly installed. These are (1) the septic tank and leaching field, (2) septic tank and evapotranspiration bed, and (3) aerobic treatment units with surface discharge or discharge into leaching fields. Each of these types of systems incorporates special design and operational considerations into their successful application. In order to better understand the functioning of these home wastewater systems, a field and laboratory study was undertaken to determine home wastewater characteristics and to relate these to the success of the treatment and disposal techniques.

Flow Characteristics

The objective of the first phase of the study was to determine the water use characteristics of individual homes. Several homes were selected for study, with families of different sizes (2–5 persons) and age groupings. Income levels were similar to those of families residing in suburban mountain residential developments utilizing individual disposal systems. The homes selected were connected to the city water and sewer systems. The water use pattern throughout the day was determined by attaching chart-recording devices to the water meters of the homes in the study and, at the same time, maintaining a log of water use by function in each residence. In this way the average water use characteristics were correlated with measured flow data.

Table 1
Individual Water Use by Function

Appliance	*Gallons per Use*	*Uses per Person per Day*	*Gallons per Person per Day*
Toilet	4.1	3.6	14.7
Sink	1.7	4.5	7.6
Garbage disposal	2.1	0.4	0.8
Bath or shower	27.2	0.32	8.7
Dishwasher	7.0	0.15	1.1
Washing machine	38.6	0.30	11.6
Total			44.5

Six distinct water uses were identified in the study. The distribution of average water use within these categories is summarized in Table 1. The total water use of 44.5 gpcd seems somewhat low but is within the range reported by other authors (Weibel et al., 1949). In developing a time distribution for the total water use, it was observed that hourly water use patterns are somewhat variable among homes and dependent on the life style of the occupants. For the families

Table 2
Strength of Household Wastes (mg/l)

	BOD	*COD*	*SS*	*MBAS*	PO_4	*Temp* °*C*
Toilet						
Feces (gm/cap/day)	5	34.2	21.5	—	—	—
Urine (gm/cap/day)	2.1	18.4	0	—	—	—
Paper (gm/cap/day)	—	18.5	15.0	—	—	—
Total (mg/l)	124	1300	650	—	—	21
Sink	172	578	228	4	0.4	43
Garbage disposal	4065	11780	6672	0	24	25
Bath or shower	168	309	76	57.5	0.08	36
Dishwasher	38	120	3	0.1	96	55
Washing machine	285	566	85	41.5	66	47
Average (flow weighted)	252	950	412	23	20	37
no./cap/day	0.10	0.35	0.16	0.01	0.01	

studied in this investigation an average water use distribution had the characteristics shown in Figure 1. The typical bimodal curve resulted, with peak flows highly dependent on the bath and shower habits of the families.

Strength Characteristics

A second phase of the study involved evaluation of the waste strength characteristics for the six major types of household wastewaters identified. Samples of each waste were collected and tested for BOD, COD, suspended solids, MBAS, total phosphate, and temperature. The results of these tests are shown in Table 2. Individual samples of feces, urine, and toilet paper were tested, and the average strength of toilet wastes was estimated. The amount of human wastes was esti-

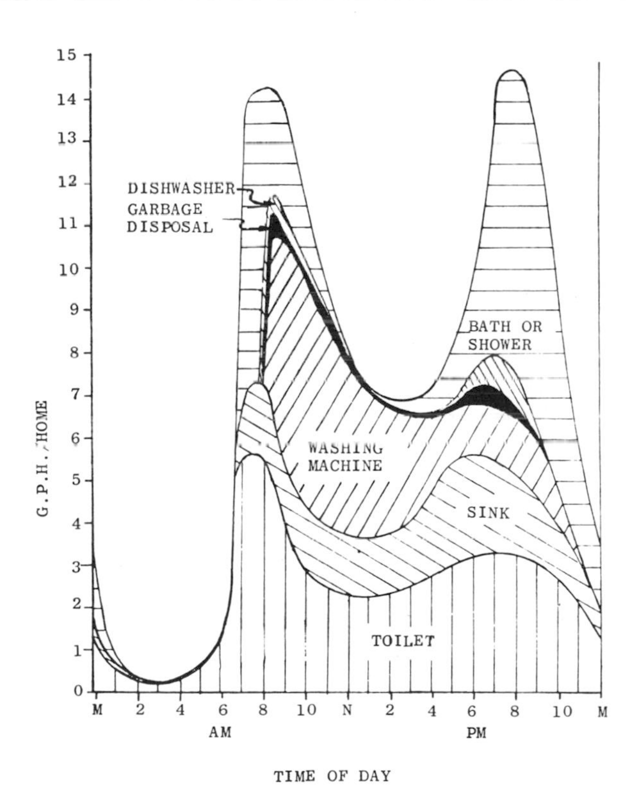

1. Daily household water use.

mated from Figure 2. Data of this type are very approximate. The plot is based on limited data from the literature (Metcalf and Eddy, 1916; Ehlers and Steel, 1965; Hawk, 1947; Rafter and Baker, 1894) and several individual measurements. The average waste strength values obtained in this study relate well to those in the literature (Zanoni and Rutkowski, 1972; Galonian and Aulenbach, 1973), although they were determined in a quite different manner.

When waste strength data are combined with flow data, loading curves of the type shown in Figure 3 for hourly COD variations result. This particular bimodal curve is strongly influenced by toilet wastes and garbage disposal wastes. The COD loading contribution from each major water use category is summarized together with similar data for suspended solids and flow in Figure 4. It can be noted that toilet and garbage disposal wastes account for about three quarters of the COD and suspended solids but comprise only about one third of the flow. Although it is not graphed, it can be noted that most of the phosphate is due to the washing machine wastes and the large majority of the MBAS is introduced with the bath and washing machine wastes.

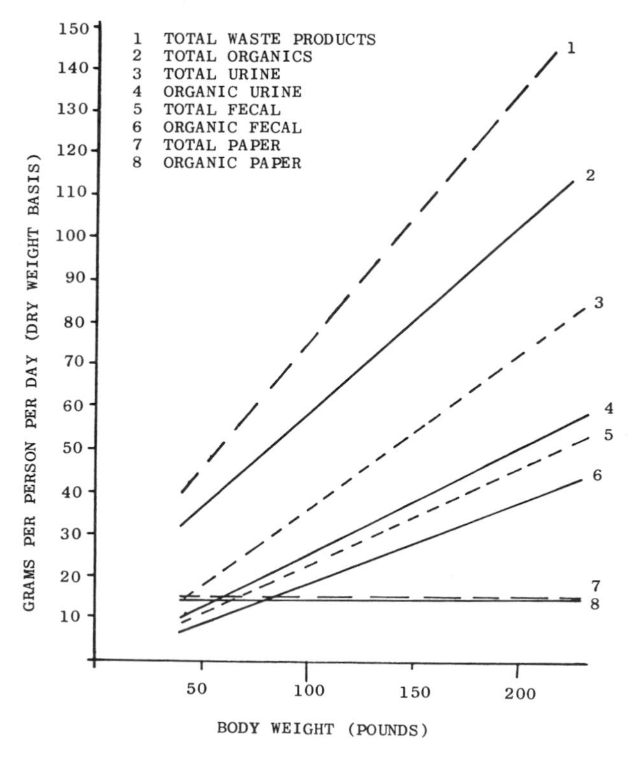

2. Approximate values utilized for total and organic solids estimates for human wastes.

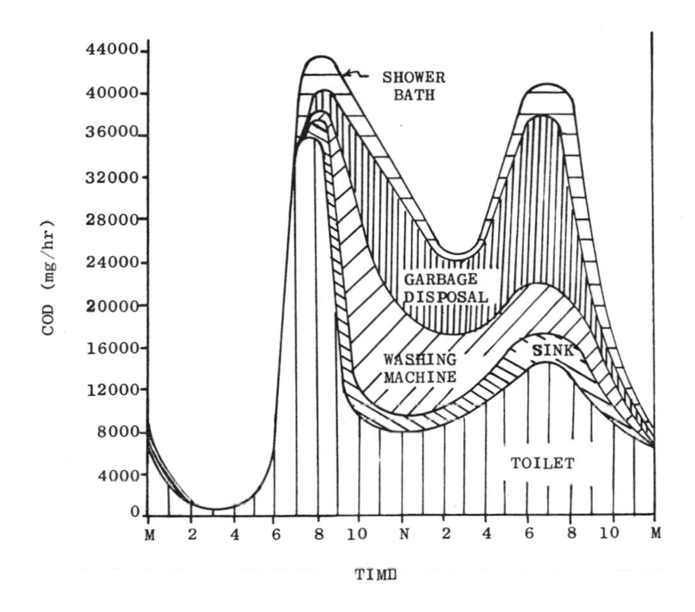

3. Hourly COD profile.

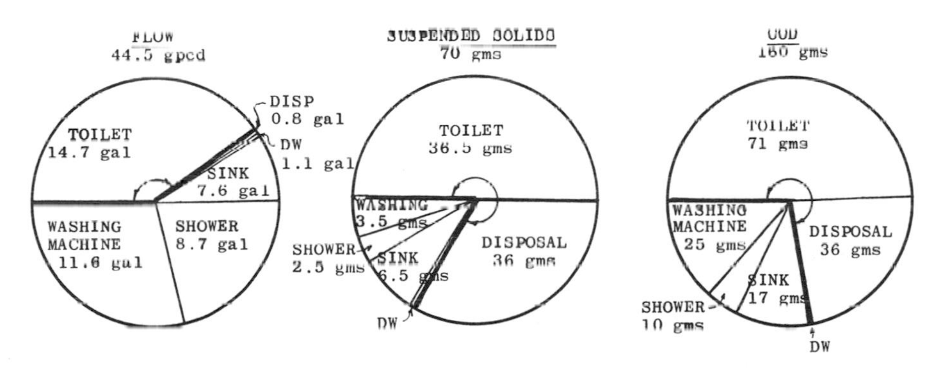

4. Distribution of flow and pollution sources.

Treatment Alternatives

Septic Tanks

The use of the septic tank extends for more than a century from the original unit designed by Mouras in 1860 and patented in 1881. The purpose of the septic tank is to settle particulate matter and provide flotation for greases and fats. The unit is of very simple construction as shown in Figure 5. The first step of anaerobic digestion takes place

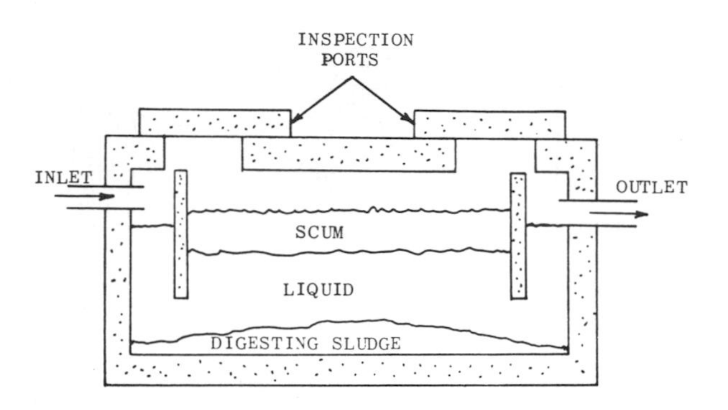

5. Septic tank.

in the tank with biological conversion of suspended organics to soluble organic acid. In this way the wastes are solubilized before entering the percolation field. Most of the biological decomposition of organic matter takes place in the leaching field. Excellent descriptions of the details of this process have been given by other researchers (McGauhey and Krone, 1967; McGauhey and Winneberger, 1964; Weibel et al., 1955; Bouma et al., 1972; Goldstein et al., 1972). The required size of the percolation field is usually determined from an *in situ* percolation test (U.S. Public Health Service, 1958). This method of testing has some major limitations, and it is the opinion of these authors that the test should be used only for the binary decision as to whether or not a site is suited for the use of a leaching field. If the percolation rate is exceedingly fast ($>$ 1 inch in 5 minutes), there is a strong possibility of local well water pollution with improperly treated effluent. On the other hand, if the percolation rate is less than one half inch in an hour, there is an indication that expanding clays could make the site unsuitable for wastewater disposal. For sites where the soil conditions give percolation test results between these limits, the septic tank and leaching field will probably be as effective as other types of disposal systems. Typical cost ranges for septic systems in Colorado designed according to U.S. Public Health Service *Manual of practice* no. 526 are shown in Table 3.

System failure with septic tanks and leaching fields typically involves either pollution of wells or surfacing of odorous effluent. The latter is usually caused by a clogged leaching field. Adequate cleaning, primarily for the removal of grease, is required about once every two years to prevent this clogging.

Several conditions can make a site unsuitable for the use of a septic tank and leaching field. These include loose soils and fissured rock

Table 3
Treatment Cost—Septic Tank System for 20 Year Life, Cost of Tank = $370

Soil Permeability	*Ft² Absorption Area Required 4 Bedroom*	*Absorption Field Costs*	*Total Annual Cost*	*Total Annual Cost/ 1000 gal.*
Good	280	$ 300–450	$ 80	$0.90
Fair	500	$ 500–750	$100	$1.15
Poor	1330	$1330–2000	$200	$2.20

Cleaning cost $15/year.
Capital recovery at 20 yr. and 7½% interest rate.

conditions, impermeable soils, shallow soil mantle above bedrock, and high water tables. When these situations exist, alternative wastewater disposal systems must be designed.

Evaporation-Transpiration Beds

Evaporation-transpiration beds following a septic tank have been used in Colorado as a solution where conditions will not permit the use of leaching fields. The systems are designed as shown in Figure 6 and in most cases the criteria of Bernhardt* have been utilized. If the unit is to be successful as an evaporation-transpiration system, imported select medium-fine sand must be used as the bed media. The sand size must be selected carefully. The pores must allow for the water to be lifted to the surface by capillary rise so that evaporation can take place. The sand utilized must be small enough to lift the water by capillary action but large enough to provide for adequate hydraulic conductivity of water up to the surface. It has been found that sand in the D_{50} size range of 0.12–0.18 mm seems to be satisfactory. A uniformity coefficient of four or less is desirable. The bed layout should allow for the sand to furrow down into the gravel to provide a wick for raising the water. Windblown deposits and gravel washings have been found to be adequate materials for evaporation-transpiration bed construction.

Experience with these systems is limited, and as a result firm performance and cost data have not been developed. Installation cost of the evaporation-transpiration bed is in the range of $2000. To date, after a few years of application, no serious complaints seem to exist.

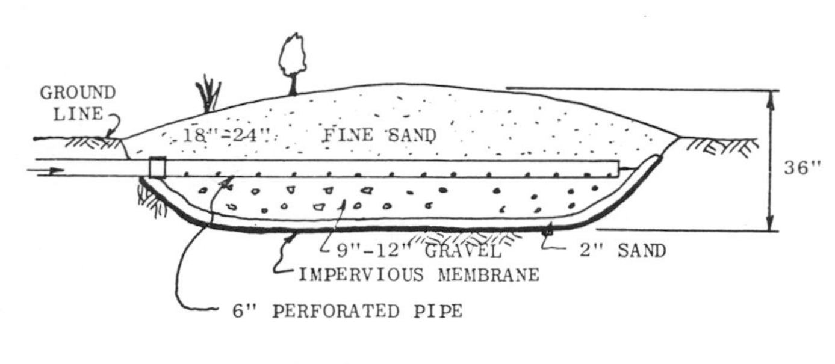

APPLICATION RATES

	SEPTIC	AEROBIC
SUMMER	0.1 GPD/FT^2	0.2 GPD/FT^2
WINTER	0.02	0.06

INCREASE 50% FOR FULL VEGETATION

0.1 GPD/FT^2 = 5 FT/YR

6. Evaporation-transpiration bed.

Aerobic Treatment Units

Aerobic treatment units have been proposed for applications where septic tank leaching fields are deemed unsuitable. These systems normally discharge directly onto the ground or into shallow, rock-filled trenches. The number of installations of this type has been growing rapidly in Colorado as indicated by the data in Figure 7 for one representative county.

The typical aerobic unit, as shown in Figure 8, is usually constructed to function in a manner similar to a small extended aeration activated sludge treatment plant. Some units have presettling chambers, although many do not. Prefabricated fiberglass installations are common, but when units are constructed in place, the building material is usually concrete. Design detention time in the aeration chamber is in the range of one to two days. Aeration is generally provided with compressed air and aided by mechanical stirring. There is usually a small final settling tank with ports or slots for gravity sludge return to the aeration chamber.

A cost analysis for aerobic systems is shown in Table 4. Total cost for capital recovery, operation, and maintenance is typically on the

*A. P. Bernhardt, 1972; personal communication.

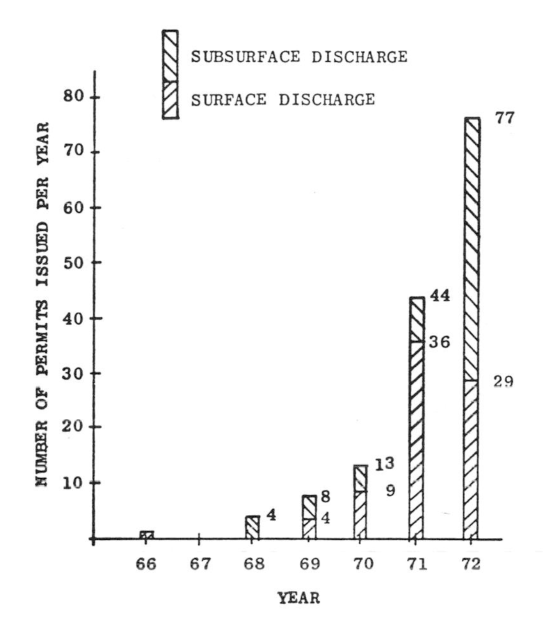

7. New permits for aerobic units in a Colorado County.

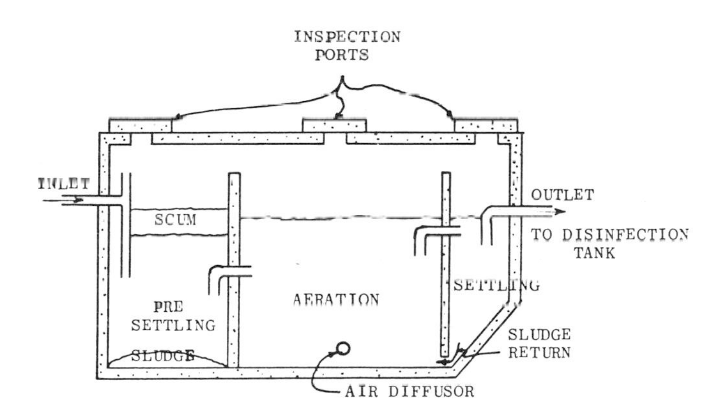

8. Typical aerobic unit.

order of three to four dollars per thousand gallons of wastewater treated.

In an effort to evaluate aeration unit performance, several installations have been field sampled and tested by students in laboratories at the University of Colorado. Effluent BOD_5 and suspended solids were measured from grab samples of units that were in normal house-

Table 4
Treatment Costs of Aerobic Systems

	Annual Cost $/Yr.	*$/1000 Gal.*
Purchase and installation $1000–$1600		
Amortization	100–160	$1.40
Electricity	25–100	0.70
Service	30– 50	0.50
	155–310	$2.60
Disinfection initial cost $150–250		
Amortization	15– 25	$0.25
Chemicals and maintenance	20– 50	0.40
	35–110	$0.65
Subsurface filter $350–500		
Amortization	35– 50	$0.50
Operation	5– 15	0.20
	40– 65	$0.70
Leach field $300–2000		
Amortization	30–170	$1.00
Operation	5– 15	0.15
	35–215	$1.15

Capital recovery in 20 years at 7½% interest.

hold use. These data were compiled to reflect the average performance of units maintained and operated by homeowners. The results of the tests were compared with values observed by local health departments during routine monitoring and were found to have a very close correlation. A summary of the data is shown in Figure 9, with pollutional parameters of BOD_5 and total suspended solids (TSS) plotted as a function of frequency of occurrence in the effluents. It can be noted that the BOD_5 and TSS values correspond on essentially a one-to-one basis and that the median BOD_5 and TSS for all of the units grouped together is approximately 150 mg/1. This average performance is below discharge standards for effluents released on the ground surface.

Two reasons have been identified for the disappointing performance. One important factor is neglect by the homeowner. Many owners will not accept any maintenance responsibility for their sewage system, and therefore a successful unit must be highly reliable with minimal maintenance. Another consideration, one that is substantiated by these studies, is the adverse effects introduced by surge

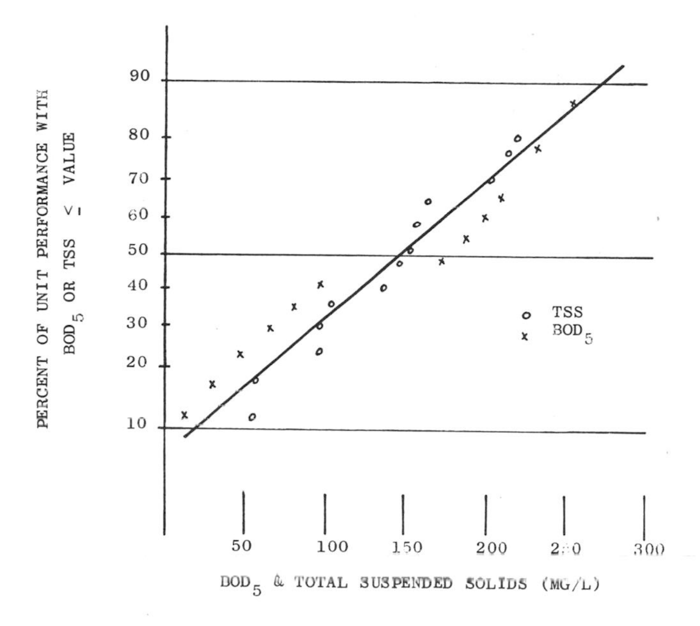

9. Spot check effluent quality of several field units.

flows. The units observed in this study had secondary settling units with volumes in the range of 30 to 100 gallons. These units are sufficiently small to be adversely affected by the surge flows which characterize individual home discharges. An evaluation of maximum surge flows has been developed from the water meter charts employed in the initial portion of the study. The results of this analysis are shown in Figure 10. It can be noted that under normal home use conditions a maximum quantity of about 60 gallons will be surged into the unit in a time period of seven to thirty minutes. Field observations have shown that such hydraulic loads will not allow any substantial concentration of organisms to build up in the return sludge. (In contrast, flow variations have little effect on septic tanks because of the extended hydraulic detention periods.)

In many of the field units tested, the suspended solids concentration in the effluent was essentially the same as in the aeration chamber. This indicates that the units functioned more as aerated ponds than extended aeration units. The combination of the pollutional load surges demonstrated in Figure 3 and the rather short detention time provided for plain aeration treatment may be the major determinant of treatment efficiency for these units. This is especially true of surface installations operating in the wintertime. Flow surges have also

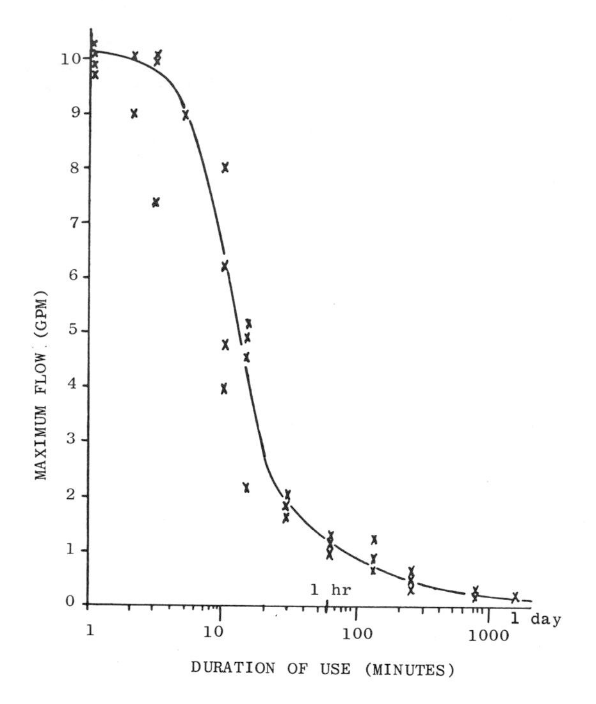

10. Duration–maximum intensity flow curve for individual home water use.

been shown to affect the reliability of the effluent chlorination operation. Effluent discharged during low flow periods tends to be overchlorinated, and that discharged during a flow surge is chlorine deficient.

Conclusions and Recommendations

1. Many septic systems and aerobic units have functioned poorly in Colorado. In some cases, they have caused dangerous pollution of wells and surface waters. In order to alleviate some of these problems, new technology is needed in individual home wastewater disposal systems as well as a better understanding and improvement of the present systems.

2. Central sanitary water and sewer systems should be required in areas with dense population or soil conditions that are not well suited for leaching fields.

3. Surface disposal of effluents from individual home systems

should be avoided except when done under carefully controlled conditions.

4. Significant flow lag and surges occur in home wastewater discharges. These factors have a significant negative impact on the operation of settling tanks, particularly the small secondary settling tank employed with some aerobic treatment units.

5. Three-fourths of the pollutional load from individual homes is contained in about one third of the total effluent volume. This fraction also contains nearly all of the pathogens. Segregation of wastes with separate disposal of the toilet-garbage disposal fraction should be given more consideration. Aerobic treatment, chlorination, and surface irrigation may be very satisfactory for disposal of the shower, sink, and washing machine wastes.

New research should be directed at defining processes specifically for treatment of the toilet-garbage disposal waste fraction. It may be possible to reduce water use in the toilets of homes with individual systems to one-half gallon to one gallon per use. On this basis approximately five to ten gallons of concentrated toilet wastes would be produced per day. Investigations are needed to determine the best method for providing storage, treatment, and transportation of this waste for disposal in an environmentally safe area. This may include such processes as wet oxidation, dry incineration, and aerobic or anaerobic digestion.

References

Bouma, J., Ziebel, W. A., Walker, W. G., Olcott, P. G., McCoy, E., Hole, F. D. 1972. *Soil adsorption of septic tank effluent*. Extension information circular 20. Madison: University of Wisconsin Extension Service.

Bernhardt, A. P. 1972. Small wastewater units for soil infiltration and evapotranspiration. Paper read at the Suburban Sewage Disposal Conference of the Michigan Department of Health, January 1972, Lansing, Michigan.

Ehlers, V. M., and Steel, E. W. 1965. *Municipal and rural sanitation*. New York: McGraw-Hill.

Galonian, G. E., and Aulenbach, D. B. 1973. Phosphate removal from laundry wastewater. *Journal of Water Pollution Control Federation* 45: 1708–17.

Goldstein, S. N., Wenk, V. D., Fowler, M. C., Poh, S. S. 1972. *A study of selected economic and environmental aspects of individual home wastewater treatment systems*. Washington, D.C.: MITRE Corporation publication no. M72-45.

Hawk, P. B. 1947. *Practical physiological chemistry*. New York: Blakiston Company.

McGauhey, P. H., and Krone, R. B. 1967. *Soil mantle as a wastewater treatment system*. Berkeley: University of California SERL report no. 67-11.

McGauhey, P. H., and Winneberger,

J. H. 1964. *Causes and prevention of failure of septic tank percolation systems*. Washington, D.C.: FHA technical studies report no. 533.
Metcalf, L., and Eddy, H. P. 1916. *American sewerage practice*. New York: McGraw-Hill.
Rafter, G. W., and Baker, M. N. 1894. *Sewage disposal in the United States*. New York: Van Nostrand.
United States Public Health Service. 1958. *Manual of septic tank practice*. Washington, D.C.: USPHS publ. no. 526.
Weibel, S. R., Bendixen, T. W., Coulter, J. B. 1955. *Studies on household sewage disposal systems, part III*. Washington, D.C.: USPHS publ. no. 397.
Weibel, S. R., Straub, C. P., Thonan, J. R. 1949. *Studies on household sewage disposal systems*. Washington, D.C.: Federal Security Agency, USPHS.
Zanoni, A. E., and Rutkowski, R. J. 1972. Per capita loadings of domestic wastewater. *Journal of Water Pollution Control Federation* 44: 1756-62.

III
Groundwater Problems

10. Groundwater Pollution—Problems and Solutions

Jay H. Lehr

Surface water pollution caught up with the public consciousness in the early 1960's, and by the beginning of the next decade action was at last under way.

Groundwater, always the forgotten stepchild of the hydrologic cycle, continued without protection of words or statutes virtually up to the present. Now at last people have begun to awaken to the inevitable fate awaiting groundwater which has already overtaken surface water.

The first major symposium addressed to groundwater quality problems was held in August 1971 in Denver, Colorado. It was sponsored jointly by the Environmental Protection Agency and the National Water Well Association and produced a major manuscript on the subject of groundwater quality. As the director of that symposium, I have chosen to draw freely from its results in order to construct this brief statement on the state of the art, on the problems and potential solutions at hand. The following causes of ground water pollution will be discussed:

(1) Well injection
(2) Solid waste disposal
(3) Septic tank inadequacies
(4) Aquifer transfer
(5) Chemical pollution

Well Injection

Deep well injection poses the greatest potential detriment to our water supplies, but also stands the best chance of being controlled, because of its well publicized lethal potential. The greatest danger comes when the well bore or the formation is plugged for any of the following reasons:

(1) Injection of fluid containing suspended solids which are often chemical precipitates;
(2) Accumulation of corrosion products;
(3) Precipitation within the formation as a result of incompatibility of the injected fluid with the connate water of the formation;
(4) Injection of insufficiently treated water resulting in bacteriological plugging;
(5) Swelling of mineral constituents in the formation.

Most states have not yet acted effectively to prevent these problems, but soon to be enacted federal regulations will force all states to close our subterranean doors to injection projects endangering our vast and valuable groundwater supplies. Of approximately a dozen states that have already taken the initiative in controlling injection programs, Illinois and Colorado stand as good examples. The Illinois statute illustrates one method for laying down requirements, and the Colorado statute illustrates the complex information which can reasonably be required to ensure protection of water supplies. Illinois requirements have been summarized as follows by H. F. Smith (1971).

(1) All zones to be considered for disposal must contain brine waters having over 10,000 mg/1 total dissolved solids. This equates to a one percent solution of total dissolved solids. Fresh water is now defined as water of 10,000 mg/1 or less total dissolved solids. A figure of 5000 mg/1 or less had been used until recently to define fresh water, but this was revised upward to preserve groundwaters that may be usable if and when current experimental methods of desalination are made practical for the production of potable water from brackish waters.

(2) There must be an effective and adequate impermeable barrier overlying the disposal zone to prevent upward migration of either the wastes or displaced brine waters into fresh water zones.

(3) The well bore must be double cased and the annular space grouted to a point 200 feet below the lowest fresh water zone. The outer casing is to be set into the next lower suitable rock barrier.

(4) The industry must indicate the character and volume of waters to be injected into the deep well. Compatibility of the wastes with the formation fluids must be indicated.

(5) Operational injection pressures to be used must be indicated in the plan documents. Injection pressures that cause fracturing of the rock formations in the injection zone have not been permitted.

(6) Detailed information about the proposed surface injection equipment must be provided as a part of the plan documents.

(7) The industry is required as a condition of the permit to submit daily injection report records each month to the Illinois Environ-

mental Protection Agency. These operating reports must include the character and volume of the waste and show the injection pressures.

(8) Currently the Illinois Environmental Protection Agency is reviewing the necessity for requiring the installation of observation wells to detect the escape of any fluids from the disposal zone.

While these conditions are both practical and reasonable, few new wells have qualified in Illinois.

The Colorado Geological Survey reports that even fewer wells have passed Colorado's standards because they rarely ever get past the application stage. The applicant in Colorado has to include a legal description and a map of the area within a two-mile radius of the proposed well showing all the surface and mineral owners. It has to show all test holes and penetrations, description and depth of all the formations penetrated by any well or mine on the map, local topography, industry, fish and wild life, description of the mineral resources believed to be present in the area, and the probable effect of the system on the mineral resources. The applicant must have a geologic structure map showing stratigraphic sections, plus a geologic cross-section drawn through the area and the proposed well, a description of all water resources, both surface and subsurface, within the zone of influence or possible influence of the system, a classification of those waters, and a map showing vertical and lateral limits of surface and subsurface supplies. There must be a description of the chemical, physical, radiological, and biological properties and characteristics of waste to be disposed of in the system, and treatment proposed for such waste including copies of all plans and specifications of the system and its appurtenances. A statement of all sources and procedures relied upon for information must be set forth in the application. If the system is to be an injection well, he has to have potentiometric surface maps of the disposal formations and those formations immediately above and below. Copies of all drill stem tests, installations, and data used in making the maps must be included. He must state the present and potential uses of fluids from the disposal or affected formation and the volume, rate, and injection pressure of the fluid to be injected. He must include the following geological and physical characteristics of the injection interval and the overlying and underlying impermeable barrier: thickness, areal extent, lithology (that is, grain mineralogy, type mineralogy, matrix, amount and type of cement, clay content and clay mineralogy), effective porosity, and the permeability, both vertical and horizontal, with mechanical logs and formation tests. He must furnish the coefficient of storage of the aquifer, the amount and extent of natural fracturing, the extent and effects of known or suspected faulting along with its location, the extent and effect of natural solution channels, the fluid saturation in the formation, the formation fluid chemistry with indicated local and regional variation, the temperature of the formation and how determined, the formation and fluid pressures, fracturing ingredients, osmotic characteristics of the rocks and fluids comprising and contiguous to the reservoir, an indication of the effect of the injected wastes on the contiguous formation in the event of leakage, diffusion and dispersion characteristics of wastes in the formation fluid and the effect of gravity segregation, and the compatibility of injected wastes with physical, chemical, and biological characteristics of the reservoir. And, of course, he must have the normal engineering data on the well as detailed for the geology. Not only has Colorado had few wells drilled for waste disposal, it has not even had many applications completed.

It should be very clear that waste disposal is a sensitive situation which under the very best of circumstances can produce potential problems. Deep well waste injection is going to be used for a long time. It can be a beneficial operation but only under the most stringent requirements and restrictions.

Solid Waste Disposal

Of no lesser threat to our groundwater is the pollution originating from the leachates of decomposed solid waste on the land surface as in the case of open dumps, solid waste composting sites, industrial refuse, and treatment plant sludge. But perhaps the worst of all are the antiseptically named but not so well operated "sanitary landfills." ASCE (1959) has defined sanitary landfill as a method for disposing of refuse on land without creating nuisances or hazards to public health or safety by utilizing the principles of engineering to confine the refuse to the smallest practical area, to reduce it to the smallest practical volume, and to cover it with a layer of earth at the conclusion of each day's operation or at such more frequent intervals as may be necessary.

In practice, sufficient water frequently passes through the decomposing mass to leach out the pollutants and convey them to the groundwater sources. The average site investigation today is absolutely minimal. The tools that are available are not being used. Most communities becoming involved in a landfill will operate it for ten or twenty years at a cost in excess of a million dollars annually for a municipality with over 250,000 people (Kaufman, 1971). A million dollars a year is a capital investment that should be handled carefully with long-term goals. For instance, infiltration can be controlled. There can be a tile system put in as the refuse goes. It can be incremented over the years as the landfill expands. After the infiltration is reduced, either by control of runoff or proper vegetative cover or a very tight asphalt cover for parking lots, tile systems should be used to collect leachate. If infiltration is still a problem, diversion wells can be put in. The type of cover material and degree of compaction are extremely important. If most of the surface precipitation is graded away from the landfill site and the top layer is relatively impervious, the degradation rate is cut to the very minimum, and the impact of leachate production will be small.

There is also a general tendency to locate landfills in quarries, ravines, and valleys that are very important reservoirs of groundwater recharge. There is a trend simply because these areas, just as they are, are worthless. A restoration plan with pretty pictures which shows usable land being developed from some of these sites is very attractive to the city fathers and political leaders, particularly from a tax base standpoint. Zanoni (1972), in a critical review of sanitary landfill literature and procedures employed by 21 states, suggests certain steps that a regulating agency can take to minimize problems in this area:

(1) Have available a geologist on the staff, preferably one trained in the area of hydrogeology, to assist in the sanitary landfill site selection processes within the state.

(2) Accumulate geological data within the state and broadly outline areas considered to be either good or poor potential landfill sites.

(3) Require more hydrogeologic and hydrologic field data for questionable sites. The burden of proof should be placed on the landfill operator or owner.

(4) Be very cautious about the approval of the ground disposal of industrial wastes. An up-to-date file should be maintained on various types of industrial wastes and their degradation properties.

(5) Consider the use of groundwater monitoring wells in those cases where some doubt exists as to future effects of a particular operation.

(6) Slow down the refuse degradation process by minimizing water percolation through the refuse mass. Slowing down degradation provides more time for leachate attenuation; longer times provide the most effective safety measure.

(7) Do not discourage novel methods of collecting and treating refuse leachates for certain installations where proper monitoring and control can be exercised.

(8) Set the distance between a point of water use and the site of a

sanitary landfill as long as possible in order to have the built-in safety factor of greater time. Figures of 500 to 1000 feet are not unrealistic. Tremendous variations in the hydrogeology surrounding each site preclude the establishment of a uniform distance.

(9) Encourage the practice of regional or district approaches to solid waste collection and disposal.

(10) Prohibit the use of abandoned rock, gravel, or sand quarries as sites for the disposal of refuse of any type. Standing water in such depressions is usually nothing more than a visible direct link to the groundwater supply. The leachate attenuation mechanism under such conditions is completely lost.

Septic Tanks

A third and always volatile problem for groundwater is the existence of an inadequately constructed or improperly placed septic tank system. Leach bed technology and soil absorption capacities are only now receiving the attention that this universal waste disposal program has long deserved.

Septic tanks and other domestic waste disposal systems such as aerobic tanks can operate so as to pose no danger to the water source buried beneath them. To achieve such a compatible situation, soil conditions must be adequate, leach beds must be constructed with care, the water table must be a safe depth below the tank and leach bed, and water wells should be located beyond the area of influence of leachate movement. A most comprehensive work on upgrading septic tank technology authored by Goldstein and Moberg (1973) holds hope for the future protection of our groundwater.

Aquifer Transfer

Of rare concern to the average citizen but of continuous consternation to the groundwater geologist and well-drilling practitioner is aquifer pollution from interaquifer transfer through inadequate or abandoned wells. It is common knowledge that water quality varies with depth and aquifer. It is always the desire to develop water from the highest quality aquifer while sealing off those with inadequate or poor quality.

For this reason water well casings are grouted into the ground and, under ideal conditions, screened only at the depth of the best aquifer. Wells which are improperly grouted may allow water of poor quality to move into an aquifer with better quality water and deteriorate it. Similarly, when the grout is inadequate, polluted surface water may run down the annular space around the casing into the well. It is, therefore, imperative that well construction techniques be regulated stringently.

Additionally, states should enforce proper procedures for abandoning useless wells so that a rusted-out casing or open hole will not become a continuous avenue for aquifer transfer or surface water entry.

Chemical Pollution

Leakage, spilling, and planned and unplanned percolation of a variety of pollutants from the ground surface or near surface down to underground water occur continuously from a variety of sources rarely considered and almost never controlled.

These sources include (1) petroleum contamination from leaking service station tanks and above-ground spills of gas and oil, (2) agricultural feedlot pollution where animal wastes concentrated in a small area are eventually washed into the ground, (3) tremendous volumes of salt thrown on icy winter roads which nearly all ends up in our precious aquifers, and (4) industrial wastes of wide varieties commonly buried in the ground and dumped in liquid waste disposal pits or evaporation pits.

William H. Walker (1973) has documented in precise detail some of these shocking disposal methods which have brought about serious incidents of pollution. Walker stated correctly that one of the primary hazards of surficial burial of solid and liquid toxic waste is that with time the exact location of many such sites are forgotten, so that future adverse effects from the waste may not be recognized or attributed to their actual source until after serious physical harm.

Industrial and chemical wastes of the varieties cited may be a greater long-term hazard than human waste. Micro-organisms from human waste can be effectively removed by filtration through only a short travel distance in granular earth material. Most chemical compounds, however, are carried without alteration or absorption through the groundwater system and into both ground and surface reservoirs.

Chemical pollutants normally move quite slowly through the

ground. Rates of one foot per day in permeable sand and gravel material and one foot a month or less in tight shales and clays are the common range of velocities. When water-supply wells do not occur near the source of pollution, the source may go undetected and the surface or groundwater supply may finally become polluted years after the toxic materials first entered the ground. The long period of time over which the pollution has already been moving makes it nearly impossible to control after it reaches the aquifer.

Thousands of toxic chemicals are now being discarded into the environment by the various means already mentioned. Many in low concentrations in drinking water have adverse effects which are not recognized as classic symptoms and may be attributed to a secondary cause. Walker has pointed out the possibility that many unusual deaths have been caused by toxic chemical poisoning but were never diagnosed as such because the doctor or coroner failed to recognize symptoms of such poisoning. Considering the questionable qualifications of some elected coroners in light of the following newspaper article, it is easy to understand why a proper cause of death diagnosis may not always be assured.

TWO SEEKING FORD COUNTY CORONER POST

> Paxton. A race between two Piper City residents is the only county office contest to be decided by Ford County voters Nov. 5. Coroner R. L. Hayslette, Republican incumbent is opposed by Robert G. Bradbury, Democrat. Hayslette, who is completing his first term, operates a radio and television shop, and he and his wife also are in the catering business. Bradbury is an employee of the Illinois Division of Highways. He holds a private pilot's license and flies his own airplane.

It is regrettably true that water samples from most domestic groundwater supplies are not analyzed on a regular basis even for bacteria or simple chemical pollutants such as nitrate, let alone for hazardous contaminants such as viruses, chlorinated hydrocarbons, polychlorinated biphenyls, cyanide, organophosphates, and heavy metals. Therefore, any discernible surficial toxic chemical source should be considered guilty of contamination until proved innocent, rather than innocent until proved guilty, as is now widely the case.

Groundwater contamination from natural causes generally is accepted as uncontrollable. Many urge a similar resignation to their waste disposal methods and argue for them as time-honored practices which pose no real or immediate hazard to public health. Groundwater contamination case histories make evident that this philosophy and the waste disposal practices it encourages must be rejected if our groundwater resources are to be adequately protected from future contamination by man.

References

American Society of Civil Engineers, Committee on Sanitary Landfill Practice of the Sanitary Engineering Division. 1959. *The sanitary landfill.* Manual of Engineering Practice no. 39. New York: ASCE.

Goldstein, S. N., and Moberg, W. J. Jr. 1973. *Wastewater treatment systems for rural communities.* Washington, D.C.: Commission on Rural Water.

Kaufman, Robert. 1971. Discussion at The National Ground Water Quality Symposium, Denver, Colorado, August, 1971.

Smith, H. F. 1971. Subsurface storage and disposal in Illinois. *Ground water* 9: 20–28.

Walker, W. H. 1973. Where have all the toxic chemicals gone? *Ground water* 11: 11–20.

Waltz, J. P. 1972. Methods of geologic evaluation of pollution potential at mountain homesites. *Ground water* 10: 42–49.

Zanoni, A. E. 1972. Ground-water pollution and sanitary landfills—a critical review. *Ground water* 10: 3–16.

11. Nitrate Contamination of the Water-Table Aquifer by Septic Tank Systems in the Coastal Plain of Delaware

John C. Miller

Introduction

Increased population growth in Delaware will result in a greater demand for potable water supplies. At present a large percentage of the public water supply is obtained from deep artesian aquifers. However, future growth will necessitate greater use of the water-table aquifer as the artesian aquifers are pumped beyond safe and economical limits. Currently, Delaware is using only 13 percent of the water from its water-table aquifer, which can yield up to 356 million gallons per day to high-capacity wells. This abundance of water represents 66 percent of the groundwater resources of Delaware.

The water-table aquifer is vulnerable to contamination by objectionable, possibly toxic fluids and water-soluble solids placed on or within the soil. The permeability of the sands that compose the water-table aquifer is such that contaminants can easily move toward public and private wells.

Chemical analyses in the files of the Delaware Geological Survey, reports from private water-analysis laboratories, and comments by officials of other governmental agencies indicated that many private water supplies in Delaware contain abnormally high amounts of nitrate. Inspection of water analyses on file at the Delaware Geological Survey revealed that 25 percent of the shallow wells (less than 50 feet deep) in the state yield water with nitrate levels above 20 ppm (parts per million). Although the U.S. Public Health Service upper limit for nitrate in drinking water is 45 ppm, the Delaware Geological Survey considers the 20 ppm level to be an indication of the beginning of nitrate contamination of the water-table aquifer. (Woodruff, 1970 believes that uncontaminated water of the water-table Columbia Formation contains less than 10 ppm nitrate.)

Although there have been no confirmed reports of fatal infant methemoglobinemia (nitrate-poisoning) in Delaware, there has been one suspected case (D. K. Harmenson, Chief, Bureau of Environmental Health, Delaware State Department of Health, personal communication, 1970). This occurred in 1969 in a small housing development near Millsboro served by a single shallow, driven well. Analyses for two sampling dates at this site indicated that the water contained 158 and 163 ppm nitrate. However, there may have been other factors contributing to the death of the infant.

Because of the number of wells that contain water with a nitrate content greater than 20 ppm and the death possibly due to high-nitrate water, investigations to determine the causes of nitrate contamination were undertaken in two suburbanized areas. The following is a summary of the findings of this study (Miller, 1972) with an expanded discussion of certain aspects.

Physical Environment of the Study Areas

Glasgow Area

The first study area is located in New Castle County on the Atlantic Coastal Plain of Delaware about two miles southeast of Glasgow, Delaware (Figure 1). The area was selected because of the extremely high water table, periodically waterlogged and poorly drained soils, and reports of overflowing septic-tank systems during rainy months. Precipitation in the area averages 45 inches per year. The lack of topographic relief favors slow overland movement of surface water.

Four soils series (Matthews and Lavoie, 1970) are present in the Glasgow study area, and they are classified by the Soil Conservation Service, U.S. Department of Agriculture (1971), according to suitability for on-lot disposal of effluent from septic-tank systems: (1) the Elkton silt loam, covering 25 percent of the area, has severe limitations because of a seasonally high water table and poor permeability within the first four feet; (2) the Fallsington loam and sandy loam, 30 percent of the area, has severe limitations because of a seasonally high water table; (3) the Matapeake silt loam, 5 percent of the area, has slight to moderate limitations; and (4) the Woodstown loam, 40 percent of the area, has moderate limitations because of a moderately high seasonal water table.

The Glasgow area is underlain by the Pleistocene-age Columbia Formation which, in this area, consists of 20 to 30 feet of clayey and silty sands with thin layers of clean water-bearing sands. Well owners find it necessary to use large-diameter dug wells in order to obtain sufficient water. The Columbia Formation is underlain by clays of the Cretaceous-age Potomac Formation.

The study area is about one mile square. Homes are located on

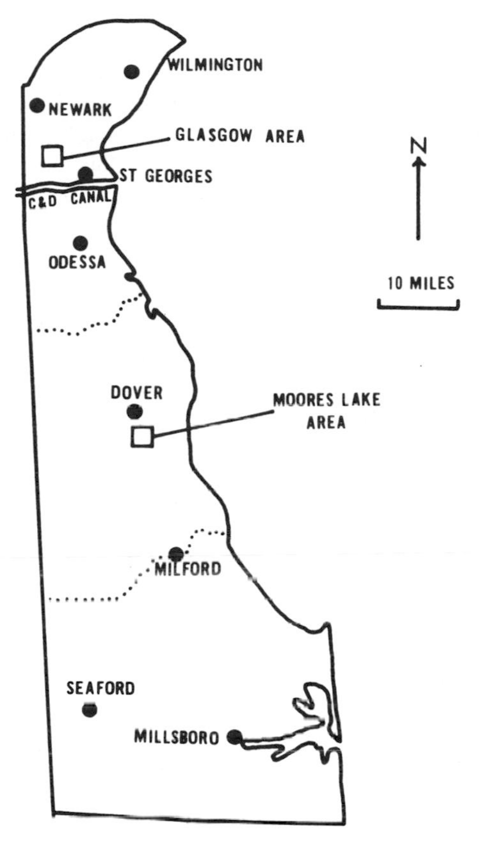

Figure 1.

one-quarter to one-half acre lots; approximately 200 people live in the area. The homes are 2 to 12 years old with shallow driven or dug wells (13 to 35 feet deep) and septic tank systems.

Moores Lake Area

The Moores Lake area, in Kent County, is also located on the Atlantic Coastal Plain of Delaware, south of Dover, Delaware (Figure 1). This area was selected because of reports of nitrate levels which double and triple the drinking-water limits. Precipitation in the area is about 46 inches per year. The topographic relief in the area is moderate, but overland runoff is slight, because of the permeable nature of the soils.

The entire study area is covered by loams and sandy loams of the

Sassafras series: deep, well-drained soils on uplands. The soils are derived from predominantly sandy Pleistocene sediments. The Soil Conservation Service (Matthews and Ireland, 1971) classifies soils of the Sassafras series as having only slight limitations for septic-tank systems because of the low water table and high permeability.

Relatively clean sands and silty sands of the Pleistocene Columbia Formation (15 to 45 feet thick) underlie the area. These sands, serving as the aquifer for local wells, are underlain by alternating sequences of Miocene-age clays and medium- to coarse-grained sands which some homes have tapped as a source of water. These lower sands, the Cheswold and Frederica artesian aquifers, are used by nearby cities and industries because of their lower iron content and higher pH. However, wells in the Columbia Formation (the water-table aquifer) will easily yield 35 gallons per minute with little drilling expenditure, and therefore these sands are used by the developer as the most economical source of water.

The study area is about 0.75 square mile in size. Homes are on one-quarter to one-half acre lots; approximately 2,000 people live in the area. The homes are 2 to 15 years old and have wells 35 to 70 feet deep.

Sampling and Analysis

Samples for nitrate analysis were taken from kitchen faucets and outside spigots and analyzed at the Delaware Geological Survey laboratory within 6 to 8 hours of collection. Analyses were made with a nitrate specific-ion electrode. A discussion of the analytical technique is given in the author's earlier report (Miller, 1972). The specific conductance of the samples (micromhos/cm at 25°C) was measured at the same time. A check on instrumentation and technique was supplied by the Delaware Division of Environmental Control by means of analyses of duplicate samples.

Results of the Analyses

Glasgow Area

In the study area near Glasgow some 37 wells were sampled. Not all wells were sampled each time, as the study area was enlarged after the first sampling date. The range of nitrate content for 93 samples was

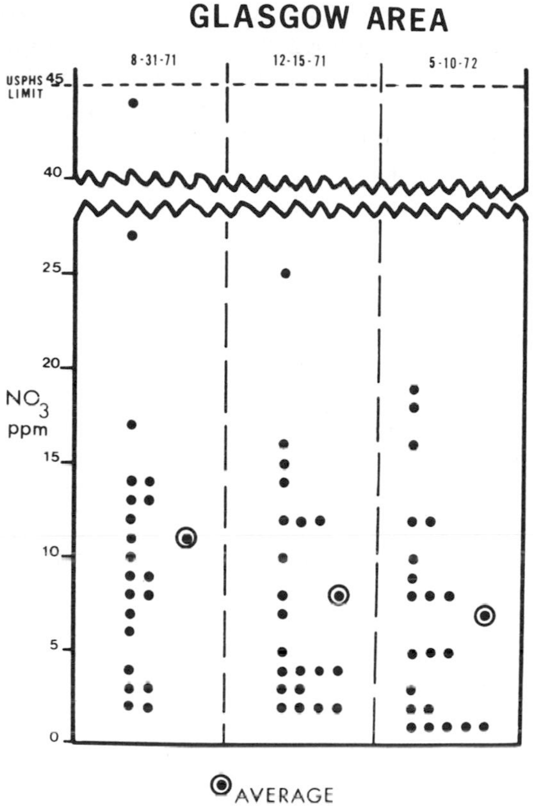

Figure 2.

0.6 to 51 ppm. The results for 21 of these wells which were sampled each time are shown in Figure 2.

The extremely low nitrate levels in most of the Glasgow area can be attributed to the fact that the combination of high water tables and impermeable, waterlogged soils favors local reducing conditions under which the nitrogen compounds of the septic-tank effluent are never oxidized to nitrate but remain as ammonia or organic nitrogen.

The averages of the 21 wells sampled seem to indicate a slight seasonal variation in nitrate level, decreasing from 11 to 7.8 to 6.9 ppm with rising water table. The specific conductance does not seem to exhibit this variation and remains at an average of about 138 micromhos/cm all year.

A number of wells in the area were reported to be contaminated by coliform bacteria at various times (source: New Castle County Health

Unit independent samplings). This is attributable to the high water table and impermeable, waterlogged soils that cause septic-tank systems to overflow and permit effluent to move across the land surface and down the outside of the casings of the large-diameter dug wells.

Under similar physical conditions in other parts of Delaware it is likely that the results will be the same: low nitrate, overflowing septic-tank systems, and bacterial contamination of private water supplies.

Moores Lake Area

During the study period a total of 20 wells was sampled in the Moores Lake area. Because the study area was enlarged after the first sampling, not all wells were sampled each time. The range of nitrate content for 49 samples was 22 to 136 ppm. The results for 12 of these wells that were sampled on each of the sampling dates are shown in Figure 3.

The extremely permeable soils of the Sassafras series favor the movement of oxidized nitrogen compounds down to the water table as nitrates. It is doubtful that this situation will become better; nitrate concentrations will probably increase as the area becomes more populated.

Figure 4 shows groundwater levels in an observation well (Jd42-3, maintained by the U.S. Geological Survey, Dover, Del.) located just 2 miles southwest of the Moores Lake area. The variations in water level are probably similar to those within the area of the wells sampled; that is, highs and lows probably take place at the same time. The magnitude of the fluctuations of groundwater level in the two areas is probably different. The water table in the Moores Lake area is 15 to 25 feet below land surface, whereas the water level in well Jd42-3 is usually about 6 feet from land surface.

The progressive decrease in nitrate and specific conductance can be attributed to simple dilution by rain falling on the site, percolating into the ground, and mixing with the groundwater at the water table. The chart of the amount of groundwater essentially in storage indicates the likelihood that the nitrate concentrations have been and will be at future dates much higher than observed during the three sampling periods.

Pathogenic organisms were quite successfully removed from septic tank effluent by adsorption on the soil despite the high nitrate levels (one to three times the health limit). The Delaware Division of Environmental Control in independent samplings reported that none of the wells was contaminated. However, there was some concern on

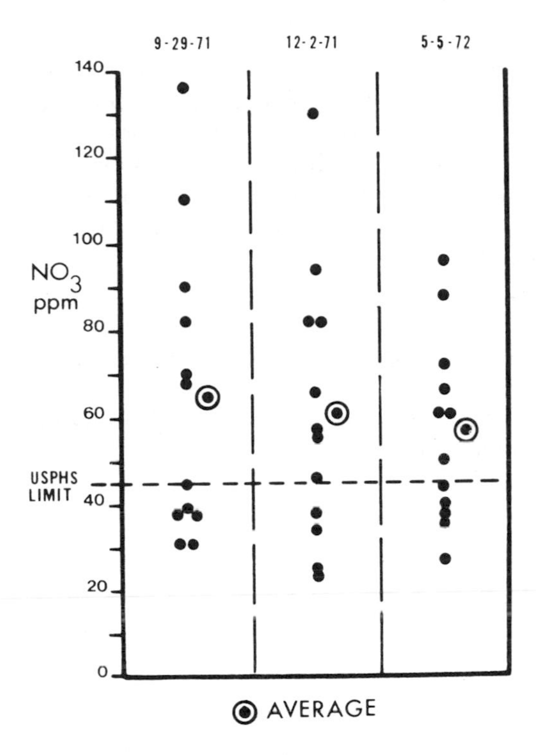

Figure 3.

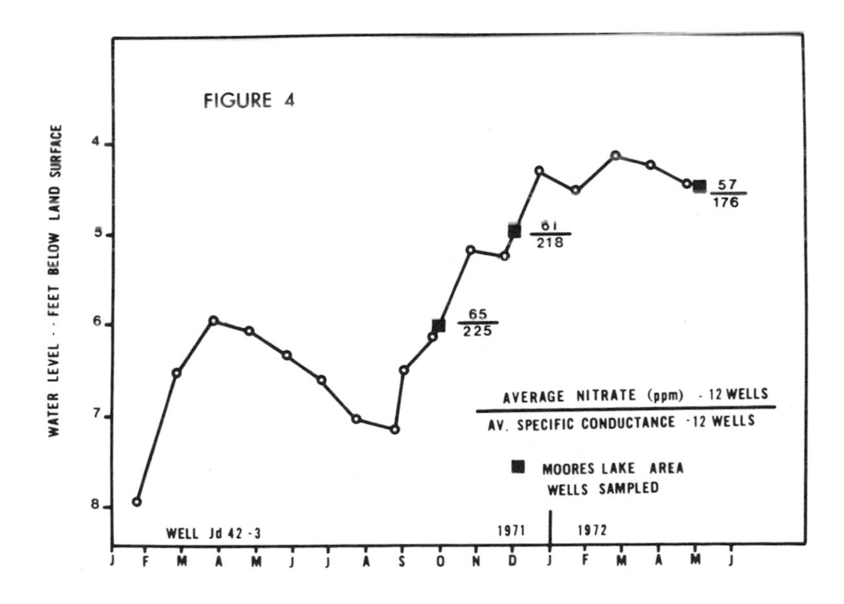

Figure 4.

their part that the extremely high permeability of the Sassafras series soils and the underlying sands would favor the movement of viruses.

Other parts of Delaware with population densities similar to those of the Moores Lake area and with permeable soils and low water table will be subject to nitrate contamination of the groundwater.

A basic conclusion is that the standard percolation test is not a suitable means for determining the acceptability of a site for disposal of septic-tank effluent. Percolation tests in the Glasgow area were conducted during dry periods: the result is overflowing septic-tank systems during wet periods. And, in the Moores Lake area, percolation tests were considered successful because of the rapid infiltration, which may filter pathogenic organisms but also favors oxidation of nitrogen compounds to nitrate.

Definition of Contaminant-Prone Areas

As was pointed out earlier, the water-table aquifer is vulnerable to contamination by objectionable, possibly toxic fluids or dissolvable solids placed on or within the soil. The most vulnerable areas will be those underlain by the most permeable earth materials.

In order to anticipate areas of potentially high nitrate concentration due to present or proposed population densities using septic tank systems, a map of Delaware was prepared which showed areas most susceptible to groundwater contamination by bacteria from overland movement of septic tank effluent. These maps were compiled from hydrologic atlases prepared by the U.S. Geological Survey which show the maximum high-water table below land surface. The boundary between the two types of contamination was taken as the five-foot contour on the maximum high water table (Figure 5). When the water table is generally less than five feet from land surface, bacterial contamination from overflowing tile fields and septic tanks is likely. When the water table is greater than five feet from land surface, nitrate contamination of the groundwater is possible. It must be emphasized that these maps are to serve as guidelines for planning and should be used in combination with soils maps and on-site investigations to evaluate an area. Precautions will have to be taken in soil map interpretation. When the soils are considered to have slight limitations for septic tank systems, they should be reinterpreted to actually have severe limitations.

There were several reasons for selection of the five-foot contour

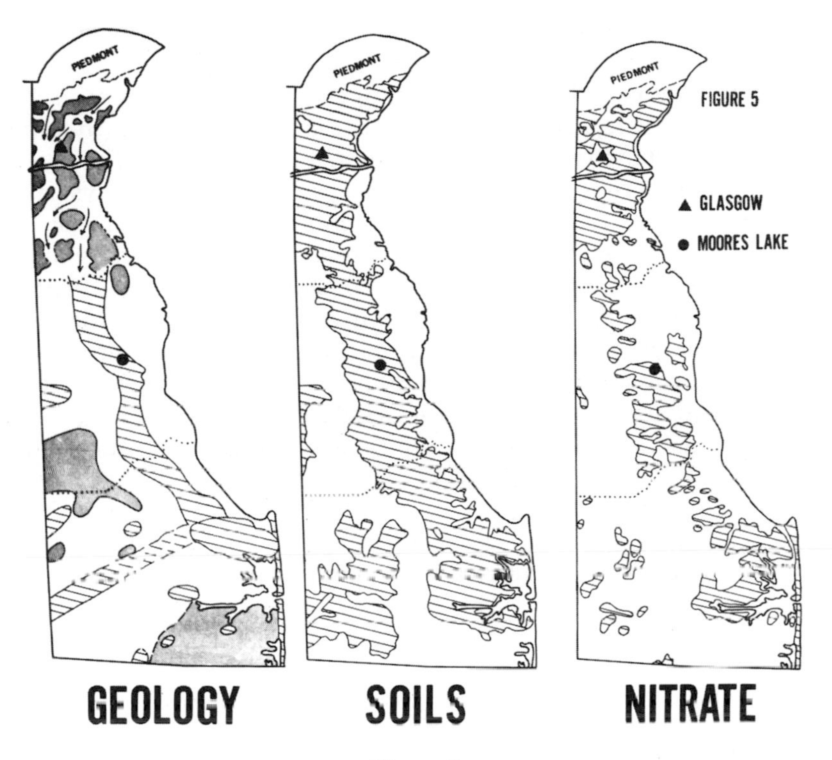

Figure 5.

line on the maximum high water table as the boundary between the two types of contamination. It was felt that conditions favorable to a low water table would also be favorable to rapid drainage of groundwater toward nearby streams. In fact, the areas of highest base-flow of streams in Delaware are located within the area where the water table tends to be more than five feet from land surface (Johnston, 1973). Additionally, a strong similarity was noted between the nitrate-prone areas and those containing soils classified by the Soil Conservation Service, U.S. Department of Agriculture (1973) as predominantly well drained. The soils map in Figure 5 (lined pattern) shows soils that are considered to offer only slight restrictions to the suitability of sites for septic tanks. Soils of lesser permeability (shown in white) are located in those areas that correspond to areas considered prone to contamination by bacteria due to overflowing septic tanks.

The subsequent inspection of geologic maps of Pleistocene sediments prepared by the Delaware Geological Survey (Jordan, 1964 and Spoljaric, 1967) provided further argument for the usefulness of the nitrate/bacteria map. The geologic maps showed the same patterns noted on the nitrate/bacteria map and the soils map. A slightly modified version of these two maps is shown in Figure 5. Basically, the surface of Delaware is covered by an extensive blanket of clays, silts, sands, and gravels of Pleistocene Columbia Formation periglacial deposits. These deposits, representing the sediments carried by ancient streams, are the actual control on both the types of soils formed and the type of pollution that will result from the use of septic tanks in highly populated areas. As shown on the map, the northern part of Delaware is traversed by deep (up to 80 feet) channels of these streams which contain deposits of permeable sands at present, with associated flood plains and islands that contain materials of lesser permeability (Spoljaric, 1967). The middle and lower portions of Delaware, although mapped by a second geologist (Jordan, 1964), show similar features: a long, southeastward-trending major river channel phase (lined pattern) with materials of lesser permeability representing the flood plain portion of the river (in white) out to either side of it. At a point half-way into the southernmost county of Delaware this major channel phase stops and is cut by a southwest-northeast ancient shoreline composed of permeable beach and dune sands. Northwest of the beach-dune line and also in the extreme southeastern portion of Delaware are areas of very low permeability (gray pattern) resulting from finer-grained materials, clayey and silty sands, representing old estuarine and lagoonal sediments.

All of these patterns seen on the geologic map are reflected by the soils map and by the nitrate/bacteria map. Permeable geologic materials will yield permeable soil materials and will favor a low water table and hence the formation of nitrates from septic-tank effluent. Geologic materials of lower permeability will yield poorly drained soils, which will favor waterlogged conditions and a high water table and will therefore result in overflowing septic tanks and tile fields.

A recommendation of the earlier report (Miller, 1972) was that a survey of private water supplies should be conducted throughout the Atlantic Coastal Plain of Delaware where septic tanks are known to be in use. Particular emphasis should be given to areas within the nitrate-prone area. Residents with high-nitrate wells should be cautioned and advised to seek alternate sources if they have infants. Preliminary samplings at two sites within this area (St. Georges and Odessa, Figure 1) show that the nitrate content of the groundwater is one to two times the drinking water standard.

Conclusions and Recommendations

The results of this study applied to other areas of glacial and fluvial sediments outside of Delaware, utilizing the fields of geology, pedology, and hydrology, will undoubtedly be of assistance in locating or preventing areas of high-nitrate ground water. A general rule to be followed for estimating contamination of the water-table aquifer in areas of permeable earth materials is that fluid or soluble contaminants will move, particularly if there are nearby centers of pumping.

Areas such as Moores Lake face the need for an immediate resolution of the problem. In this area, fortunately, there are deeper aquifers separated from the contaminated water-table aquifer by thick clays. The nitrate level of the deeper artesian aquifer is about 1.5 ppm at the site of contamination, which indicates that the contaminants are restricted to the water-table aquifer. Septic tanks should be phased out as public sewerage enters the area. Eventually the water-table aquifer should be purged of nitrate through natural dilution by downward-percolating rainwater. No estimate of dilution time is available as yet. Future residential developments should not utilize septic tank systems. Alternate means of domestic sewage disposal should be considered.

If septic tank systems are used indiscriminantly around the edge of larger cities in Delaware, the groundwater within the Pleistocene Columbia Formation, the water-table aquifer, will be so contaminated by nitrates that it will not be suitable for use by the larger communities as their water needs increase and wellfields move outward. In Dover, Delaware, where water levels in the Cheswold and Piney Point artesian aquifers are declining as a result of heavy pumpage in the Dover, Dover Air Force Base, and Cambridge, Maryland areas, the community will eventually have to resort to wellfields in the Columbia Formation. Such a city as Dover will be surrounded by a ring of nitrate-contaminated water (Figure 6) that it cannot use. It will then be necessary for the city to extend costly water lines out to uncontaminated areas.

The data within this report support the long-held and little-applied belief that the use of septic tanks in high density suburban developments has detrimental effects on ground water. The septic tank was not intended to be used in this manner but in areas of lower population density away from cities. If the trend toward suburbs with septic tanks is continued, it is very likely that the health effects from such contamination will soon be seen.

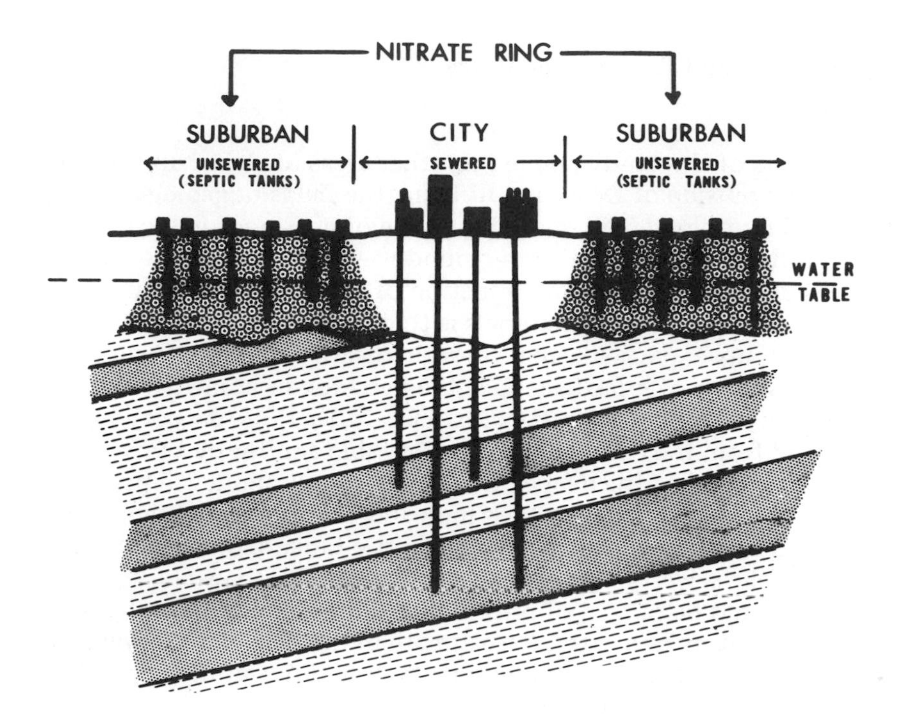

Figure 6.

References

Johnston, R. H. 1973. *Hydrology of the Columbia (Pleistocene) deposits of Delaware: an appraisal of a regional water-table aquifer.* Delaware Geological Survey, Bulletin no. 14.

Jordan, R. R. 1964. *Columbia (Pleistocene) sediments of Delaware.* Delaware Geological Survey, Bulletin no. 9.

Matthews, E. D., and Lavoie, O. L. 1970. *Soil survey of New Castle County, Delaware.* Delaware Agricultural Experiment Station & Soil Conservation Service, United States Department of Agriculture.

Matthews, E. D., and Ireland, W., Jr. 1971. *Soil survey of Kent County, Delaware.* Delaware Agricultural Experiment Station and Soil Conservation Service, U.S. Dept. of Agriculture.

Miller, J. C. 1972. *Nitrate contamination of the water-table aquifer in Delaware.* Delaware Geological Survey, Report of Investigations no. 20.

Soil Conservation Service, U.S. Dept. of Agriculture and Delaware Agricultural Experiment Station. 1971. *Soil limitations for on lot disposal of effluent from septic systems, New Castle County, Delaware.*

Soil Conservation Service, U.S. Dept. of Agriculture. 1973. *The first state resource conservation and development plan (Delaware).* Dover: Soil Conservation Service.

Spoljaric, N. 1967. *Pleistocene channels of New Castle County, Delaware.* Delaware Geological Survey, Report of Investigations no. 10.

U.S. Public Health Service. 1962. *Public Health Service drinking water standards.* U.S. Dept. of Health, Education, and Welfare Publication no. 956.

Woodruff, K. D. 1970. *General ground-water quality in freshwater aquifers of Delaware.* Delaware Geological Survey, Report of Investigations no. 15.

12. Virus Movement into Groundwater from Septic Tank Systems

Otis J. Sproul

Viruses can be recovered from any water that has been subjected to viral contamination. In situations where wastewater is to be discharged to the local environment, e.g., one's dooryard, as with the septic tank system, the concern of the homeowner should be obvious, especially if his water supply is a private well only a few feet from the septic tank system.

Obviously, however, people have not been as concerned as they should be, for the literature is replete with instances of viral contamination of well-water supplies (Mosley, 1967; Berg, 1973). Viruses from these supplies are routinely involved in outbreaks of infectious hepatitis and gastroenteritis.

It is the object of this paper to discuss the methods of predicting the capacity of a septic tank–soil absorption system to remove viruses and to develop criteria to assess this capacity.

Characteristics of Viruses

Viruses are very small particles characterized by their inability to reproduce outside a living cell. The polioviruses, for example, are on the order of 0.02 microns in diameter. Most viruses have a protein coat which surrounds the nucleic acid, the infectious part of the virus. The reactions of viruses in the environment are characteristic of the protein in the coat. Predictions of virus behavior in water can often be made solely from a knowledge of protein chemistry.

Virus infections in humans can be produced by the ingestion of very low numbers of viruses. Katz and Plotkin (1967) have shown that only 1 or 2 PFU or $TCID_{50}$ (PFU-plaque forming units; $TCID_{50}$ -tissue culture infectious dose to infect half the tubes) is capable of producing infections. We should be concerned, therefore, with the presence of only one virus particle in water.

Virus Density and Virus Removal in Septic Tanks

Clarke et al. (1964) estimated that the level of viruses in domestic wastewater was about 7000 PFU/liter. Berg (1973) reported recoveries of 32 to 107 PFU/liter in the midwestern United States during September and November 1970. Nupen and Stander (1972) reported peak virus concentrations of 270,000 to 2,147,000 $TCID_{50}$/liter in settled wastewater in South Africa.

Little information is available on virus removal in septic tanks. Laboratory studies by Clarke et al. (1961) showed minor removal of poliovirus type 1 after three hours of sedimentation of domestic wastewater. Plant scale data by Malherbe and Strickland-Cholmley (1967) show an inconsistent and undependable removal of viruses. In some cases, viruses increased in numbers after sedimentation, probably because of the breaking up of fecal material and liberation of the viruses to the bulk solution. One should note, however, that the methods of virus sampling in the field normally include preliminary sample cleanup which discards the suspended solids and examines only the water phase for its virus content. Viruses are embedded in these solids, and the raw sludges which result from the solids contain very large numbers of viruses and other pathogens.

Factors Affecting Virus Removal in the Soil

Virus removal or inactivation in the soil after discharge from a septic tank may be effected by any or a combination of the mechanisms below:

(1) Adsorption
(2) Bacterial enzymatic attack
(3) Natural dieoff

Dilution, while not a method of removal or inactivation, will reduce the concentration of viruses and hence the likelihood that any particular glassful of groundwater will contain viruses. Of these possible removal mechanisms, the most important is adsorption. To maximize the removal of viruses in the ground, then, one should promote those conditions which make adsorption possible or which promote the rate at which it occurs.

The mass transfer equation indicates that the important elements to consider in the adsorption reaction are as follows:

(1) Approach velocity
(2) Specific surface, surface area per unit volume
(3) Packing of adsorbing media
(4) Rate constant *k* which in turn is a function of the surface characteristics of the soil and of the viruses
(5) Reaction rate constant *r*

The reaction rate constant *r* is necessary because viruses are nonconservative pollutants that are subjected to bacterial attack and to natural dieoff.

Attempts to develop a rational model of virus removal for field use have not been successful, because soil and wastewater characteristics vary widely. Investigations have, therefore, proceeded on an empirical basis and have not examined the removal system in toto. The following areas have been studied by a number of investigators for virus adsorption under absorption trench and groundwater conditions:

(1) Type of subsoil
(2) Degree of saturation of soil
(3) Flowrate or application rate of water
(4) Water quality
(5) Adsorption characteristics of different viruses

Type of Subsoil

Virus adsorption in the soil occurs in part because of an interaction between chemical groups on the virus protein coat and groups on the adsorbing surface. The opportunity for this adsorption to occur is obviously greater when more groups in the soil are available for reaction. It follows, therefore, that finer soils have a greater adsorption capacity than coarser soils. Viruses are not removed from groundwater after it has reached fractured ledge, since there are very few fine materials present in the fissures.

Robeck et al. (1962) showed a trend toward better removal of poliovirus type 1 on 2 feet of 0.27 mm (effective size) Ottawa sand than from a 0.78 mm sand with a surficial flow rate of about 1.55 ft/day. The poliovirus was suspended in clean tapwater at concentrations of from 10,000 to 30,000 PFU/m1. Virus removals under these conditions were in excess of 97 percent. Upflow experiments with two-foot beds of a 0.18 mm Newton sand and a 0.38 mm Chillicothe sand showed only an occasional virus from the fine sand but usually 8 PFU/m1 from the coarser sand. Saturation of the sand with viruses was not reached in the seven months of the test. It was speculated that since no saturation was reached, a virus dieoff occurred which was sufficient to expose new adsorption sites. No evidence was presented to show this or the capacity of the sand to adsorb viruses.

Eliassen et al. (1967), using the T2 bacteriophage and Monterey sand with effective sizes from 0.12 mm to 0.50 mm, determined that sorption capacity measured in a static batch test had no relation to the sand size. When these sands were used in columns with continuous throughput, however, the finer sands exhibited better virus-removal characteristics.

A number of investigators have shown that soils with lower pH reactions with water have a higher adsorption potential. Hori et al. (1970) found immediate breakthrough of poliovirus type 2 with a Tantalus soil but not with Wahiawa and Lahaina soils. The tantalus soils had a pH of 7.2, while the Wahiawa and Lahaina soils had pH values of 5.0 and 5.3 respectively. Wentworth (1969) determined that similar amounts of T4 bacteriophage were adsorbed onto four different silicate minerals at any pH when the surface area of minerals was that which would give the same acidity. The weights of olivine, kyanite, microcline, and actinolite necessary to give the same acidity to the surface area differed by factors of up to 2. Drewry and Eliassen (1968) indicated that decreasing the pH of the soil reduces its negative charge and decreases the repulsion between the virus and the soil.

Soils with higher fractions of silt and clay particles than sand have higher capacities for virus removals. Eliassen et al. (1967) showed that their typical sand had a retention capacity for the T2 bacteriophage of 10^8 viruses per gram of sand at linear velocities of 6 to 7 feet per day. Their soils at linear velocities of about 1.8 feet per day had retention capacities of 7.6×10^9 to 2.0×10^{10} viruses per gram. The T2 virus was used in concentrations of 10^7 to 10^8 PFU/ml. The soil columns with depths of about 12 inches retained 99 percent or more of the viruses charged to the column.

Degree of Saturation of Soil

Data are not available to compare the removal of viruses on the same soil under saturated and unsaturated conditions. It is known from the wastewater treatment literature that viruses are removed better when there is a greater amount of biological activity, i.e. activated sludge vs. trickling filtration. Better penetration of oxygen into the wastewater near the ground surface would occur under unsaturated conditions. This in turn would permit more biological activity and consequently better virus removal.

Flowrate of Water

All studies that have investigated the effect of flowrate show decreased virus removal under higher flowrates. Table 1, adapted from data of Robeck et al. (1962), demonstrates this well. Duplicate runs at flowrates in excess of 1.6 feet per day gave widely varying results for virus removal. More consistent results were obtained with rates less than this. Eliassen et al. (1967) working with the T2 bacteriophage also showed better removals on sand at lower flowrates. They also concluded that the virus saturation breakthrough (the time required for the effluent virus concentration to equal the influent concentration) varied directly with the natural logarithm of the filtration rate.

Wellings et al. (1973)* have found viruses in a full-scale irrigation system after filtration of an activated sludge effluent through 5 feet of a sandy soil in St. Petersburg, Florida. Examination of the effluent being spray-irrigated determined that there were about 2 viruses/liter present and that the BOD removal in the plant was about 90 percent. The examination results obtained at a surficial water application rate of 0.13 feet per day are shown in Table 2. Three viruses were isolated by the membrane adsorption procedure from the 115 gallons sampled. The actual number of viruses present was larger than this, since the concentration and isolation techniques used would not affect a complete recovery. More recent preliminary work has failed to find viruses at a flowrate of 0.024 feet per day (Wellings, 1973).

Water Quality

The principal water-quality parameters that affect the soil virus adsorption process are the pH, ionic strength, and concentration of organics. The adsorption of viruses to silts and clays is particularly sensitive to the inorganic ion concentration. Carlson et al. (1968) have shown that adsorption of T2 bacteriophage and poliovirus type 1 on clays was less than 30 percent or so in distilled water conditions but reached a maximum of 80 to 99 percent when the suspension was 0.03 M to 0.1 M in sodium. Only one tenth of these concentrations were required for similar results with calcium. All of their results indicated that the primary impact of the added cations was to cause a proper charge distribution leading to a clay-cation-virus bridge after which the virus was removed with the clay. Competition for the adsorption sites was demonstrated with several proteins and with

*Wellings amplified these findings in a personal communication to the author (1973).

Table 1
Poliovirus 1 Removal at Various Flowrates[a]

Media	*Surficial Velocity Ft/day*	*Percentage of Removal*
California Dune Sand		
Eff. size: 0.28 mm	2.7–5.3	> 99.99
Depth: 1.25 feet		
Ottawa Sand		
Eff. Size: 0.28 mm	1.6	> 98
Depth: 2 feet		
"	6.7	22–96
"	385–1155	1–50

[a]Data adapted from Robeck et al. (1962).

domestic wastewater. Ten ml of domestic wastewater per 50 mg/1 of clay reduced the adsorption of T2 virus from 93 to 81 percent. Eliassen et al. (1967) also found reduced adsorption of the T2 virus to their soils when the inorganic concentration was reduced in the suspending medium. They indicated that this effect was most pronounced on soils with lower clay contents.

Lo and Sproul (1974) presented the data in Table 3 for the adsorption of poliovirus 1 on a 6.5 inch column of the silicate mineral microcline (effective size = 0.15 mm) at a surficial flowrate of two feet per day from an artificial natural water and from an activated-sludge effluent. Breakthrough of virus on the clean-water column did not occur until the sixth day of the run, but only two days were required from breakthrough with the activated sludge effluent. The

Table 2
Viruses Obtained after Five Feet of Filtration of St. Petersburg's Activated Sludge Effluent at a Rate of 0.13 Feet per Day[a]

Date	*Sample Size Gallons*	*Viruses Isolated*	
		Number	*Type*
1-24-73	15	0	
2-02-73	5	2	Reovirus type 1
2-14-73	15	0	
2-15-73	30	0	
2-27-73	50	1	Not identified

[a]Data from Wellings et al. (1973).

Table 3
Poliovirus 1 Adsorption on a 6.5 inch Column of Microcline[a]

	Natural Water		*Activated Sludge Effluent*	
Day	*Viruses in Influent*	*Viruses in Effluent*	*Viruses in Influent*	*Viruses in Effluent*
1	28×10^6	0	28×10^6	0
2	51×10^6	0	49×10^6	25×10^3
8	204×10^6	115×10^3	161×10^6	600×10^3

Surficial application rate = 2 feet/hour.
Effective size = 0.15 mm.

[a]Abstracted from Lo and Sproul (1974).

numbers of viruses in the effluent from the activated-sludge effluent column did not increase as rapidly as that from the clean-water column. This indicated an initial and continued penetration through the column, perhaps after adsorption on the suspended solids in the feed solution or displacement on the adsorption sites by the organics. The suspended solids in the effluent from the latter column were 5 mg/1 at day one and increased to 9 mg/1 on day seven.

Adsorption of viruses to soil particles is higher under lower pH conditions. As explained earlier, lower pH values reduce the magnitude of the negative charge on the soil particles. Lower pH values also reduce the negative charge on the virus particle by decreasing the ionization of the carboxyl group and increasing the ionization of the amino groups. Hori et al. (1970) and Eliassen et al. (1967) have noted these effects in their virus adsorption work.

Adsorption Characteristics of Viruses

Viruses have different proteins in their protein coat, which would be expected to have different adsorption characteristics. Additionally, the bacteriophages frequently have more complicated structures than animal viruses and would be expected to have different adsorption characteristics. The T2 bacteriophage, for example, has a head, a tail, a baseplate at the tail, and tail fibers attached to the baseplate. The adsorption of this virus takes place at the baseplate after proper orientation of the tail fibers. Hori et al. (1970) noted a decreased adsorption of the poliovirus type 2 as compared to that obtained by Tanimoto et al. (1968) for the T4 bacteriophage under similar adsorption conditions for Wahiawa and Lahaina soils.

Discussion of Data

The distinguishing feature of the data from the literature is that viruses are always seen in the effluents from soil and sand columns. It is true that these soils were challenged with higher virus concentrations than would exist in the "real" situation and that the columns seldom exceeded two feet in depth. However, the field experience in St. Petersburg,* with a relatively low application rate and low influent virus numbers but with viruses present after five feet of filtration, does not encourage confidence in the ability of the soil environment to reduce virus numbers to zero under marginal conditions. This is especially significant since these groundwaters are used for drinking-water supplies without disinfection.

Long filtration distances will affect virus removals. Ward (1965) in a full-scale test by the swab-detection technique could find no viruses after filtration of a completely treated effluent through 200 feet of an 8 to 12 foot layer of sand and gravel. The test used 160,000 gallons of effluent containing 500 $TCID_{50}$ /ml of poliovirus type 3.

Methods to Assess Virus Removal

In the soil absorption trench system the only variables that can be controlled are the percolation rate, soil type and depth, depth to impervious layers, and water-application rate. Criteria for maximizing the capacity of these variables to remove viruses are given in Table 4. The values are about the same as used at present with the exception of the application rate for water. The currently used rate of about 0.13 feet per day (1 gpd/sf) is in excess of what would be desirable where the soil is underlaid by fractured ledge. Rates up to 1 gpd/sf would be satisfactory where the flow distance through the soil is much in excess of 10 feet.

Any set of standards is of little value if improperly applied. The agency administering the program must vigorously pursue the application of its standards to prevent viral contamination of groundwater supplies. Most of the cases of virus disease transmission by groundwater appear to have been caused by gross violations of existing standards and could have been prevented by adherence to even the most casual standards.

*F. M. Wellings 1973; personal communication.

Table 4
Criteria to Promote Virus Removal in the Soil

Type of subsoil	Deep soils most desirable. Sandy soils are effective but require greater depths for good removals. Fractured ledge gives no removal.
Depth of soil	Minimum of 5 to 10 feet to ledge or impervious material.
Degree of saturation	Minimum of 5 feet to groundwater table.
Application rate of water	Maximum rate of 0.05 to 0.10 feet per day (0.4 to 0.7 gpd/sf).
Water quality	Little control possible. Clean water best.
Adsorption characteristics of viruses	No control possible

Research

The development of a rational mass transfer model to predict virus removal would be useful. Because of the great differences in the boundary conditions which exist from one soil to another, the model's greatest use would be in basic studies rather than in routine fieldwork. Only two or three of the more than 100 known enteroviruses have been examined for the potential of soil systems to remove them. Additional work is required to determine the soil adsorption potential for many more of the other viruses.

The use of the hundred-foot separation distance between a well and the absorption trench is based on little other than folklore and extrapolation of results using two-foot columns of soil. Laboratory and field studies are needed to verify the relationships for a variety of field conditions.

References

Berg, G. 1973. Microbiology—detection and occurrence of viruses. *Journal of Water Pollution Control Federation* (annual literature review) 45: 1289.

Berg, G. 1972. Reassessment of the virus problem in sewage and surface and renovated waters. Paper read at Sixth International Water Pollution Research Conference, 18-23 June 1972, Jerusalem, Israel.

Carlson, G. F., Jr., Woodard, F. E., Wentworth, D. F., Sproul, O. J. 1968. Virus inactivation on clay particles in natural waters. *Journal of Water Pollution Control Federation* 40: R89-R106.

Clarke, N. A., Berg, G., Kabler, P. W., Chang, S. L. 1964. Human enteric viruses in water: source, survival and removability. In *Advances in water pollution research*, vol. 2, pp. 523-536. New York: Macmillan Company.

Clarke, N. A., Stevenson, R. E., Chang, S. L., Kabler, P. W. 1961. Removal of enteric viruses from sewage by activated sludge treatment. *American journal of public health* 51: 1118-29.

Drewry, W. A., and Eliassen R. 1968. Virus movement in groundwater. *Journal of Water Pollution Control Federation* 40: R257-R271.

Eliassen, R., Ryan, W., Drewry, W., Kruger, P., Tchobanoglous, G. 1967. *Studies on the movement of viruses in groundwater.* Final report to the Commission on Environmental Hygiene of the Armed Forces Epidemiological Board. Stanford: Stanford University Press.

Hori, D. H., Burbank, N. C., Jr., Young, R. H., Lau, L. S., Klemmer, H. W. 1970. *Migration of poliovirus type 2 in percolating water through selected Oahu soils.* Honolulu: University of Hawaii, Water Resources Research Center technical report no. 36.

Katz, M., and Plotkin, S. A. 1967. Minimal infective dose of attenuated poliovirus for man. *American journal of public health* 57: 1837-40.

Lo, S. H., and Sproul, O. J. 1974. Virus adsorption from natural water and wastewater by silicate minerals. In preparation.

Malherbe, H., and Strickland-Cholmley, M. 1967. Quantitative studies of virus survival in sewage purification processes. In *Transmission of viruses by the water route*, ed. G. Berg, pp. 379-388. New York: John Wiley and Sons.

Mosley, J. W. 1967. Transmission of viral diseases by drinking water. In *Transmission of viruses by the water route*, ed. G. Berg, pp. 5-25. New York: John Wiley and Sons.

Nupen, E. M., and Stander, G. J. 1972. The virus problem in the Windhoek waste water reclamation project. Paper read at Sixth International Water Pollution Research Conference, 18-23 June 1972, Jerusalem, Israel.

Robeck, G. G., Clarke, N. A., Dostal, K. A. 1962. Effectiveness of water treatment processes in virus removal. *Journal of American Water Works Association* 54: 1275-90.

Tanimoto, R. M., Burbank, N. C. Jr., Young, R. H., Lau, L. S. 1968. *Migration of bacteriophage T4 in percolating water through selected Oahu soils.* Honolulu: University of Hawaii, Water Resources Research Center technical report no. 20.

Ward, P. C. 1965. Santee filtration study. Sacramento: California Department of Public Health, Bureau of Sanitary Engineering, unpublished report.

Wentworth, D. F. 1969. Virus adsorption as related to surface properties of silicate minerals. Master of science thesis, University of Maine.

Wellings, F. M., Lewis, A. L., Mountain, C. W. 1973. Virus studies in a spray irrigation project. Paper read at Institute of Food and Agricultural Sciences Second Annual Wastewater Workshop, May 1973, Gainesville, Florida.

13. Engineering Economics of Rural Water Supply and Wastewater Systems

Michael D. Campbell and Steven N. Goldstein

PART I. ELEMENTS OF SYSTEM DESIGN

A Rural Problem

Millions of Americans, especially in remote, economically depressed regions, e.g. "Appalachia" and rural minority centers, do not have safe drinking water or sanitation facilities. Consider the following facts from the studies which have been done:

(A) Some 75 percent of the population is served by public water systems, but a recent Public Health Service survey of selected systems found that 41 percent of these systems failed to meet PHS Drinking Water Standards. Only 50 percent of the systems serving fewer than 500 people met even the minimum standards (McCabe et al., 1970).

(B) In 1971, 70 percent of the population was on central sewer systems and 92 percent of these were provided with some sewage treatment, but only 54 percent were estimated to have adequate treatment. Americans without sewer service or with inadequate sewage treatment thus number nearly 65 percent of the total population. The dominant portion of the unsewered population can be classified as rural, and poor sewage treatment is also more common in rural areas than in large cities (Wenk, 1971).

(C) A 1969 study by the Farmers Home Administration (FmHA) identified over 30,000 communities with populations below 5,500 which needed assistance in either building a water system for the first time or improving an inadequate one. A similar number of communities needed assistance for sewer systems. The study did not cover communities considered by FmHA to be unsuitable for central systems (U.S. Department of Agriculture, 1969).

(D) Surveys which have been done of individual water supplies suggest that they are worse than community supplies. Recent field studies in three southern states found a substantial majority of the individual supplies contaminated, in some cases as high as 90 percent. Nearly one quarter of the American population, most of them in rural areas, relies on individual water supplies (U.S. Environmental Protection Agency, 1972). From these figures and the studies of rural community water supplies, an estimated 20 to 30 million Americans in rural areas are drinking unsafe water.

These and other data lead to one conclusion: existing water and sanitation facilities in the United States are inadequate, and it is rural America that is hardest hit. This is due partly to weaknesses in the present system and partly to the nature of rural life itself (Morgan and Cobb, 1973).

Pollution and population dispersal are the main causes of the rural water problem. Most of the nation's streams are now polluted by human and industrial wastes, and groundwater aquifers, which are the major source of water in rural areas, have also been adversely affected. As a result, an ample and sanitary supply of water can be obtained only by drilling, pumping, treating, piping, and storing. But the scattered rural population, especially those living outside any incorporated municipality, cannot usually be reached by central water systems.

Weaknesses in the National Delivery System

While the national delivery system for rural water and sanitation has reached many rural residents in the past, the increased pollution of water supplies, the need for more complex and expensive facilities, and the general shift of human and financial resources away from rural areas have made this system progressively less satisfactory in recent years. The following are major weaknesses in the present system:

(1) *Policy and Priority*. There is no coherent and adequately supported national commitment to the provision of basic sanitation services, both water and wastewater, for rural residents or to assistance for areas most in need.

(2) *Financing*. Subsidized and nonsubsidized financing is not available in many rural areas to residents who need it most.

(3) *Development*. The limited availability of public and private developers prevents many rural residents from assembling and managing the resources required to provide sanitation services.

(4) *User Support*. Users, who generally want adequate domestic sanitation services, are not aware of the steps involved in obtaining and sustaining the service and the role they must play in this endeavor.

(5) *Technology*. Design and construction of facilities is not sufficiently directed at meeting technical problems in rural areas, although most common problems can be overcome at a reasonable cost with existing technology.

(6) *Operation and Maintenance*. Inadequate attention to operation and maintenance has often meant that services which have been

established could not be maintained over the long period (Zimmerman, 1973).

The Responsibility for Action

In order to develop a national commitment to good water and sanitation facilities the responsibilities of the public and private sectors must be determined. The private sector does play a major role in water and sewerage in the United States, especially in the development of groundwater technology, sewer treatment, etc. In addition, about 15 percent of all water companies are private, profit-making concerns. They serve approximately 15 million people mostly in rural areas.

Profit-making interests have not been able, however, to extend service to all rural residents. Scattered, low-income rural families cannot afford a central distribution or collection system, nor even individual facilities constructed by private contractors. According to the Office of Economic Opportunity, low-income families cannot pay water bills or meet loan payments for water supply or sanitation facilities much in excess of $7.00 per month.

What seems to be required is some form of public subsidy for the construction and operation of facilities. This kind of public assistance is now so common that it is no longer considered a public subsidy. Certainly schools and roads for all citizens would have been impossible without governmental support.

Local governments, in one way or another, operate the vast majority of the nation's water systems. State governments, through health departments and planning bodies, provide research, regulation, and sometimes financing. The federal government, long involved with research in water and sanitation through such agencies as the U.S. Public Health Service, the U.S. Office of Water Resources Research, the U.S. Geological Survey, and the U.S. Environmental Protection Agency, has been an important source of funding or significant technical assistance for water and wastewater projects.

There is substantial precedent for public support of water supply systems. Indeed, between the early 1940's and 1968, the U.S. Agency for International Development contributed close to one billion dollars to the development of water supply systems in foreign countries (McJunkin, 1969). Within the United States, however, there has been no public effort of a scope and cohesiveness to match the effort which has been made abroad.

Weaknesses in Funding Policy

Although financing by the Farmers Home Administration has significantly advanced the quantity and quality of rural water facilities, it is not unfair to point out that their eligibility requirements for individual loan programs (numbers 502 and 504) have kept funds from reaching the rural poor. These people sometimes lack clear title to their property and often cannot meet loan payments when the life of the loan is a short ten years. A persistent shortage of grant funds, recently deteriorating into complete stoppage, has also limited the agency's ability to respond to low-income residents.

Even more importantly perhaps, grant and loan funds for the community systems have by and large been extended only to "central" systems with one water source and treatment facility, mostly on the grounds that efficiency and continuity of operation are best assured by such systems. Rural communities or clusters of houses not meeting FmHA criteria for central systems have thus been excluded.

Policy for the Future

Public assistance to those who need help and are willing to help themselves hardly needs a justification; nor should there be any misunderstanding about the extent to which low-income residents help themselves. Under the previous FmHA water association plan, where systems were financed by a combination of 50 percent grant and 50 percent loan funds, users in the end would pay about 79 percent of the total cost of the system (initial capital plus operating expenses) or 62 percent of the present value of the total system cost. Loan funds should have lower interest rates and longer repayment periods so that low-income families can use them.

In any case, future funding agencies should be less restrictive about the kinds of water systems they will finance. The difficulties involved in reaching low-income residents by traditional central systems have led in recent years to attempts to devise a new approach based on the experience of rural electric cooperatives. This approach makes possible the kind of effort which is broadly endorsed in the American ethic, a combined public-private effort.

The public-private approach is being tested in various parts of the country by the National Demonstration Water Project (NDWP). NDWP has developed a number of model projects all designed to demonstrate the effectiveness of local organizations in providing water and sanitation facilities for rural residents.

NDWP has also established a national clearinghouse for rural water information. Developmental work for the NDWP program has been funded by OEO. FmHA has provided much of the money for construction of water supply and sewage facilities. EPA is now granting funds for construction of sewage facilities. Private social and technical research organizations representing the groundwater industry and various community organizations have participated in the development activities.

The original NDWP project sprang from the effort of low-income residents in a five-county area around Roanoke, Virginia, to obtain adequate water supplies (*Water Well Journal*, 1971). With funding from the U.S. Office of Economic Opportunity, NDWP established a series of separately incorporated water companies and trained residents to operate the companies as nonprofit associations. After a considerable struggle, the necessary approvals were obtained and financing secured from the Farmers Home Administration for the construction of facilities. Several companies are now in full operation and others are in various stages of development. Water is now being supplied to residents who never before had an adequate water supply.

With the Roanoke experiment a success, NDWP is testing this model in various areas with different problems. A project now under way in Logan, West Virginia, includes wastewater facilities as well as water. Figures 1 and 2 are typical of the West Virginia project area. Another model project in Beaufort-Jasper counties of South Carolina is an attempt at a cooperative arrangement with a local municipality in one area and a water system which included fire protection in another. NDWP is working on a field demonstration project in conjunction with the National Rural Electric Cooperative Association (Zimmerman and Cobb, 1972).

The NDWP Approach

The precise choice of implementing organization is not the crucial issue. What is important is that the present lack of an organization to implement a program of rural water development is a serious weakness in the present national delivery system. To meet this weakness in national delivery, NDWP adopted certain concepts in the development of its field projects (Zimmerman and Cobb, 1972; Campbell and Lehr,

Figure 1.

1973a; Goldstein and Moberg, 1973). For example, governmental agencies have tended to feel that only a central water source or a central treatment facility could provide satisfactory service and quality (Andrews, 1971), because central *management* has been critical to the performance of these facilities. NDWP has felt, however, that central management need not be tied to central systems in the physical sense. NDWP early recognized the possibility of employing central management for a number of water sources and sewage disposal facilities. Wells and small treatment plants could serve a varying number of people all as part of the "system" in the sense of management.

In NDWP experience, the local conditions and their respective impact on the specific project are critical to ultimate configuration of the system. These conditions or field parameters fall into three broad categories:

(1) Geographical (topography, population density, surface reservoir proximity, suitability, etc.)
(2) Hydrological (surface water quality and availability, groundwater quality and availability, etc.)
(3) Political (state and federal regulatory agency attitude, etc.)

Figure 2.

In one NDWP project geographic factors such as frequent bedrock exposure and a clustered population density in isolated areas motivated a design which reduced the extent of distribution line construction between each isolated group of homes as much as possible. Areas of low relief having unconsolidated sediments at the surface require less capital for distribution-line construction.

With regard to the hydrologic parameter, local water availability and quality are reviewed with respect to either a surface-water or groundwater source. Since a groundwater source is usually favored on economic grounds, its quality becomes a significant factor (Campbell and Lehr, 1973; Mirshleifer et al., 1960; Bourcier and Forste, 1967). If previous test drilling and production analyses indicate that significant treatment will be required to remove iron, manganese, etc. from the groundwater, the use of an available surface-water source may be more economical. And sometimes it is less expensive to construct a water reservoir with its attendent treatment plant than a well system drawing from the groundwater reservoir.

The third field parameter which affects the design of the system is the political factor. Problems often arise when a specific design is presented to the regulatory agency for approval. One agency historically prohibits PVC pipe. Another may require specific well sizes. Most states apparently require system design based on a per capita usage of 100 gpd, an unrealistically high figure even for many affluent suburban homes and one which places an unduly high construction burden on the rural community. A strong tendency exists for regulatory agencies to favor systems that are overdesigned and economically unsuited to the project's local conditions. This tendency most likely stems from the agencies' concern for the system's longevity and operational simplicity. The common preference for a central high-capacity well system is based more on operational control considerations than on engineering grounds.

Local Parameters Translation

There are four generally accepted system types or alternatives for obtaining a community water supply in rural areas:

A. Treatment of raw surface water, e.g. small surface reservoir, river, etc.
B. Purchase of treated surface or ground water, e.g. extension of existing water lines, etc.
C. Construction of a single high-capacity well system, e.g. one well, central treatment plant, extensive distribution system.
D. Construction of multiple or "cluster" well system, e.g. more than one well, additional treatment plants, less extensive distribution systems.

NDWP field affiliate projects have so far employed all but Alternative A. In the NDWP approach the relative impact of the local parameters is translated into these possible system types and the *total* cost of each of these systems is compared by means of the following equations:

$$S_R = P_W + T_{CR} + D_{CR} + 0_{CR} + M_{CR} \text{ (Raw Surface Water System)}$$

$$S_T = P_P + D_{CE} + M_{CE} \text{ (Purchased Water System)}$$

$$S_C = W_{CC} + T_{CC} + D_{CC} + O_{CC} + M_{CC} \text{ (High Capacity Well System)}$$

$$S_M = \sum_{i=1}^{n} (W_{CM_i} + T_{CM_i} + D_{CM_i} + O_{CM_i} + M_{CM_i}) \text{ (Medium-Low Capacity Well Systems)}$$

Where: S_R, S_T, S_C and S_M = Total Cost of System Over Project Life

P_W = Pumping Plant Construction Cost

P_p = Purchased Water Cost Estimation Over Project Life

$T_{CR, CM, CC}$ = Treatment Plant Construction Cost

$D_{CR, CE, CC, CM}$ = Distribution System Construction Cost

$W_{CC, CM}$ = Well System(s) Construction Cost

$O_{CR, CC, CM}$ = System(s) Operation Cost Over Project Life

$M_{CR, CE, CC, CM}$ = System(s) Maintenance Cost Over Project Life

Each factor on the right side of the equations can be evaluated in terms of the effect of every significant local parameter on total system costs. Furthermore, comparison of equivalent factors, e.g. maintenance costs for a central-well system and a multiple-well system, can be made.

Central-Well Systems Versus Cluster-Well Systems

As previously mentioned, the single high-capacity or central-well system has advantages of central management, efficiency, and continuity in system operation, particularly if the central management is a governmental body. The principal drawback of this system, however, is that it cannot be extended to many scattered rural residents except at a prohibitive cost.

This drawback has led NDWP to the development of the cluster-well alternative. Several wells of medium to low capacity are constructed. Each well serves a small cluster of homes, but the multiple system of wells and low capacity treatment plants is centrally managed. Both low construction cost and efficient operation may thus be achieved. In Figure 3 a number of cluster well systems are shown, all of which are under central management.

The distance between homes to be served is the key factor when comparing the costs of the central-well system with the cluster-well system. There is a point beyond which it becomes more economical to construct a second well than to lay pipe to connect a distant house

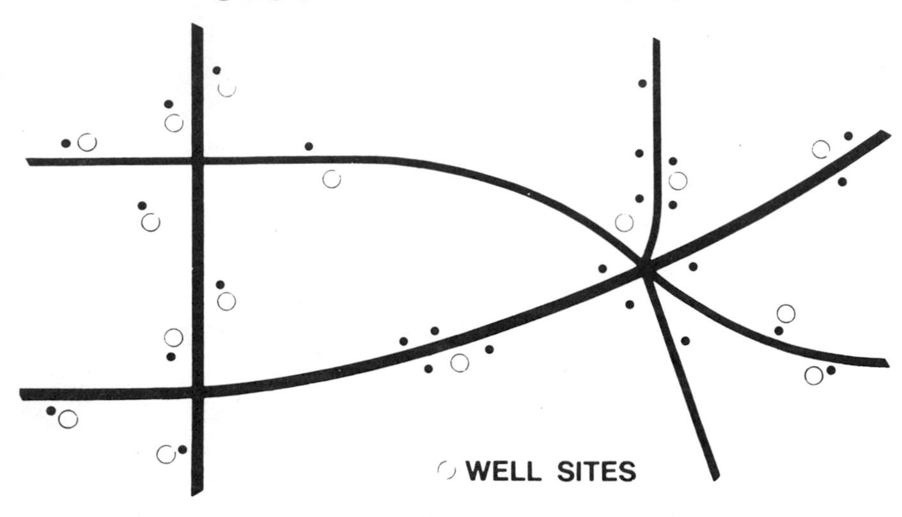

Figure 3.

to an existing well. NDWP has developed the following equation for comparing these costs:

$$\frac{S_c}{S_M} = C_R$$

or:

$$\frac{W_{CC} + T_{CC} + D_{CC} + O_{CC} + M_{CC}}{\sum_{i=1}^{n} W_{CM_i} + T_{CM_i} + D_{CM_i} + O_{CM_i} + M_{CM_i}} = C_R$$

Where: C_R = Total System Cost Ratio

S_C = Total System Cost of High Capacity or Central Well System

S_M = Total System Cost of Medium-Low Capacity or Multiple Well System

Central Well Extensions

$$P_C > W_{CM} + T_{CM} + D_{CM} + O_{CM} + M_{CM}$$

Where: P_C = Total Interconnecting Distribution Cost (in Place)

After translating the effects of local field parameters into estimated dollars, estimated cost components are entered into the respective expressions and the equations solved for total systems, i.e. S_C for the central system and S_M for a multiple system. If the total system cost ratio is unity (one) or less, the central well system should be the superior design for the particular project area and vice versa. In the lower part of the preceding equation, a useful relationship is shown. If the following is valid, then one or more outlying wells are justifiable:

$$P_C > (W_{CM} + T_{CM} + D_{CM} + O_{CM} + M_{CM})$$

Or: $P_C > S_M$

Where: P_C = Total Interconnecting Distribution Cost (In Place)

If P_C is greater than S_M, a cluster system is more economical. If P_C is less than S_M, an extension of the existing system is more economical.

Operation and Maintenance

The pipe-laying cost is not the only critical factor. The effect of operation and maintenance (O&M) must also be carefully estimated. Although O&M costs are difficult to calculate, they can be approximated with far greater accuracy than is currently attempted by many consulting engineers, who often resort to convenient, but inappropriate rules of thumb. Under sponsorship of the NDWP and the Commission on Rural Water two guides are in preparation for the operation, maintenance, and management of support companies for rural utilities.* These guides should be available for public dissemination by late 1974.

Local field parameters dictate the scope of operation and maintenance required for the system. In one area well incrustation and/or corrosion may be a problem. The water supply may require an iron treatment plant. Some systems even require additional treatment to remove other dissolved minerals. As treatment needs increase, and with them both initial construction costs and the attendant operation and maintenance costs, the suitability of treatment-plant consolidation also increases (Campbell and Lehr, 1973b).

* Michael D. Campbell and Jay H. Lehr, 1974, *Rural water systems operation and maintenance: A guide for the engineer and operator* (Houston: Commission on Rural Water, National Water Well Association Research Facility); Edwin L. Cobb and Steven N. Goldstein, 1974, *Managers' guide for the support of rural water-wastewater systems* (Washington, D.C.: Commission on Rural Water).

NDWP Focus on Operation and Maintenance

No matter how well and how inexpensively water and wastewater systems are constructed, the rural problem cannot be solved unless these facilities are adequately operated and maintained. Probably the biggest mistake made by the U.S. Agency for International Development in financing water development in foreign countries was ignoring maintenance needs. After facilities were constructed, the developers simply went away, leaving the residents to shift for themselves. Without any real local maintenance, the systems deteriorated, wells failed, treatment plants malfunctioned, distribution lines broke, etc. The lesson was not that development will not work in rural areas but that development must include a strong awareness of operation and maintenance.

Whether future rural water systems are administered by municipalities, public service districts, or nonprofit corporations, the administering authority must have enough funds and expertise to keep the system running effectively and efficiently. Wells and treatment plants must be inspected and repaired, meters read, bills collected, records kept, and loan payments made. The officers of a water company themselves must be able to carry out all its functions from seeing that wells operate at peak efficiency and remain uncontaminated to making sure that taxes are paid. These tasks will not be easy where users' incomes are low and where reliable public services have not been a part of the heritage of the area. The problem can be tackled either by training the members of each individual water company to perform the necessary work or by developing a separate support company to manage operations for all companies in a project area.

At the present time, maintenance provisions for rural water systems are generally poor. There is inadequate inspection of community water systems, and individual facilities are rarely checked at all. System designers usually underestimate the maintenance requirements during the life of a loan made for the construction of facilities. Reports and other devices for evaluating the performance of water systems are spotty. It is not surprising that small rural water companies have obtained a reputation for performing badly.

No attempt has been made here to discuss all the technical problems and developments that relate to water and wastewater systems. A few of these problems will be elaborated in Part II of this presentation. It is clear in any case that the major weaknesses of the national delivery system for rural areas are not solely technical. Indeed there are *no* technical problems that cannot be solved with the proper political and financial backing. The major needs, commitments and funding, can be met only by state and federal government. System development, centralized management, economically sound engineering, and strong operation and maintenance are features the National Demonstration Water Project is attempting to implement in its field projects. As the present NDWP projects age, their performance will test the effectiveness of the NDWP approach.

References

Andrews, R. A. 1971. Economies associated with size of water utilities and communities served in New Hampshire and New England. *Water resources bulletin* 7: 905–912.

Bourcier, D. V., and Forste, R. H. 1967. *Economic analysis of public water supply in the Piscatagua River watershed, part I: an average cost approach*. Durham: University of New Hampshire, Water Resources Research Center, bulletin no. 1.

Campbell, M. D., and Lehr, J. H. 1973a. *Engineering guide for rural water systems development*. Washington, D.C.: Commission on Rural Water.

Campbell, M. D., and Lehr, J. H. 1973b. *Water well technology*. New York: McGraw-Hill.

Goldstein, S. N., and Moberg, W. J. 1973. *Wastewater treatment systems for rural communities*. Washington, D.C.: Commission on Rural Water.

McCabe, L. J., et al. 1970. Survey of community water supply systems. *Journal of American Water Works Association* 62: 670–687.

McJunkin, F. E. 1969. *Community water supplies in developing countries: a quarter-century of U.S. assistance*. Chapel Hill: University of North Carolina Press.

Mirshleifer, J., et al. 1960. *Water supply: economics, technology and policy*. Chicago: University of Chicago Press.

Morgan, M. E., and Cobb, E. L. 1973. *Water and wastewater problems in rural America*. Washington, D.C.: Commission on Rural Water.

U.S. Department of Agriculture. 1969. *Rural water and sewer needs—1969–1970*. Washington, D.C.: FmHA.

U.S. Environmental Protection Agency. 1972. *Georgia-Tennessee water supply survey*. Washington, D.C.: Water Supply Division, EPA.

U.S. Public Health Service. 1963. *Public Health Service: background material concerning the mission and organization of the Public Health Service*. Washington, D.C.: USDHEW, USPHS.

Water well journal. 1971. A breakthrough on rural water district planning. *Water well journal* 25: 25–28.

Wenk, V. D. 1971. *A technology assessment methodology*. Washington, D.C.: Mitre Corporation.

Zimmerman, Stanley. 1973. *Program design, functional organization and basic management program*. Washington, D.C.: Commission on Rural Water.

Zimmerman, Stanley, and Cobb, E. L. 1972. *Guide for the development of local water projects*. Washington, D.C.: Commission on Rural Water.

13. Engineering Economics of Rural Water Supply and Wastewater Systems

Michael D. Campbell and Steven N. Goldstein

PART II. APPLICATION OF ECONOMIC CRITERIA TO THE EVALUATION OF PROJECT FEASIBILITY (A CASE STUDY)

Problem Statement and Study Objectives

The subject of this case study is the Big Creek Public Service District's proposed water supply and wastewater collection and treatment systems in Logan County, West Virginia. The engineering objectives were: to design domestic water supply and wastewater treatment facilities serving the 250 buildings in the district, to design a system meeting all applicable quality standards both for supply and treatment, to incorporate simplicity, durability, reliability, and maintainability into the design and materials, and to provide first-class service at a total monthly cost that the residents could reasonably afford to pay. In order to accomplish the objectives, several alternative approaches for both water and wastewater systems were evaluated and compared.

Generally speaking, a water and wastewater system for a small community widely dispersed over inhospitable terrain would be prohibitively expensive to set up and would probably not be maintained in proper operating condition for any appreciable period. As previously stated in Part I, the National Demonstration Water Project approach is to provide for centralized management of the utility, including operation and maintenance, while relegating decisions about centralizing the physical facilities to considerations of economic and technical feasibility. A major constraint is that the services must be provided at a price that is within the ability of all subscribers to pay, and this requires that full advantage be taken of all available grant and favorable loan programs.

Description of the Big Creek Public Service District

General Description

The service area of the Big Creek Public Service District consists of eight square miles of steep slopes with level areas in the Guyandotte River and creek bottoms (Figure 1). The 250 buildings to be served include convenience commercial buildings such as a neighborhood sundries shop, small motel, and gasoline station, a grammar school at the mouth of Big Creek Hollow, and several churches.

No industrial hookups are included, nor are any anticipated in this essentially residential public service district. The area currently has neither central water nor sewer services. Some residences have no source of water on the premises. The median family income is below $4,000, and 75 percent of the families have an income below $6,000. Because of the scarcity of flat land with access, there are not 400 suitable building lots available in Logan County for relocating the families who will be displaced when a nearby highway is constructed. About 25 additional building lots could be developed if sanitary (water and sewer) services were available to qualify them for development grant funds.

Population and Existing Facilities

Demographic and housing characteristics of the community and existing facilities are summarized in Tables 1 to 4. The detailed data were collected during the early phases of the project by the Guyandotte Water and Sewer Development Association and are based on detailed information supplied by 154 families. The statistics of Table 1 do not reveal a number of harsh realities in the Big Creek Public Service District. For example, while about half of the families appear to be served by either septic tanks or cesspools, the soil in the area is generally ill suited to subsurface effluent disposal and ponded, daylighted septic effluent is not an uncommon sight. Many sanitary disposal facilities are located dangerously close to wells. Many wells are shallow dug wells which intersect a generally high water table. The quality of water from drilled wells is variable and often not suitable for drinking, and many families with wells haul in drinking water from outside sources. It can be conservatively estimated that close to half of the sewage runs off with only minimal treatment to the creeks which feed the Guyandotte River. Thus existing water and wastewater services for the Big Creek Public Service District must be

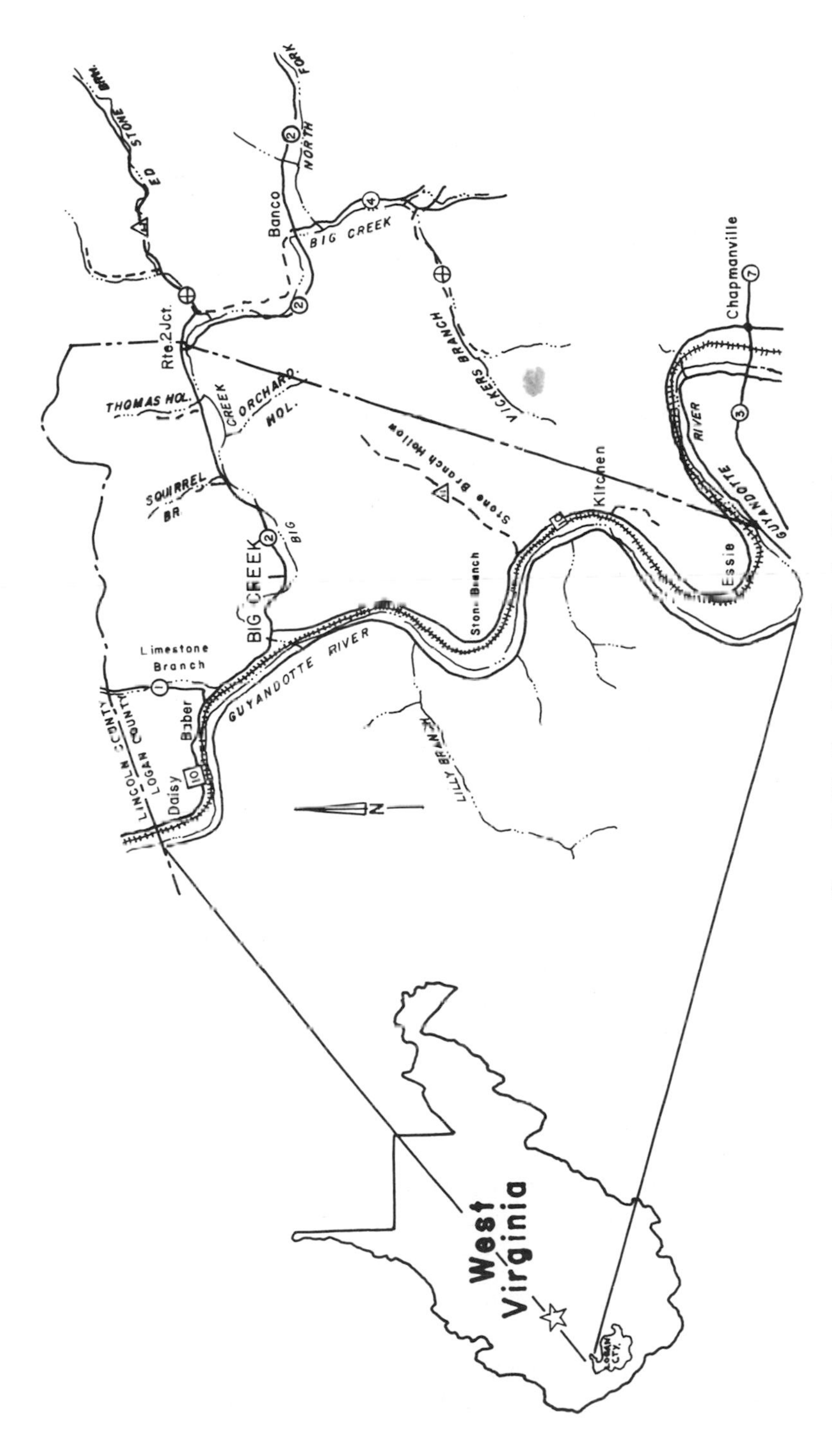

Figure 1. Big Creek Public Service District.

Table 1
Population Size and Household Characteristics

Number of people currently in service area (Estimate based on 251 houses and 3.43 people/house county-wide occupancy rate from 1970 census)	860
Design Population	950[a]
Number of children enrolled in Big Creek Elementary School	142
Number of families responding to detailed household survey (100% of sample)	154
Owner occupied homes	69%
Renter occupied homes	31%
Families using well water	82%
Families using cisterns for collection and storage of drinking water	1%
Families without wells	10%
Families using outdoor privies	44%
Families with bathrooms	56%
Families with septic tank systems	47%
Families with cesspool systems	3%
Families discharging raw sewage to creek	6%

[a]1%/year compound growth for 10 years—1983 design year.

Table 2
Distribution of House Size

Rooms per home	1	2	3	4	5	6	7	8	9	10	11
% homes in category	0	2	3	34	29	19	5	3	1	3	1

considered almost totally inadequate in an age of technological enlightenment and public health awareness. There are few areas that would present a greater challenge to the ingenuity of the project developer or system designer than this one.

Physical Features of the Big Creek Public Service District and their Effect on System Design

Topography

The topography of the Big Creek Public Service District is characterized by significant relief. Land flat enough for homebuilding is

Table 3
Distribution of Children Living at Home

Number of children living at home	0	1	2	3	4	5	6	7	8-12	13
% families in category	44	18	17	6	6	5	1	2	0	1

Note: 28% of sampled population were children.

Table 4
Distribution of Family Income Levels

Family Annual Income	*Percent of Families*
$ 300-3000	39
3000-4000	20
4000-6000	16
6000-8000	15
8000-10000	4
10000-12000	6
over 12000	0

located adjacent to creek bottoms in the hollows and along the Guyandotte River. The secondary roads are mainly dirt surfaces and deeply rutted from heavy truck traffic.

Geology

The Big Creek Public Service District area is underlain by alternating intervals of sandstone, shale, coal, and associated lithology. The relatively shallow sandstone beds are the major aquifers of this area. Abandoned coal mine shafts along with active coal mining and highly ferruginous sandstone aquifers present water-quality problems to any groundwater development program.

Hydrology

Rarely is a well found which meets the standards for chemical content of the West Virginia State Department of Health. Iron concentrations vary from 0.05 to 9.75 mg/l, and manganese from a trace to 0.7

mg/l. The pH varies from 7.4 to 8.4, though at one well the pH dropped to 4.6 and the manganese concentration reached 3.7 mg/l. Small amounts of H_2S are detected in most wells. The chloride concentration ranges from a trace to 125 mg/l. In cases where a salty interface is penetrated at an elevation of about 550 feet, the chloride concentration reaches as high as 1675 mg/l.

Records of groundwater development in Logan County show that the yields of standard vertical drilled wells of 8 to 10 inch diameter range from 50 to as much as 300 gpm. Yields of 100 and 200 gpm are actually common in properly developed wells where the saltwater interface is relatively deep.* In the Big Creek Area, however, controlled pumping and relatively low-capacity wells will most likely be necessary to prevent saltwater coning from below.

The Guyandotte River, which will be the recipient of discharged effluent, has a mean flow of about 1,500 cubic feet per second (cfs) with a maximum of over 16,000 cfs and a minimum of 89 cfs.

Soils

The soils in the area are generally unsuited to septic tank–soil absorption systems because of at least one of many limiting factors: excessive slope, shallow water table, shallow layer of impermeable material (mainly bedrock), and periodic flooding. For example, near the creek bottom, the soil may be of sufficient depth and permeability for septic tanks but poorly drained because of a shallow water table.

Governmental and Administration Relationships

The Big Creek Public Service District was established on February 5, 1973, by the County Court of Logan, West Virginia, under the provisions of Chapter 16, Article 13a, Section 2 of the West Virginia Statutory Code to provide water supply and wastewater treatment services to the residents within the boundaries. As a municipal body, the Big Creek Public Service District qualifies for state and federal construction grant and loan funds. Administratively and legally, the Big Creek Public Service District is the applicant for all grant and/or loan funds, the mortagee, and the owner and operator of all water and wastewater facilities that are proposed in this report. The com-

*Wilmoth Benton, 1972; personal communication.

munities within the service area are all unincorporated, and many are unnamed.

The critical importance to project development of working through existing political institutions and agencies cannot be overemphasized. In this regard, both the Big Creek Public Service District and its agent, the Guyandotte Water and Sewer Development Association, enjoy excellent and cordial working relationships with all levels of local and state government.

Water Supply System

Basic Objectives

The design objective is to provide safe drinking water for domestic purposes in adequate quantities for the existing community of about 860 people with allowance for moderate growth to 950 people over a decade.

Description and Evaluation of Alternatives

The alternatives for the water supply system in the preliminary engineering study are as follows:

Alternate 1: Centralized treatment of Guyandotte River water. A central water treatment and distribution system would withdraw 100,000 gallons per day of water from the Guyandotte River. In addition to chlorination costs of about $.02/1,000 gallons, other physical and chemical treatment costs of about $.07/1,000 gallons are anticipated for a total treatment cost estimate of $.09/1,000 gallons. This is by far the highest treatment cost of any of the alternatives—which is to be expected, since the other alternatives all involve the use of groundwater.

Alternate 2: Purchase of treated water from the town of Chapmanville. The purchase of water from Chapmanville would necessitate a water main connection to Chapmanville and a distribution system for the project areas. The assumed water demand per household is 4,000 gallons per month (Table 5). Using the basic rate for water purchase from Chapmanville of $126.40 for the first 100,000 gallons per month and $90.00 per 100,000 gallons per month thereafter and the state design standard of 100 gcpd and four people per house, the

Table 5
Cost Comparison of Water Supply System Alternatives

	Alternate 1 Guyandotte River Water Central System	*Alternate 2 Chapmanville Purchase Central System*	*Alternate 3 Single Well Central System*	*Alternate 4 (Recommended) Five Cluster Well Systems*
Total first costs	$496,024	$388,188	$526,886	$368,152
First cost annualized at 5%, 38 yr. (.05928)	29,404	23,012	31,234	21,824
Annual labor costs	6,604	5,010	6,610	8,210
Annual nonlabor costs	10,862	15,703	8,307	9,543
Total annual cost (estimate)	46,876	43,725	46,151	39,577

Basic Assumptions:
- Skilled operator and assistant at $4.00/hr., including fringe benefits.
- Bookkeeper/Administrative Assistant at $2.50/hr., including fringe benefits.
- Meter Readers, pump-house checkers, part-time, at $2.00/hr.
- 4 hr./wk. skilled labor or 1 hr./wk. skilled plus 6 hr./wk. unskilled labor per well treatment-storage facility, on the average.
- Meter-reading monthly at average rate of 6 per hour.
- Maintenance of equipment at 2.5% of capital cost.
- Distribution pipe maintenance of $2000/year labor, based on $.04/ft for Alternate 4.
- Treatment chemicals for ground water at $.02/1000 gal; $.09/1000 gal for surface water.
- Electric power at $.025/kilowatt hour.
- System components capitalized at 6% (for replacement) at reasonable lifetimes.

annual estimated cost of water purchase from Chapmanville would be $32,832 (compared to $11,280 based on 4,000 gallons per month per connection).

Alternate 3: Central water supply and distribution system based on single well field. A single well field would be drilled and an iron (and perhaps manganese) removal and chlorination water treatment plant installed. This alternative would include a 100,000 gallon storage facility (standpipe), which, based on 4,000 gallons per connection per month (or 35,140 gallons per day for 251 connections), would provide almost three days' storage capacity. The cost of the storage facility is estimated to be $45,000 installed, which is included in the cost estimate for Alternate 3 in Table 5. (The technical feasibility of this alternative is clouded by uncertainties about the saltwater interface.)

Alternate 4: Five individual water systems. Five individual (cluster) systems, each composed of a well field, treatment facility, storage tank and distribution system, would be installed. They would provide for the various communities a total of 105,430 gpd, which satisfies the full 100 gpcd design standard.

Cost Comparison

It can be seen from Table 5 that Alternate 4 (five separate systems) has the lowest first cost, lower by $20,000 than its nearest competitor, purchase from Chapmanville; but that would amount to a savings of only $.04 per month per customer. The choice should be based on the annual costs of operation and maintenance (including a reserve for equipment replacement, repair, and overhaul) and the annual loan payment on debt services as well as first costs. Annualizing total first costs with debt service figured on the full cost of the system shows the total burden on society, irrespective of who pays for what. The costs of the alternatives are presented in Table 5.

Alternative 4, the five-cluster system as conceived, not only has the lowest first costs, but also the lowest total annual costs. Even though the annual operation and maintenance costs of Alternate 4 are higher than the O&M costs of Alternates 1 and 3, the high debt burden of the latter two dominates the annual payment. Alternate 2, water purchase from Chapmanville, looks attractive as a second choice, but it should be recalled that the cost is estimated on the basis of 4,000 gallons per connection per month (or about 40 gpcd).

Uncertainties

This preliminary appraisal is highly dependent on the suspected but unknown yields and quality of groundwater supplies. If the groundwater should require extensive treatment, the costs will change. Test wells must be drilled to establish the quantity and quality of the groundwater.

Cost Estimates

The estimated total in-place costs and O&M costs are given in Table 6. The engineering fee estimate is based on a resident construction

Table 6
Total In-Place Costs and O&M Costs

In-Place Costs		*O&M Costs*	
Construction cost	$290,405	Utilities (electric power, heat)	$ 718
Land acquisition	9,500	Chemicals and other consumable supplies	763
Engineering fees	29,041	Equipment overhaul/repair/replacement	6,062
(Includes resident construction engineer; approximately 10% of construction costs)		Service equipment (truck)	1,500
		Purchase of services	—
Legal fees	5,808	Rental (provided at wastewater treatment plant site)	No charge
(Approximately 2% of construction costs)			
Interest during construction	4,357	Insurance	500
(1% per month for three months on half of total construction cost)		Subtotal, nonlabor O&M	$ 9,543
Contingencies	29,041	Field personnel (chief operator, assistant, meter readers)	6,960
(Approximately 10% of construction costs)		Office personnel	1,250
		Subtotal, labor	$ 8,210
Total In-Place Cost	$368,152	*Total Operative Costs* (excluding debt service)	$17,753

engineer whose services will be provided by the Guyandotte Water and Sewer Development Association. The estimate of 10 percent for contingencies includes miscellaneous small fittings and similar small cost items as well as uncertainties related to the difficult terrain and geological conditions. Approximately $2,000 annually was projected for repairs to the water distribution system. This estimate was by analogy with the $.04/foot estimate for sewer maintenance costs. Since the cost of pipeline maintenance and repair will be almost entirely labor, it is grouped with the labor costs.

Wastewater System

Basic Objectives

The State Board of Health has stated: "On and after July 1, 1970, the date these regulations become effective, every dwelling or establishment whether publicly or privately owned where persons reside, assemble, or are employed, shall be provided with toilet facilities and a sanitary system of sewage or excreta disposal.... The use of a cesspool as a means of sewage disposal is prohibited."

Consequently, from the standpoint of both the community's health and enacted legislation, the existing sanitary facilities of the project area are inadequate.

Wastewater disposal involves two basic functions, collection and treatment. Consideration of the requirements for waste collection via a system of laterals and trunk line sewers readily indicates that most of the smaller hamlets are too widely dispersed to be economically served by a *conventional* gravity system terminating at a single central treatment facility. These limitations suggest that meeting the design objectives will involve trade-offs between central treatment and collection costs.

Description and Evaluation of Alternatives

The following alternatives for the wastewater system were examined in the preliminary study:

Alternate 1: Centralized. This consists of a single 120,000 gpd rotating biological surface bio-disc treatment plant served by a combination of gravity and pressurized sewer systems. This configuration is

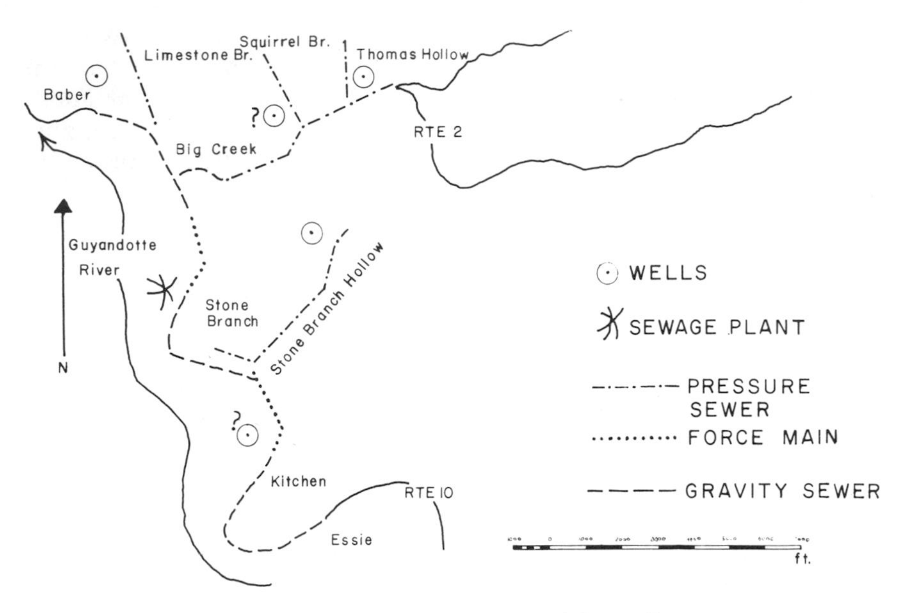

Figure 2. Schematic Representation of Sewer System.

depicted schematically in Figure 2. The centralized approach is made economically feasible by the use of new pressure sewer technology.

Alternate 2: Decentralized with eight treatment plants. This approach uses many smaller sewage plants (eight extended aeration plants ranging between 500 and 30,000 gpd capacity) served by individual cluster collection systems. Central management would ensure proper operation of all the plants.

Alternate 3: Decentralized with sixteen treatment plants. At an earlier stage of the design investigation, clustering to as many as sixteen individual extended aeration treatment plants was examined. These included several "individual home aerobic" plants which would serve as few as one or two connections. Consultation with the Environmental Health Service of the West Virginia State Department of Health indicated that decentralization to this extent would be most impractical to operate and maintain.

Cost Comparison

The full, unsubsidized first cost is annualized to remove any bias from arbitrary allocations of grant, loan, and community funds. Recurring annual expenses of operation and maintenance are separated into labor, including field work and office work, and nonlabor, including utilities, equipment repair and replacement, rents, service vehicles, and outside services such as sludge removal. The alternative approaches to the wastewater collection and treatment system are presented in Table 7. Alternate 1, the central system, is preferred. Even with the obstacles to central sewers, the use of pressure sewers makes first costs lower for Alternate 1 than for Alternate 2 (eight plants).

The eight individual plants can treat sewage as effectively as the single plant, if they receive the necessary attention. Performance would be degraded mainly by the greater variations in loading that could be expected in smaller systems as well as by the overcapacity design that would result from the 100 gpcd design loading requirements, but small, batch-treat, extended aeration plants are available which could accommodate variable loadings. Water quality in the creeks would be somewhat degraded with the decentralized plants. Given the steep terrain, however, groundwater recharge from effluent discharged to the creeks would be minimal. In summary, the superiority of the central system is mainly on the basis of annual cost; in other respects the two systems would be about equally effective.

Table 7
Cost Comparison of Wastewater System Alternatives

	Alternate 1 (Recommended) Central	*Alternate 2 Eight Plants*	*Alternate 3 Sixteen Plants*
Total first costs	$751,425	$763,629	$700,630
First cost annualized at 5%, 38 yr. (.05928)	44,544	45,268	41,533
Annual labor costs	7,356	9,838	16,238
Annual nonlabor costs	12,977	16,729	16,729
Total annual cost (estimate)	64,877	71,835	74,500

Basic assumptions:
- Skilled operator and assistant at $4.00/hr., including fringe benefits.
- Bookkeeper/Administrative Assistant at $2.50/hr., including fringe benefits.
- 4 hr./wk. skilled labor for each small wastewater plant on the average.
- 12 hr./wk. skilled labor for single large wastewater plant.
- Sludge handling and disposal at $25/1000 gallons, about 95,000 gal/year.
- Sewer annual O&M at $.04/ft. gravity; $.06/ft. pressure or force mains.
- Chlorine at $.02/1000 gallons.
- Electric power at $.025/kilowatt hour.
- System components capitalized (for replacement) over reasonable lifetimes at 6%.

Detailed Description of Recommended Wastewater System Approach

The wastewater treatment plant is fairly new to the U.S., but it is based on designs well tested in Europe. The plant uses plastic media disks which rotate on a horizontal shaft, half-submerged in a basin of wastewater and half-exposed to the air. This arrangement obviates the need for control of mixed liquor solids. There are no air compressors or air diffusers. When underloaded with respect to nominal design capacity, the plant responds by giving more complete treatment. For example, at the required 120,000 gpd design loading, the type of plant envisioned would reduce the BOD_5 by 93 percent. At an anticipated loading of less than 60,000 gpd, the same plant will reduce BOD_5 by 97 percent, according to manufacturer-supplied design curves (*Bio-Surf Design Manual*). As a further illustration of the importance of this property, the final stage disks in the plant are covered with nitrifying bacteria which oxidize ammonia, thereby considerably lowering the nitrogenous oxygen demand on the receiving waters.

Chlorination of the effluent will most likely be accomplished by means of a powdered or compressed powder-tablet form of chlorine.

Pressure Sewers

Pressure sewers either create positive pressures at each house with pumps or a negative pressure in the system with a central vacuum pump. The sewers range in size from about 1¼ to 4 inches, and they are normally made of plastic pipe. Pumping station costs are about $250 to hook up a house to a gravity sewer and around $1,000 to hook up to a pressure sewer.

Pressure sewer systems can either grind up the wastewater contents and convey a pressurized slurry, or they can use a modified septic tank to settle the solids and pump relatively clear septic tank effluent. In either case, the sewers must be designed so that a scouring velocity of about two fps (feet per second) will be achieved at least once a day in order to flush out settled solids and keep the lines from clogging. The requirement is much more critical with systems that convey ground-up sewage than with systems which convey septic tank effluent, especially since the sewage in the former contains greases. In terms of initial costs, the pump and septic combination will cost about the same or perhaps slightly less than the grinder pump. Since sewage is partially stabilized (i.e. BOD is reduced) and settled in the septic tank,

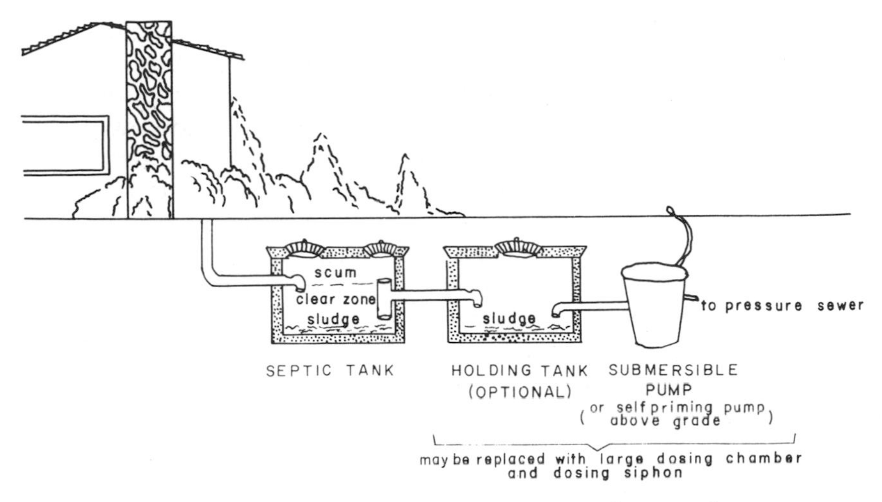

Figure 3. Septic Tank and Pump Preceding Pressure Sewer.

the treatment plant that receives septic tank effluent will be faced with a lower organic and solids loading than one which receives grinder pump effluent, although septic effluent will require more aeration. The septic tanks will, of course, have to be pumped out every few years, and the pumpings will require sanitary disposal.

A cost comparison between pressure and gravity sewers is determined by considerations of population density and population size and by local parameters such as prices, topography, and state and local codes. Gravity sewers have considerable excess capacity built into their design and therefore are to be preferred in certain locations where significant growth is imminent. If a community is composed of both densely and sparsely populated areas with low growth rates, the most economical solution may be to serve the former with gravity sewers and the latter with small-diameter pressure sewers.

The system recommended for the Big Creek Public Service District will employ a combination of gravity sewers along the main highway, connecting force mains where necessary, and pressure sewers in the sparsely populated hollows that have inhospitable geology for gravity sewers. The septic tank with pump version (Figure 3) is recommended to minimize problems of grease clogging of the lines. A holding tank of volume equal to the septic tank will be interposed between the septic tank and the pump. The holding tank will fill from the top (as a septic tank), but it will drain from the bottom (as a bathtub), so that it will normally empty. Its function is solely to provide two to three days' worth of effluent storage in the event that a prolonged power failure, blockage, or other emergency makes the pressure sewer inoperative. Wherever possible, several houses have been clustered onto a single septic (collector) tank, and the four-inch house sewer has been extended out beyond the normal house-sewer hookup point to make this possible. The septic tanks and excess house sewers will be the property of the Big Creek Public Service District, and they will be located on land either owned by the Public Service District or transferred to the Public Service District under easement. Therefore, the entire pressure collection system should qualify as an allowable cost for Environmental Protection Agency grant funding.

The sewers will be of PVC plastic with solvent welded joints or materials of equivalent performance to minimize infiltration. Infiltration specifications will be written into the construction bid package, and the sewers will be tested for watertight integrity before acceptance.

Inflow will be minimized by strict sewer use regulations prohibiting the hookup of household storm drains, downspouts, etc. Manhole covers will either be sealed or raised in areas subject to flooding. One exception will be that the backwash brine from the domestic water treatment units at the pump houses will be connected to the sewers, but through a flow equalization tank with small orifice to prevent shock loading of salts at the wastewater plant.

Operational Performance and its Effect on the Receiving River

The anticipated performance of the waste treatment plant well exceeds 90 percent reduction of BOD_5, with 93 to 97 percent expected. Ammonia removals of about 80 percent are expected.

Flows of the Guyandotte River vary from 89 to 16,200 cfs. Assuming an effluent from the Big Creek Public Service District of 125,000 gpd, 25 mg/l BOD (based on a conservative 90 percent removal and 250 mg/l influent), and zero dissolved oxygen (again, conservative), the dilution at low flow will be

$$\frac{125{,}000}{52{,}518{,}208^*} = .002 \text{ or } 500:1$$

The change in BOD will then be

$$\text{BOD} = \frac{25 \text{ mg/l}}{501} = .05$$

or the BOD of the stream will be raised only from 3.9 to 3.95 under

*89 cfs = 52,518,208 gpd.

worst conditions. The reduction in dissolved oxygen will be approximately

$$\text{D.O.} = \frac{10.1}{501} = .02 \text{ mg/l}$$

and the stream D.O. would be lowered from 10.1 to 10.08, a negligible decrease. Coliform count of the river is already extremely high. At a 500:1 dilution the maximum addition from sewage plant effluent will raise the coliform count from 800 to 820. Thus the treated wastewater will have no significant effect upon the quality of the receiving river. When completed in 1976, the R. D. Bailey Dam upstream of Big Creek will even the flows and should therefore reduce even further the impact of Big Creek sewage plant effluent.

Cost Estimates for Wastewater Treatment System

A construction cost estimate for the treatment plant is given in Table 8. The estimate is based in part on manufacturer-supplied information

Table 8
Construction Cost Estimate for Treatment Plant (950 Design Population, 120,000 gpd)

Life station preceding plant	$ 5,500
Primary treatment screen	3,500
Flow equalization tank (4 hr. flow = 20,000 gal.)	10,000
Feed mechanism (bucket)	1,000
Shaft with biological surfaces (20 ft. 4 element, 64,900 ft.^{2} of surface will give 93% BOD removal when loaded at 120,000 gpd, 97% at 60,000 gpd for fresh sewage, and 92% removal for 24 hr. septic tank effluent)	
Scoop clarifiers—two 100 ft.^{2} area each, at $4,500	9,000
Sludge and primary screenings storage tank (at 3 ft.^{3}/population equivalent = 21,300 gal.)	10,000
Tankage for shaft, 35 yd.^{3} concrete at $200	7,000
for clarifiers (20 yd.^{3} each), 40 yd.^{3} at $200	8,000
Chlorinator	2,000
Basic plant cost, subtotal	$ 77,500
Transportation of plant components to site, 1500 mi. at $2	3,000
Installation and connecting utilities to plant at 20% of basic cost	15,500
Site preparation, including landfill to bring plant above flood level, access road, grading, landscaping, etc.	10,000
Shell enclosure around plant, including basic water laboratory, work space, and small office area	14,500
Total Treatment Plant Construction Cost	$120,000

Table 9
Estimated In-Place Costs for Wastewater Collection and Treatment System

	Collection System	*Treatment System*	*Total*
Construction cost	$478,630	$120,500	$599,130
Land acquisition	6,500	5,000	11,500
Engineering fees (includes resident construction engineer; approx. 10% of construction costs)	47,863	12,050	59,913
Legal fees (approx. 2% of constructions costs)	9,572	2,410	11,982
Interest during construction (1% per month for three months on ½ of total construction cost)	7,180	1,807	8,987
Contingencies (approx. 10% of construction costs)	47,863	12,050	59,913
Total In-Place	$597,608	$153,817	$751,425

for a typical design which is believed representative of the cost of the final design. Estimated total in-place costs are given in Table 9. All operation and maintenance costs are summarized in Table 10. The annual cost of equipment repair and replacement was estimated by capitalizing over assumed equipment lifetimes at 6 percent. Pump lifetimes of seven years (replacement costs of $400), lift station life of 20 years for the structures (replacement at $3,000), seven years between pump overhauls (at $200), and 30 years of life for components of the treatment plant that might need repair or replacement (estimated at $30,000) were assumed. The overhaul, repair, and replacement annual costs for the 26 pumps, five lift stations, and treatment plant amount to $5,511.

In addition, sewer repair, based on an estimate of $.04/foot for gravity sewers and $.06/foot for pressure sewers, amounts to $2,906 annually, but most of the cost is associated with labor or outside services rather than materials. The sewer repair cost was estimated for gravity sewers by dividing national average per capita sewer O&M costs by national average lengths of sewer per capita for appropriately sized communities (Goldstein and Moberg, 1973). An arbitrary 50 percent was added to the estimate for pressure sewers in anticipation of greater O&M costs.

Table 10
Annual Operation and Maintenance Costs for Wastewater System

Utilities (heat and power for treatment equipment, pumps, and lift stations)	$ 1,836
Chemicals and other consumable supplies	830
Equipment overhaul/repair/replacement	5,511
Service equipment (truck)	2,300
Rental (rights-of-way)	500
Insurance	500
Subtotal, nonlabor O&M	$13,844
Field personnel (chief operator, assistant)	6,106
Office personnel	1,250
Subtotal, labor	$ 7,356
Total Operating Costs (excluding debt service)	$21,200

Estimated Costs and Funding for the Entire Project

Tables 11 through 13 show in-place costs, O&M costs, funding, and an annualized cash flow model for the recommended alternatives, the five-cluster water system and central wastewater system, in the project. The indebtnedness must be limited to about $70,000 if the monthly payment is to be in the $14 to $15 range per connection, which represents about 5 percent of the median family income in the service area. The annualized cash flow model (Table 13) shows that the monthly payment can be kept down to about $14 with 250 connections. If enough users, e.g. the country club or motel, exceed the 1,000 gallon minimum consumption base, the anticipated break-even rate may be reduced.

Conclusions and Discussion

The water supply system that is likely to be superior, both technically and economically, will be comprised of between three and five local "cluster" systems each of which will have a well, treatment plant, one or more storage facilities, and distribution lines. A preliminary estimate of the first costs for the water system amounts to about $1,472 per connection.

The recommended wastewater system will have a centralized collection system terminating at a single nominal 120,000 gpd treatment plant that will discharge wastewater (receiving up to 97 percent

Table 11
Recapitulation of Total In-Place and O&M Cost Estimates

	Water	*Wastewater*
Construction	$290,405	$599,130
Land acquisition	9,500	11,500
Engineering fees	29,041	59,913
Legal fees	5,808	11,982
Interest during construction	4,357	8,987
Contingencies	29,041	59,913
Total In-Place Cost	$368,152	$751,425
Utilities	$ 718	$ 1,836
Chemicals	7,763	830
Equipment repair and replacement	6,062	5,511
Service equipment	1,500	1,500
Purchase of services	—	—
Rental	—	500
Insurance	500	500
Subtotal, nonlabor annual costs	$ 9,543	$ 12,977
Field personnel	6,960	6,106
Office personnel	1,250	1,250
Subtotal, labor annual costs	$ 8,210	$ 7,356
Total Annual Operation and Maintenance	$ 17,753	$ 20,333
Monthly, for 250 Connections	$ 5.92	$ 6.78
Combined Total—$12.70/Connection/ Month		

BOD removal and chlorination) into the Guyandotte River. The hollows will employ septic tanks from which a settled effluent will be pumped under pressure in small-diameter pipes to the gravity sewers in the adjacent roadside communities. The communities will connect to the treatment plant with pressure mains. A preliminary estimate of first cost per connection for the wastewater system is $3,006. This is only about 5 percent more than the national average cost (in 1973 dollars) for systems that serve similarly sized communities (Goldstein and Moberg, 1973). Inasmuch as the Big Creek Public Service District area presents a most difficult terrain for sewer construction, this should be a most cost-effective system.

It would seem reasonable that the water and sewer bill should not exceed 5 percent of family earnings; otherwise it would become too burdensome for the residents to pay. Based on a median family income of around $3,500 (see Table 4), the 5 percent guideline amounts to about $14.60 per month. The recommended design should provide

Table 12
Allocation of Loan and Grant Funds to Big Creek Public Service District

	Water	*Wastewater*	*Total*
EPA grant	—	553,250	553,250
State of West Virginia grant	213,250	36,750	250,000
Guyandotte Water and Sewer Development Association grant	88,575	161,425	250,000
Farmers Home Administration Water system loan	66,325	—	66,325
Totals	368,150	751,425	1,119,575

fully adequate sanitation services for just under this figure (approximately $14 per connection per month).

Major equipment failure, lack of management ability, and an inflated initial membership (which after attrition leaves too few subscribers to pay the bills) have been indicated by Peterson (1971) as the three major causes of failure in rural water systems. Management includes the routine day-to-day operation of the system with systematic attention to the details of meter reading, bill collection, and maintenance. Major equipment failure after the initial warranty period can seriously threaten the continued operation of a system. Equipment repair and overhaul, which account for about 30 percent of total O&M costs, have been included in the O&M cost estimates for this project.

The National Demonstration Water Project has developed and is now revising training materials and programs for the management, administration, and support of water and wastewater companies. Operational experience with existing companies in the Roanoke, Virginia, and Beaufort, South Carolina, affiliated field projects is being factored into the revisions.

The public service district form of organization for the local com-

Table 13
Annualized Cash Flow Model

	Water	*Wastewater*	*Total*
Debt service (.05928)	3,932	—	3,932
Operation and maintenance	17,753	20,333	38,953
Totals	21,685	20,333	42,885
Monthly, per connection (based on 250 connections)	$7.23	$6.78	$14.01

pany should obviate the problem of inflated initial membership, since all homes in the service area will be legally required to subscribe and pay their bills. The existence of sanitary facilities should also spur a modest growth rate that would compensate for any subscriber attrition. Also, the public service district, unlike a private association, will qualify for state and federal grant programs.

This study illustrates quite conclusively that the Big Creek system or any similar rural system will require large grants of initial capital to minimize the debt service load and/or some form of subsidy for operating expenses if it is to be a feasible venture. Low cost loans such as the 5 percent, 40-year construction loans from the Farmers Home Administration, while by no means an insignificant form of subsidy, do not meet the financial needs of rural communities. For comparison, it is interesting to note that were this million-dollar project to have been installed and operated as an investor-owned venture without subsidy, the monthly rates per subscriber would very likely exceed $40, well beyond the financial capabilities of the community.

References

Autotrol Corporation. Undated. *Bio-surf design manual.* Milwaukee: Autotrol Corporation, Bio-Systems Division.

Goldstein, S. N., and Moberg, W. J., Jr. 1973. *Wastewater treatment systems for rural communities.* Washington, D.C.: Commission on Rural Water.

Peterson, J. H., Jr. 1971. *Community organization and rural water system development.* State College, Mississippi: Mississippi State University, Water Resources Research Institute.

Phillips, S. A. 1972. Evaluation of the bio-disc wastewater treatment process for summer camp operation. Master's thesis, West Virginia University.

14. Improved Field Techniques for Measurement of Hydraulic Properties of Soils

Johannes Bouma

Introduction

Soil suitability for the disposal of septic-tank effluent is being determined on the basis of several criteria, one of which is the percolation rate of the soil (U.S. Public Health Service, 1967). The percolation test is useful to rank the relative capacity of different soils to accept water when applied 6 inches high in a 6-inch diameter hole. But the test does not yield a representative rate of the movement of waste liquid into soil surrounding a seepage bed or trench. For example, the critical percolation rate of 60 min/inch represents a quite high flow rate of 14 gal/square foot/day. Only 21 square feet of seepage area would be needed to dispose of 300 gallons/day if this rate were to apply. However, this percolation rate is considered to be a medium to slow one and a large septic tank leach field would be required. This discrepancy has been known to occur, of course, and the flow rates used to size required seepage areas are only fractions of 5 percent or less of the percolation rate. For example, the recommended wastewater application rate determined from a percolation rate of 60 min/inch would be 0.6 gallons/square foot/day (U.S. Public Health Service, 1967). These fractions were determined empirically by Ryan in the 1920's and have been somewhat refined since (McGauhey and Krone, 1966).

Real flow conditions around operating seepage beds in the field are often governed by processes of unsaturated flow because of crusting or clogging of infiltrative surfaces and associated ponding of effluent in seepage beds (Bouma, 1971; Bouma et al., 1972). Prediction of flow conditions in the soil requires, therefore, knowledge of the hydraulic conductivity (K) of the soil under both saturated and unsaturated conditions. K-data is physically well defined in that any K value represents the real flow rate in a one-dimensional system at a hydraulic gradient of 1 cm/cm. Flow rates (including percolation rates) in the same soil can differ at the same moisture content as gradients vary. K-values, in contrast, are considered characteristic at each moisture content and relatively constant. Moreover, use of sophisticated physical flow models, predicting movement of liquid as a function of specific physical boundary conditions, requires the availability of K-data (Hanks et al., 1969).

Significance of Unsaturated Flow

Knowledge about the unsaturated flow regime of different soils is very important in designing infiltration systems. All soil pores are filled with water under saturated conditions (when water pressures are positive or zero), and the permeability will be relatively high. As the moisture content decreases below the saturation value, the water pressure in the pores will become negative, larger pores will empty first, and the permeability will decrease because the water will move through smaller pores that offer a higher resistance. The rate of decrease is a function of the pore size distribution of the soil, and K-curves of different soil materials have, therefore, characteristic shapes. For example, Figure 1 shows K-curves for a sand, a sandy loam, a silty clay loam, and a clay. Under saturated conditions (soil moisture tension = 0), the sand with many large pores is most permeable, while the clay with dominantly fine pores has the lowest permeability. However, at a water pressure of -60 cm (moisture tension = +60 cm) indicating unsaturated soil conditions, the clay is more permeable (0.13 cm/day) than the sand (0.04 cm/day) because all large pores in the sand are filled with air at that moisture content and the few remaining small pores containing water can conduct only a limited amount. K-values decrease more strongly upon increasing soil moisture tension when there are more of the relatively large pores. The curves for the fine-porous silty clay loam and clay are particularly interesting, for they show the effects of structural porosity (large worm and root channels and cracks) in yielding relatively high K-values near saturation. However, these pores empty when tensions exceed values of approximately 10 cm, and a very sharp drop of K in this tension range, even greater than the drop occurring in sand, results.

A major decision to be made in the design of any soil disposal system is to select a compromise between a high infiltration rate in a relatively small absorption area on the one hand and a low infiltration rate in a relatively large area on the other. High flow rates may result in poor purification, due to insufficient contact between effluent and soil during the relatively short period of percolation. This problem was demonstrated during field monitoring of an experimental mound system where fecal indicators moved more than five feet through a

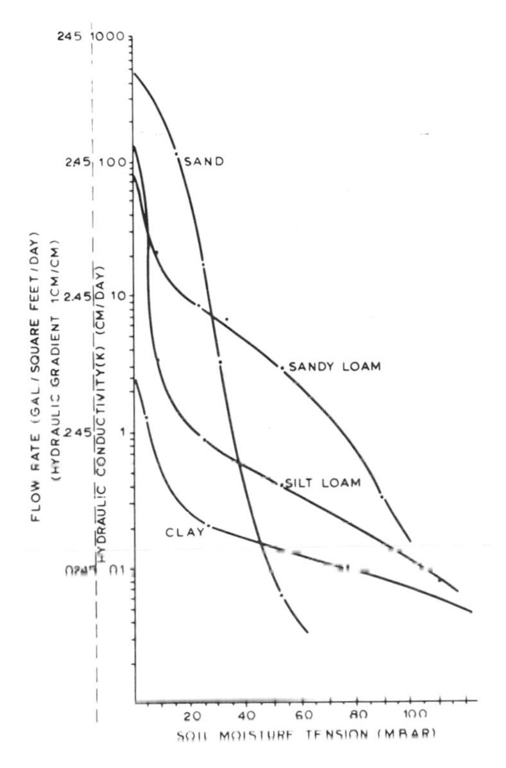

Figure 1.

sandy fill (Bouma et al., 1972, p. 205). Low flow rates, on the other hand, result in better purification because the contact with the soil particles is better and lasts longer. Removal of fecal indicators may then occur within a few inches (Bouma et al., 1972, chap. 5). These basic flow phenomena are illustrated in Figure 2, showing cross-sections of a sand at three moisture contents: saturated, very moist, and slightly moist. The increased vertical travel time for the liquid to move one foot at decreasing moisture content illustrates the essential fact that the vertical distance of percolation is not as important as the prevailing hydraulic conditions during percolation which determine the retention time and the degree of contact with the soil. These flow characteristics can be calculated only if K-data are available.

Field Methods to Measure K

Several *in situ* field tests have been developed in recent years which are preferable to laboratory methods where disturbed samples may

PLAINFIELD LOAMY SAND (C HORIZON: MEDIUM SAND)

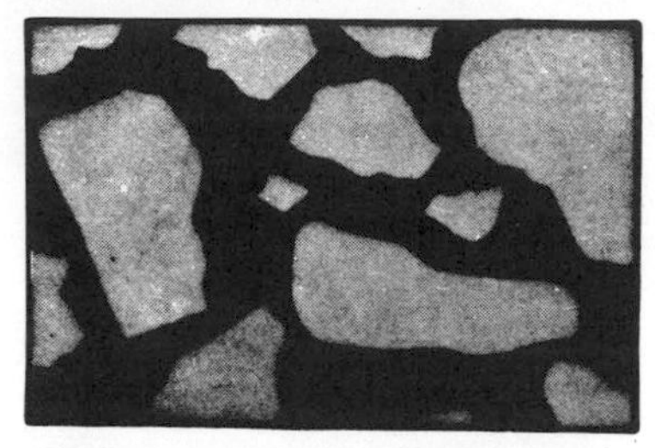

SATURATED
K = 500 cm./day

ONE FOOT (30cm) MOVEMENT IN THE SOIL IN: 33 minutes (hydraulic gradient: 1cm/cm.)

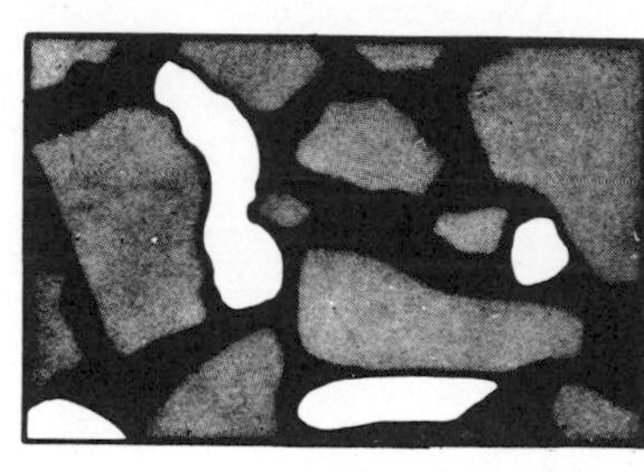

AT 30 mb. SUCTION
K = 5 cm./day

13 hours

Sand grains

Voids

Liquid

100 microns

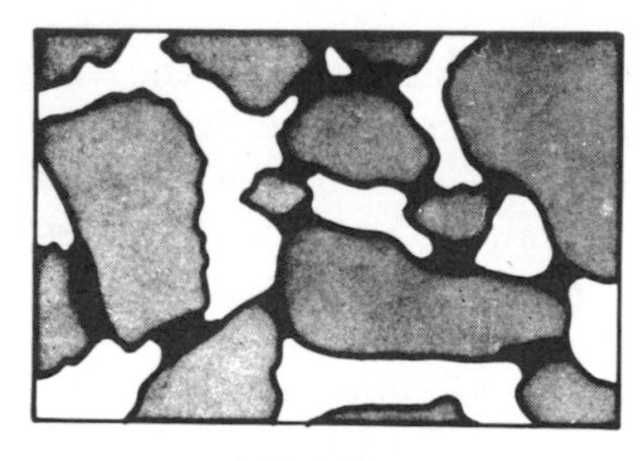

AT 80 mb. SUCTION
K = 0.1 mm/day

300 days

Figure 2.

perhaps be tested. The double-tube method can be applied to measure K of saturated soil (Bouwer, 1962; Bouma et al., 1972). The crust test was developed in Wisconsin to measure unsaturated K-values (Bouma et al., 1971; Bouma and Denning, 1972) (Fig. 3). K-values were determined from infiltration rates at unit hydraulic gradient into

Figure 3.

crust-capped soil columns 30 cm high carved out *in situ*. Matric suctions were measured with tensiometers. A crust on top of an infiltrating profile causes a potential head loss at that point. Thus if the water head over the crust is kept small, it is possible to maintain infiltration into an unsaturated column yet retain the experimental advantages

Figure 3.9.

of an easily measured inflow rate afforded by a flood infiltrometer. The use of a successive series of crusts with progressively lower resistance on the same column can give progressively higher K-values corresponding with higher water contents up to saturation. Each crust yields one point on a K-curve which gives the relation between K and soil moisture tension. The method has been applied to major soils of the state and results are used to design experimental systems for on-site liquid waste disposal. Some representative results were presented in Figure 1.

Only one other field test for measurement of K-unsat, the instantaneous profile method, is available and is currently part of the field-testing program in Wisconsin. Development of this procedure started forty years ago. A detailed description was recently given by Hillel (1972). In this method the complete soil profile is initially saturated. The decrease of the moisture content in the different horizons due to drainage is monitored as a function of time with tensiometry and the neutron probe. Evapotranspiration is avoided by covering the soil. K curves of the different horizons can be calculated with the drainage data. The instantaneous profile method is not very suitable to obtain K-values close to saturation.

Appendix. A Detailed Description of the Crust Test Procedure for Measuring Hydraulic Conductivities of Unsaturated Soil in Situ.

A horizontal plane is prepared at the desired level on the test site by using a putty knife and a carpenter's level (Figure 3.1). A cylindrical column of soil at least 25 cm high with a diameter of 25 cm is carved out from the test level downward, taking care to chip or pick the soil away from the column as the desired boundary is approached, so as to prevent undue disturbance of the column itself (Figures 3.2 and 3.3). A ring infiltrometer 25 cm in diameter and 10 cm high with a 2.5 cm wide brim at the top is fitted onto the column (Figure 3.4). The sides of the column are then sealed with aluminum foil and soil is packed around it (Figure 3.5). Complete sealing is not necessary because water will not flow from the column under unsaturated conditions. A half-inch thick acrylic plastic cover with a diameter of 12 inches (30 cm) and a thin rubber gasket glued to it is bolted to the top of the infiltrometer (Figure 3.6). An intake port and air-bleeder valve are provided in the cover.

Thin pencil-size mercury-type tensiometers attached to 1/8-inch plastic tubing (Figures 3.7 and 3.8) are placed 1 cm below the crust in the center of the column and 3 cm deeper. Careful positioning of holes in the steel infiltrometer ring and use of a small auger to prepare a tight-fitting cylindrical cavity for the tensiometer cup may be helpful in installing the cups. Stony soils present some difficulties, but successful insertion of tensiometers is usually possible after probing at several points. Recent test results have indicated that one tensiometer 3 cm below the crust may suffice.

In the first experiments with the crust-test procedure, various puddled soil materials were used for crusts (Bouma et al., 1971). Additional field experience, however, showed that some of these crusts (in particular the ones with a relatively low resistance) were rather unstable and easily disturbed because of continuous swelling of the clay particles. A different procedure was developed therefore in later experiments using dry gypsum powder thoroughly mixed with varying quantities of a medium sand. After sufficient wetting and continuous mixing, a thick paste is obtained. Then this material is quickly transferred to the prepared column and applied on top with a carpenter's knife as a continuous crust with constant thickness. Special care is taken to seal the crust to the wall of the cylinder to avoid boundary flow. Within about 30 minutes, crusts of this type harden, thereby providing a stable porous medium with a fixed conductivity value. Crust resistance can be varied by changing the relative quantities of gypsum and sand. Crusts composed only of gypsum have the highest resistance. For example, a subcrust tension of 52 mbar was induced in a sand column capped with a 5 mm thick gypsum crust with 3 mm water on top. This crust had a K_{sat} value of 0.007 mm/day. Another crust, formed from a pre-wetted mixture composed of 14 percent gypsum and 86 percent sand by volume as measured in the field using a graduated cylinder induced a subcrust tension of only 11 mbar. K_{sat} of this crust was 8.3 cm/day. The higher K_{sat} value is due to the occurrence of larger pores between the sand grains (Bouma and Denning, 1972).

A series of crusts is applied to the same column for succeeding runs. A common series of four crusts would be composed of 100 percent, 30 percent, 20 percent, and 12 percent gypsum by volume respectively. Relative volumes of gypsum can be changed as needed to obtain data points in the desired range. Each infiltration run through a particular crust yields one point on a curve of hydraulic conductivity versus soil suction (see Figure 1). The small space between the crust surface and the cover of the cylinder (C, in Figure 3.9) is kept full of water. A Mariotte device in a burette (B) maintains a constant pressure of about 3 mm water over the crust (Figure 3.9). The infiltration rate into the soil, corresponding to the rate of movement of the water level in the burette, is recorded as soon as the tensiometers show that equilibrium has been reached. Soil moisture tensions are measured in the columns and are derived from the mercury rise in 1/8-inch plastic tubes along calibrated scales (S) (Figure 3.9). The infiltration rate, when constant for a period of at least 4 hours, is taken to be the unsaturated K-value at the subcrust suction when the suction gradient is zero. (This means only gravitational flow occurs: $v/i = 1$.) In some cases a suction gradient remains at steady state conditions. Hydraulic conductivity is then calculated according to: $K = v/i$, where v=infiltration rate and i=hydraulic gradient below the crust (in such a case $v/i \neq 1$).

References

Bouma, J. 1971. Evaluation of the field percolation test and an alternative procedure to test soil potential for disposal of septic tank effluent. *Soil Science Society of America proceedings* 35: 871–875.

Bouma, J., Hillel, D. I., Hole, F. D., Amerman, C. R. 1971. Field measurement of unsaturated hydraulic conductivity by infiltration through artificial crusts. *Soil Science Society of America proceedings* 35: 362–364.

Bouma, J., and Denning, J. L. 1972. Field measurement of unsaturated hydraulic conductivity by infiltration through gypsum crusts. *Soil Science Society of America proceedings* 36: 846–847.

Bouma, J., Ziebell, W. A., Walker, W. G., Olcott, P. G., McCoy, E., Hole, F. D. 1972. *Soil absorption of septic tank effluent*. Madison: University of Wisconsin Extension Service, Geological and Natural History Survey information circular no. 20.

Bouwer, H. 1962. Field determination of hydraulic conductivity above a watertable with the double-tube method. *Soil Science Society of America proceedings* 26: 330–335.

Hanks, R. J., Klute, A., Bresler, E. 1969. A numerical method for estimating infiltration, redistribution, drainage and evaporation of water from soil. *Water resources research* 5: 1064–69.

Hillel, D., Krentos, V. D., Stylianon, Y. 1972. Procedure and test of an internal drainage method for measuring soil hydraulic characteristics *in situ*. *Soil science* 114: 395–400.

McGauhey, P. H. and Krone, R. B. 1967. *Soil mantle as a wastewater treatment system*. Berkeley: University of California SERL report no. 67-11.

U.S. Public Health Service. 1967. *Manual of septic tank practice*. Washington, D.C.: USPHS publication no. 526.

IV
Rural Water Supply Problems

15.

Rural Water Supply: How Bad a Problem?

Wilbur J. Whitsell

A survey of states conducted by the Environmental Protection Agency in 1971* revealed that one third of the state and local agencies responding gave their rural water supplies ratings of good or better, while two thirds rated them as less than good, i.e., fair to bad. Many of the shocking conditions listed below can be found in rural areas throughout the country.

1. Dug wells with a few loose boards over them and weighted down by bricks, rocks, or dirt.
2. Wells dug or drilled to shallow depths, near streams, with the water level in the well reacting immediately to changes in the levels of the stream, and the appearance of the well water matching that of the stream.
3. Springs whose flow increases rapidly, along with the turbidity, shortly after heavy rains.
4. Drilled wells with only six to ten feet of well casing standing loosely in the hole.
5. Drilled wells with no cover whatever or with burlap or rags stuffed into the top of the casing.
6. Unprotected cisterns with debris, even small animals, floating on the surface, with heavy accumulations of sediments on the bottom, and with no convenient means for draining or cleaning.
7. Springs with no enclosing fence, structure, or cover, and accessible to animals.
8. Homes without any water supply whatever of their own and to which water must be carried or hauled in open containers.
9. "Public springs" repeatedly posted by health departments as "unsafe" but to which people in the area return, destroying the signs and drinking the water.

Enough data have been collected by federal, state, and local agencies to establish beyond reasonable doubt that the problem is serious for many areas of the country. From these data one can estimate that at

*For all EPA studies and surveys cited in this manuscript, the reader is referred to W. J. Whitsell and G. D. Hutchinson (1973a) and (1973b).

least 40 percent of the individual home water supplies deliver water which does not meet the U.S. Public Health Service 1962 Drinking Water Standards (Whitsell and Hutchinson, 1973b; U.S. Public Health Service, 1962) because they contain too many coliform bacteria or too much of certain minerals or have an unpleasant taste or odor.

Sources of Water

While there is no accurate breakdown of kinds of water supply sources for the entire country, the great majority are wells. Springs, catchments with cisterns, ponds, and streams follow in that order. Wells are favored sources because water good enough in quality and quantity to satisfy most home owners can be obtained by drilling nearly anywhere.

Though springs have an image of purity in the popular mind, health agency personnel have established that spring water is frequently contaminated. Springs in limestone regions are among the worst. Here many of the springs are little more than the downstream termini for networks of channels draining water from one surface location to another. In some regions, governmental agencies suggest that springs be considered only after ruling out the possibility of a water well. Both the U.S. Department of Agriculture in its "Water Supply Sources for the Farmstead and Rural Home" (U.S. Department of Agriculture, 1971) and the U.S. Environmental Protection Agency in its newly revised "Manual of Individual Water Supply Systems" (1973) emphasize water wells as the preferred sources. This is not to say that springs cannot be satisfactory sources of water; many are. Unfortunately, however, the regions in which springs abound are also those in which their contamination may be difficult or impossible to prevent.

The old cistern enjoyed popularity many years ago before well-drilling equipment became generally available. It is still used where the cost of drilling deep wells to groundwater is quite high or where groundwater quality is so poor chemically that rainwater is a welcome replacement. There are few regions, however, where groundwater is so poor or so inaccessible that requirements cannot be better satisfied with a well. The studies conducted by engineers of the Water Supply Division of EPA show that roof catchment and cistern supply systems are likely to be contaminated with coliform bacteria. With more than 80 percent of cistern systems found contaminated, they rank right along with the old dug wells in the probability that the water will be contaminated.

Water Wells

Geology and construction are the main determinants of water supply quality. Geology controls the chemical and sanitary quality of the groundwater in its undisturbed state. Geology frequently determines the equipment and procedures to be used in drilling and completing the well, and these in turn affect the ease with which the well can be protected from contamination. Some unconsolidated formations practically guarantee a safe water supply with no special attention. Loose sandy formations overlying sand aquifers readily collapse against the well casing to form a natural seal against pollution from the surface. Consolidated formations, on the other hand, frequently allow coliform bacteria to contaminate the groundwater, and, of course, the proper sealing of wells in such formations is more difficult. Groundwater in some limestone regions may be polluted to such a depth that tapping an uncontaminated source is impossible.

As for construction methods, jetted and driven wells have compiled the best record for safety. Wells drilled with cable tool and mud rotary equipment in unconsolidated formations occupy second place. The EPA studies conducted in Georgia and Tennessee showed that nearly half of the wells drilled by any method in consolidated formations were contaminated with coliform bacteria. Those drilled in unconsolidated formations such as sand, gravel, or clay were much less likely to be contaminated.

For the rural home owner who lives in an area of relatively high groundwater levels and light, unconsolidated earth materials, the hand-bored well appears to be an attractive, inexpensive solution to his water supply problem. Unfortunately, the method requires the use of short sections of pipe, usually made of vitrified clay, and the many unsealed joints permit surface waters to enter the well. In the same EPA studies, these homemade wells had so many sanitary defects, e.g., lack of formation seals and inadequate or no well cover,—that it could not be readily determined just how contaminants were entering the wells. It is especially interesting to note that these bored wells were all constructed in unconsolidated formations, which are supposed to protect water quality. Advantageous geology cannot compensate for poor construction. Contrariwise, development of a safe water supply is nearly always possible, regardless of geology, if enough care and money are expended on construction.

In summary, careless construction certainly accounts for most contamination problems in wells completed in unconsolidated formations and probably also accounts for a large proportion of contaminated wells in the consolidated or hard-rock formations. The EPA studies in Georgia, Tennessee, and Kentucky as well as many other studies conducted by states support this conclusion.

The literature and field experience indicate the following as the most serious deficiencies in the construction of individual rural water systems:

1. Insufficient and substandard (thin-wall) well casing
2. Inadequate "formation seal" between the well casing and the wall of the bore hole
3. Poor welding of casing joints and indiscriminate perforation of casing
4. Lack of sanitary well and spring covers
5. Unsealed, jointed casing in certain types of well construction, e.g., hand-augered (bored) wells and dug wells
6. Reliance on the difficult-to-protect dug well
7. Use of well pits to protect from freezing

Water and Disease

The whole subject of how disease organisms find their way into drinking water, how they move from one area to another, how they infect people, and how people are affected by them is thoroughly documented in the literature. A few of the more interesting articles are cited in the references at the end of this paper (Weibel et al., 1964; McCabe and Craun, 1971; Schliessmann et al., 1958). Thousands of well documented cases of water-borne diseases have been published, and doubtless many thousands more unpublished reports gather dust in the files of state and local health departments. Sanitarians, epidemiologists, and doctors generally agree that for every case heard about there are many more that are never reported. The problem is greatly compounded by consumer apathy. Unfortunately, the presence of coliform bacteria cannot be detected simply by tasting and, unless the water tastes unpleasant or the quantity is inadequate, the average consumer never complains. This attitude was clearly demonstrated by the EPA Water Supply Division studies conducted in Georgia, Tennessee, and Kentucky, and there is no reason to believe that the attitude would be limited to those three states.

Chemicals in the Water

Generally speaking, most rural water supplies do not seem badly polluted by chemicals. There are some notable exceptions, but these are localized and they generally result from poorly constructed wells drawing on shallow groundwater sources in polluted areas. The most common chemical pollutant in rural areas is nitrate salts, usually from ground where livestock have been concentrated and fed for long periods of time. Second in frequency of occurrence are those cases wherein private wells are contaminated by highway salting operations and salt stockpiling. Some of the most dangerous chemicals used in rural areas are the agricultural chemicals, especially the herbicides and insecticides. So far these chemicals have rarely been found in wells, unless the wells are old hand-dug ones which have virtually no protection from surface water infiltration.

Hard water and water with too much iron and manganese account for most of the "chemical problems" in individual home water supplies, at least in the opinion of the home owners. The housewife regards iron and manganese as real nuisances in the family laundry. Special agents which chemically tie up the metal ions with a potential for staining have recently been put on the market to overcome this problem (University of Minnesota Agricultural Extension Service, 1970). Iron and manganese have no health significance except in very high concentrations, and then taste, appearance, and staining characteristics make them so objectionable that they are not likely to be drunk. Health officials are more concerned with the likelihood that residents who find mineralized groundwater disagreeable will rely on polluted sources that taste better.

Work of State, Local, and National Agencies

State and local health department and other agencies having responsibilities for water supply protection react in different ways to reports of contamination. Half of the states responding to an EPA questionnaire knew of no agency with authority to investigate the safety of individual rural water supplies. Forty percent of the states answering had no authority to impose minimum construction standards. Most of the reporting agencies stated that separate bacteriological records for individual supplies were being kept. Nevertheless, only one quarter of them knew of analyses, summaries, or interpretations of the data. One encouraging sign was that about one half of those reporting knew of plans to promote individual water supply legislation in some form.

Since that EPA survey a few states have enacted laws which regulate rural water wells in one way or another, but the purpose in controlling them seems to be to prevent their becoming sources of contamination for public water supply wells.

In July 1965 a "Model Water Well Construction and Pump Installation Law" (Groundwater Resources Institute, 1966) was recommended for adoption by the states. This model law was promoted by the Committee for Private Water Resources Protection, in turn sponsored by several influential national organizations, including the National Water Well Association and the Water Systems Council. The U.S. Public Health Service, the Conference of State Sanitary Engineers, and the American Public Health Association all endorsed the law. Of special interest is the fact that it represented the consensus of more than 100 drillers, pump installers, and suppliers from all parts of the nation. Unfortunately, very few states have adopted legislation that resembles the model law even for public water supplies.

The work being performed by the Commission on Rural Water through their National Demonstration Water Projects and related activities is described elsewhere in this text by Michael Campbell and Steven Goldstein (pp. 145–180). Of special importance is the problem of financing, operating, and maintaining rural water supply systems, which they have handled quite effectively.

The Future

There is a silver lining to the cloud of despair. We do have the tools necessary to resolve the problems. Technology is available to correct any deficiency in rural water supplies and at reasonable cost. The desire for improvement is becoming stronger every day. These two elements will surely combine to bring substantial benefits in health, economics, and service to the millions of users of rural water systems.

References

Campbell, M. D., and Lehr, J. D. 1973. *Rural water systems planning and engineering guide*. Washington, D.C.: Commission on Rural Water.

Commission on Rural Water. 1973. *Closing the rural water-sewer gap*. Washington, D.C.: Commission on Rural Water.

Goldstein, S. N., and Moberg, Jr., W. J. 1973. *Wastewater treatment systems for rural communities*. Washington, D.C.: Commission on Rural Water.

Ground Water Resources Institute. 1966. The "final" water well law and regulations. *Water well journal* 10: 3–11.

McCabe, L. J., and Craun, G. 1971. Waterborne disease outbreaks

1961–1970. Paper read at annual meeting of American Water Works Association, 13–17 June, Denver, Colorado.

Schliessman, D. J., Atchley, F. O., Wilcomb, Jr., M. J., Welch, S. F. 1958. *Relationship of environmental factors to the occurrence of enteric disease in areas of eastern Kentucky.* Washington, D.C.: U.S. Public Health Service publ. no. 591.

U.S. Dept. of Agriculture. 1971. *Water supply sources for the farmstead and rural home.* Washington, D.C.: USDA, Agricultural Engineering Research Division.

U.S. Environmental Protection Agency. 1973. *Manual of individual water supply systems.* Publ. no. EPA-430-9-73-003. Raleigh: EPA Forms and Publications Center.

U.S. Public Health Service. 1962. *Public Health Service drinking water standards.* USPHS publ. no. 956. Washington, D.C.: U.S. Gov't. Printing Office.

University of Minnesota Agricultural Extension Service. 1970. *Iron in drinking water.* Bulletin no. M-154. Minneapolis: U. of Minn. Ext. Serv.

Weibel, S. R., Dixon, F. R., Weidner, R. B., McCabe, L. J. 1964. Waterborne-disease outbreaks, 1946–1960. *Journal of the American Water Works Association* 56: 947–958.

Whitsell, W. J., and Hutchinson, G. D. 1973a. The forgotten water consumer. *Transactions of the American Society of Agricultural Engineers* 16: 782–786.

Whitsell, W. J., and Hutchinson, G. D. 1973b. Seven danger signals for individual water supply systems. *Transactions of the American Society of Agricultural Engineers* 16: 777–781.

Supplemental Bibliography

American Association for Vocational Instructional Materials. 1973. *Planning for an individual water system.* Athens, Ga.: American Association for Vocational Instructional Materials.

Bureau of Water Hygiene. 1971. *Evaluation of the Tennessee water supply program.* Atlanta: Environmental Protection Agency, Region IV.

Environmental Protection Agency. 1973. *A pilot study of drinking water systems at Bureau of Reclamation developments.* Washington, D.C.: Water Supply Division, EPA.

Environmental Protection Agency. 1972. *Evaluation of the Kentucky water supply program.* Atlanta: Water Supply Section, WSPD, EPA, Region IV.

Environmental Protection Agency. 1972. *Evaluation of the Wyoming water supply program.* Denver: Water Supply Branch, Denver Regional Office, EPA.

Environmental Protection Agency. 1971. *Sanitary survey of drinking water systems on federal water resource developments.* Washington, D.C.: Water Supply Division, EPA.

Gibson, U. P. and Singer, R. D. 1971. *Water well manual.* Berkeley: Premier Press.

Jones, E. 1971. Well construction and water quality. Paper read at the 1971 winter meeting of American Society of Agricultural Engineers, 7–10 December, 1971, in Chicago, Illinois.

LaCavera, A. J. 1969. *Water quality survey, Coweta County, Georgia.* Atlanta: Southern Regional

Education Board and U.S. Public Health Service.

Michigan Department of Public Health. 1966. *Ground water quality control act, 294 P. A. 1965, and Rules R325.1601 through R325.1722.* Lansing: Michigan Department of Public Health.

National Sanitation Foundation. 1970. *Pitless well adapters, criteria C-8.* Ann Arbor: National Sanitation Foundation.

National Water Well Association. 1971. *Water well driller's beginning training manual.* Columbus, Ohio: National Water Well Association.

Ohio Department of Health. 1973. *Water supply improvement project.* Appalachian Regional Commission final report. Columbus: Ohio Department of Health.

Oregon State Engineer's Office. 1963. *Oregon ground water laws.* Salem: State Engineer's Office.

Oregon State Engineer's Office. 1962. *Rules and regulations of the State Engineer's Office prescribing general standards for the construction and maintenance of water wells in Oregon.* Salem: State Engineer's Office.

Peppers, Larry. 1970. *Water quality survey, Haywood County, Tennessee.* Atlanta: Southern Regional Education Board and U.S. Public Health Service.

U.S. Public Health Service. 1965. *Grade "A" pasteurized milk ordinance.* USPHS publ. no. 222. Washington, D.C.: U.S. Government Printing Office.

Universal Oil Products, Johnson Division. 1972. *Ground water and wells.* St. Paul: U.O.P., Johnson Division.

Water Systems Council. 1965. *Guide to water systems construction.* Chicago: Water Systems Council.

16. Erosion-Type Hypochlorinators for the Disinfection of Water Supply and Wastewater in Low Density Areas

Eric W. Mood

Need for Water Disinfection in Low Density Areas

Recent testimony presented at Congressional hearings on the Safe Drinking Water Act (S.433 and H.R. 10955), which has been passed by the U.S. Senate and which is pending action by the House of Representatives, includes data which demonstrate that millions of residents of these United States drink water which does not conform with the standards for drinking water quality of the U.S. Public Health Service (1971 and 1962). Some of the data offered in testimony included findings made in 1969 by the Bureau of Water Hygiene, a division of the Public Health Service (McCabe et al., 1970). This study analyzed the drinking water quality of community water supplies which serve more than 18 million people and found that the water in 120 of the 969 systems that were studied exceeded the limits established for coliform bacteria. Most of these systems which did not conform to the bacterial standards served small communities.

A review of reported waterborne disease outbreaks for the decade of the 1960's reveals that during the interval 1966–70 there were an average of fourteen waterborne disease outbreaks per year, ten of which involved private or small water supply systems (Craun and McCabe, 1973). The number of persons made ill in these outbreaks averaged 93 per year for the period 1966–70. While this is not a large number of cases, it represents only those cases which were reported to the National Center for Disease Control, a division of the U.S. Department of Health, Education and Welfare.

During the twenty-five year period, 1946–70, a total of 12,620 cases of waterborne disease from among the persons served by private or small water supply systems was reported (McCabe et al., 1970). This figure accounts for 18 percent of the total number of waterborne disease cases in the U.S.A. These 12,620 cases involved 253 outbreaks or 71 percent of the total number of reported waterborne disease outbreaks for this twenty-five-year interval.

Of the reported waterborne disease outbreaks involving private or small water supply systems for the interval 1946–70, 56.5 percent involved an untreated groundwater source (e.g., wells) and 8.7 percent involved an untreated surface water source (Craun and McCabe, 1973). Of these, 9.9 percent were attributable to inadequate chlorination. In summing up these findings, it may be stated that 75.1 percent of these reported waterborne disease outbreaks involving private or small water-supply systems probably could have been prevented by the application of chlorine in sufficient quantity before the water was delivered to the consumer.

The magnitude of the health hazard associated with water delivered by community systems in low-density areas is emphasized by a recent report issued by the U.S. General Accounting Office (*Environmental Health Letter*, 1973). This report stated that "potentially dangerous water [is] being delivered to some consumers, particularly by small systems serving populations of 5,000 or less." The findings of this report were based on data from six randomly selected states, including Vermont and Massachusetts.

Water-supply problems of the U.S.A. may be divided into two major categories, with each category having distinct characteristics. One division involves the large municipal and urban-area water-supply systems wherein problems are largely associated with the quality of the water source and may not be significantly improved by present treatment methods at an economical cost. These problems include pollution with toxic metals, chlorinated hydrocarbons, and other organic chemicals (Crossland and Brodine, 1973). Solution to these problems seems to lie not in treatment but in preventing these compounds from entering the water supply source. The second division involves problems of small water supplies serving rural or low density areas. The problem of these systems is—or has been—largely a technological one concerned with practical and economical means of disinfection. Generally, the quality of raw, fresh water available to residents of low density areas is such that disinfection is the aspect of treatment basically needed. Man-made chemicals and physical pollutants which are degrading, unsafe, and hazardous elements in much of the untreated fresh-water available to urban populations do not usually pose problems in low density areas. Microbiological contaminants of human and animal origin are the common degrading elements in the fresh-water resources of rural and low density areas. Therefore, the primary need in the treatment of drinking water in these areas is usually disinfection.

Problems of Disinfection of Water in Small Volumes and/or at Low Rates of Flow

Disinfection of water in small volumes and/or at low rates of flow poses many problems. There is the problem of selecting a suitable disinfecting agent—one which is effective and efficient, nontoxic, and nonirritating or nonsensitizing at use concentrations even after many years of continuous consumption; does not give the water undesirable characteristics, e.g. of taste, odor, color, or physical appearance; is stable and safe when stored properly; and is economical, safe, and easy to use. Another problem involves the means to apply the disinfectant to the water in the proper amounts at all times when needed and under varying flow conditions. There are other related problems, but they are not of the magnitude of the two enumerated.

Chlorine, either in its elemental (gaseous) form or as a hypochlorite, is the most effective, efficient, and economical disinfecting agent currently available for water disinfection (Behrman, 1968). It is not practical to consider the use of elemental (gaseous) chlorine to disinfect water in small volumes or at low rates of flow. Chlorine gas is too hazardous for such use and the technology to apply it to small volumes of water and / or at low rates of flow has not been developed. Therefore, disinfection of small-community water supplies with chlorine is limited to the use of hypochlorites.

Other forms of chlorine have been used for select purposes. For example, many resident swimming pools in the United States are being disinfected through the use of one of the chlorinated isocyanurates. However, these newer, chlorine-releasing chemicals have not been approved by governmental authorities for the treatment of drinking water.

Hypochlorites for water disinfection are available in four common forms, namely high test calcium hypochlorite, a powder usually with 70 percent available chlorine by weight; sodium hypochlorite, a liquid usually with 5 to 15 percent available chlorine; chlorinated lime, a powder with 12 percent available chlorine and consisting of an impure solid mixture of calcium hypochlorite and calcium chloride; and lithium hypochlorite, a product which, at this time, is not recommended for use in drinking water disinfection, because of insufficient evidence that its hydrolysis products are nontoxic when consumed for a long period of time.

Hypochlorite solutions, either sodium hypochlorite or calcium hypochlorite, are generally unacceptable for disinfecting small volumes of water and / or water at low rates of flow, because solution-type hypochlorinators, to be operated efficiently and effectively, need sophisticated flow-control devices which maintain the desired rate of application for hypochlorite solution. Also, the use of hypochlorite solutions requires the regular checking of the strength of the solution as well as the periodic, but annoying and sometimes hazardous task of making fresh hypochlorite solution if either calcium hypochlorite or chloride of lime is used. In rural and other low-density areas sodium hypochlorite with a known and uniform amount of available chlorine is not usually available at an economical price.

Most solution hypochlorinators require electrical power and operate only at a constant output rate. Solution hypochlorinators that have the capabilities and capacity to adjust the output proportionally to variations of flow are very complicated and costly. These limiting factors prohibit the use of automatic solution hypochlorinators in most rural and low density areas. Hence, disinfection of drinking water and wastewater has been virtually unheard of in rural areas and small communities. Instead, residents of these areas have had to be content with drinking water that was not disinfected, running the constant risk of waterborne disease, and have had to rely on liquid waste disposal methods that did not require terminal disinfection.

Development of the Erosion–Type Hypochlorinator

Conditions are changing, and a practical, economical, reliable, efficient, and effective means for disinfecting small volumes of water and / or at low rates of flow seems to be at hand. During the past decade, stimulated by the sudden popularity of backyard swimming pools, a new type of hypochlorinator has been developed and produced in large quantities. Several erosion-type chemical feeders (also known as "flow through chemical feeders" and "tablet hypochlorinators") have been marketed as a means to apply disinfectants to swimming pool water. These feeders have been developed to apply bromo-chloro-dimethyl hydantoin ("stick bromine"); calcium hypochlorite in tablet, briquet, and cartridge forms; and chlorinated isocyanurates, also in tablet, briquet, and cartridge forms. One erosion-type hypochlorinator has been developed to disinfect treated wastewater with briquets of chlorinated isocyanurates. The extensive use of erosion-type hypochlorinators for the treatment of swimming-pool water led the National Sanitation Foundation to promulgate a set of standards for the

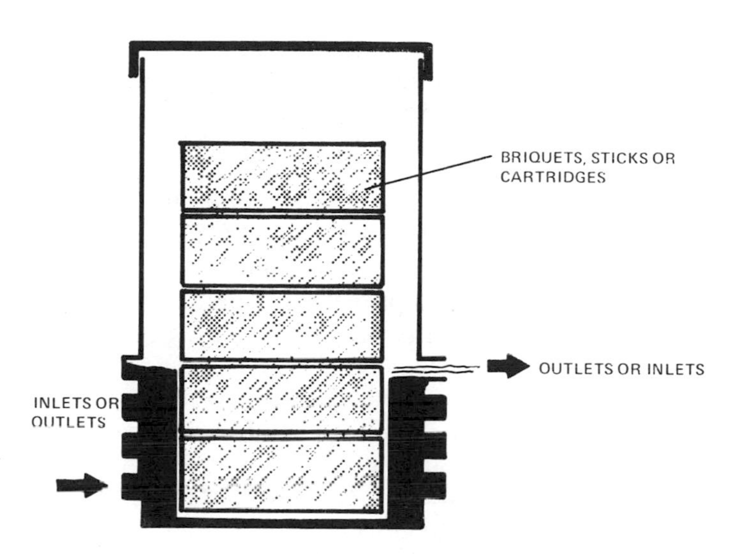

1. Diagrammatic sketch of the dissolving type of erosion feeder. This piece of apparatus was designed for use with solid disinfectants in the form of briquets, sticks, or cartridges.

design, manufacture, and operation of these devices (National Sanitation Foundation, 1973). (The erosion-type hypochlorinator has been referred to as a "flow through chemical feeder" by the swimming pool industry.)

Erosion-type hypochlorinators have assumed several forms with varying basic operating characteristics. The fundamental designs of these devices are as follows:

1. the dissolving type
2. the sloping-bottom type
3. the impingement type
4. the suspended mesh-basket type
5. the split-flow type

The basic design and operating characteristics are described in the following paragraphs.

One of the early models to be developed was the dissolving type erosion feeder (Figure 1). This probably is the simplest design of all and simply entails the placing of slowly dissolving briquets, sticks, or cartridges of a disinfectant in a container with water inlets located at an elevation different from the outlets. The usual configuration, shown in Figure 1, has the inlets at a lower elevation than the outlets. This arrangement assures a uniform depth of water in the dissolving chamber. The use of this design is limited to constant flow conditions and to a continuous and uniform rate of application of the disinfectant. Several pieces of the equipment on the market today for this design are intended only for use with briquets of a chlorinated isocyanurate.

The sloping-bottom type erosion feeder was developed as an attempt to overcome some of the problems associated with excessive submergence of the disinfectant tablets in the eroding or dissolving stream of water (Figure 2). This design provides a degree of control over the rate of application of the disinfectant by regulating the volume of water flowing into the dissolving chamber. Also, it does provide for some degree of intermittency in the operation of the unit, as the flow of water into the dissolving or eroding chamber may be shut off, thereby discontinuing the dissolving or eroding action of the incoming water. However, this design does not provide for easily calibrated adjustments of the rate of dissolving action.

A third design of an erosion-type hypochlorinator is the impingement type feeder (Figure 3). This feeder, which operates most efficiently on a solid stick or stack of briquets of the disinfectant, depends upon the eroding action of a jet of water which is directed at the lower end of the stick or briquet. It is a true erosion-type hypochlorinator, in contrast to the units previously described. For example, the sloping-bottom type of feeder seems to depend more upon the dissolving action of the incoming water than upon the eroding action of this stream. In the impingement type feeder the rate of erosion of the disinfectant may be varied within a narrow range of values by adjusting the hydraulic head on the eroding jet of water ("h" in Figure 3).

Still another type of erosion-type hypochlorinator is the suspended mesh-basket type feeder (Figure 4). This design was developed to be used with tablet-form hypochlorite. It may be used with some of the more soluble disinfectants in forms other than tablets. The rate of feed of the disinfectant is varied by adjusting the overflow level to change the degree of submergence of the mesh basket in the dissolving reservoir. In so doing, more or less of the disinfectant is exposed to the dissolving or eroding action of water.

The fifth type of erosion-type feeder is the split-flow type hypochlorinator (Figure 5). The design embraces many of the better features of the first four types of erosion hypochlorinators and incorporates some additional characteristics. The operation of this design is dependent upon splitting the flow through the hypochlorinator, with the smaller of the two flows being directed to the dissolving chamber, which, when feeding a disinfectant, is kept at a

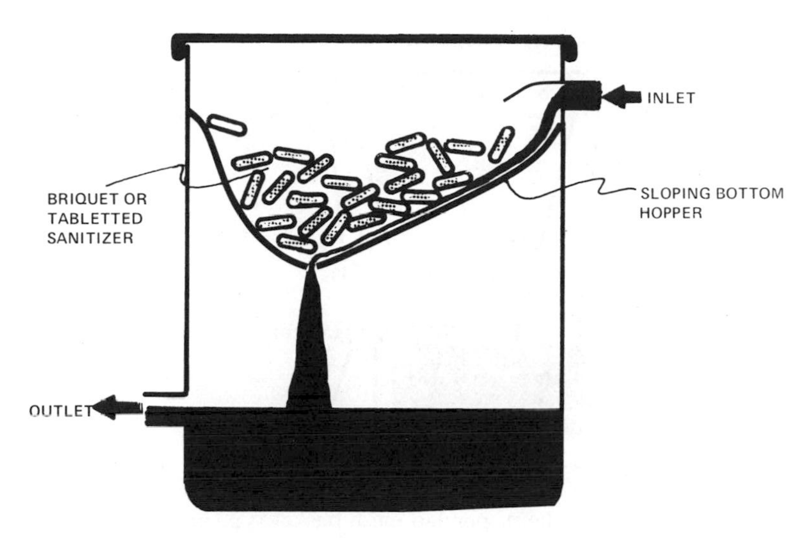

2. Diagrammatic sketch of the sloping bottom type erosion feeder for use with briquets or tablets of a disinfectant.

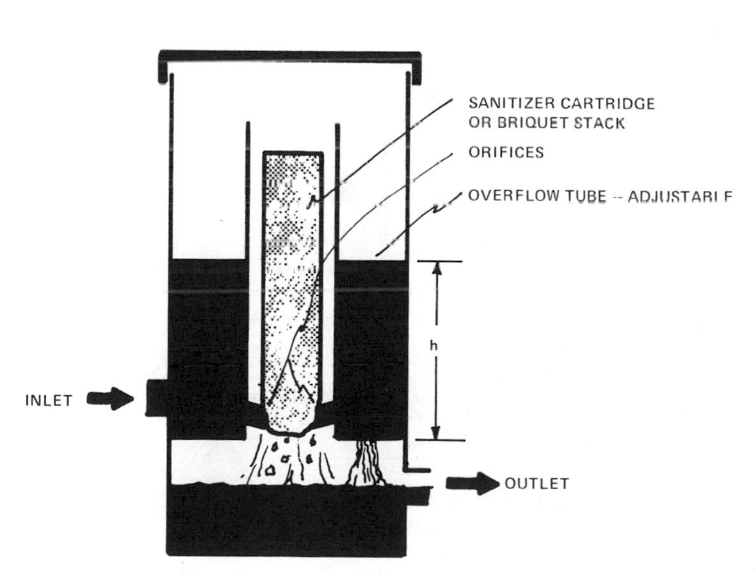

3. Diagrammatic sketch of the impingement type erosion feeder. The rate of dissolving or erosion is varied by changing the head of water "h" on the eroding stream of water.

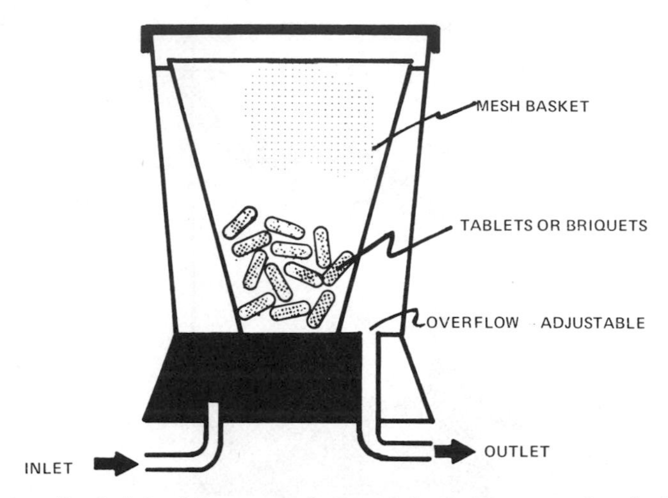

4. Diagrammatic sketch of the suspended mesh basket type erosion feeder for use with a disinfectant in the forms of tablets or briquets. Note the adjustable overflow which permits some variation on the rate of chemical application.

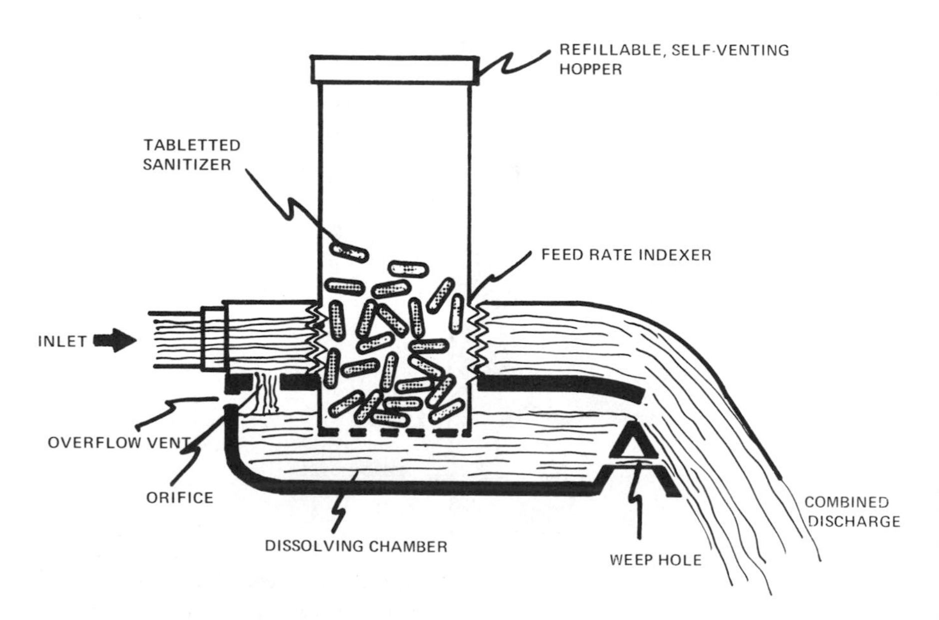

5. Diagrammatic sketch of the split-flow type erosion feeder.

constant level by a fixed overflow weir. The rate of application of the disinfectant is varied by raising or lowering the hopper containing the tablets of disinfectant and thereby changing the degree of submergence of the disinfectant in the dissolving chamber. When the inlet water is shut off, the eroding or dissolving action of water on the disinfectant is stopped as a weep hole, located at a level below the lowest submergence depth of the chemical storage hopper, drains the dissolving chamber of water.

The operating characteristics of the split-flow type erosion feeder may be summarized as follows:

1. The flow of water into the hypochlorinator is split into two streams of water, with the smaller of the two streams being diverted into the dissolving chamber.

2. During the operating phase, the depth of water in the dissolving chamber is kept at a constant value by a fixed-level overflow weir.

3. During the shut-down or nonoperating phase, a weep hole drains the water in the dissolving chamber to a level that is lower than the lowest possible position of the hopper, thereby stopping the dissolving or eroding action of water on the disinfectant.

4. The rate of application of the disinfectant may be varied by adjusting the depth of submergence of the hopper in the dissolving chamber.

Operation and Performance of a Split–Flow Type Erosion Hypochlorinator

During the past three years, studies have been conducted on the operation and performance of a split-flow type erosion hypochlorinator.* Although this unit was designed to meet conditions found in the developing nations of the world, it seems highly suited for the disinfection of drinking water and wastewater in rural and low density areas, provided that the unit may be operated with zero back pressure. Field experience with this unit has been limited largely to use in other countries; however, extensive laboratory tests and simulated in-use field evaluation have been conducted to determine the suitability of these units for use in the United States.

The units tested were designed specifically for use with high-test calcium hypochlorite tablets. In addition to conforming with the specifications previously enumerated for the split-flow type of erosion

*The unit studied was the WaterSure, Type 1, erosion-type hypochlorinator manufactured by World Water Resources, Inc., New York, New York.

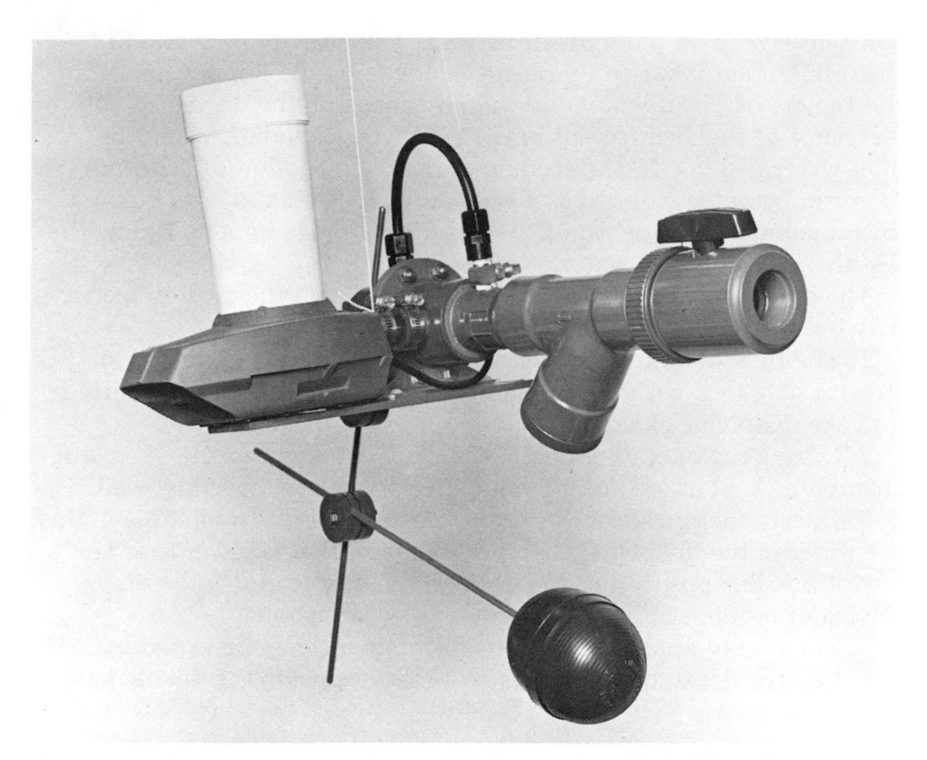

6. The split-flow type erosion feeder manufactured by World Water Resources, Inc. The connection to the untreated water source is to the right. The ball float controls the operation of the feeder.

hypochlorinator, these units were equipped with a quick-opening and closing valve actuated by a ball float (Figure 6). This valve is positioned at the influent end of the split-flow chamber with the ball float positioned near the effluent end and with the floatation unit on the surface of the treated water receptacle or reservoir.

While the quick-opening and closing valve will operate satisfactorily over a range of pressure heads from three to fifty feet, most of the laboratory and simulated field tests that were conducted in the United States were performed with pressure heads of five to ten feet. Figure 7 shows one of the arrangements of equipment for laboratory testing which simulated field conditions.

The lever arm of the ball float was positioned so that the float traveled approximately twelve inches along the vertical axis between activation and shut-off. This permitted approximately fifty to sixty gallons of water to pass through the hypochlorinator. With a three-foot head of pressure on the hypochlorinator, the length of time

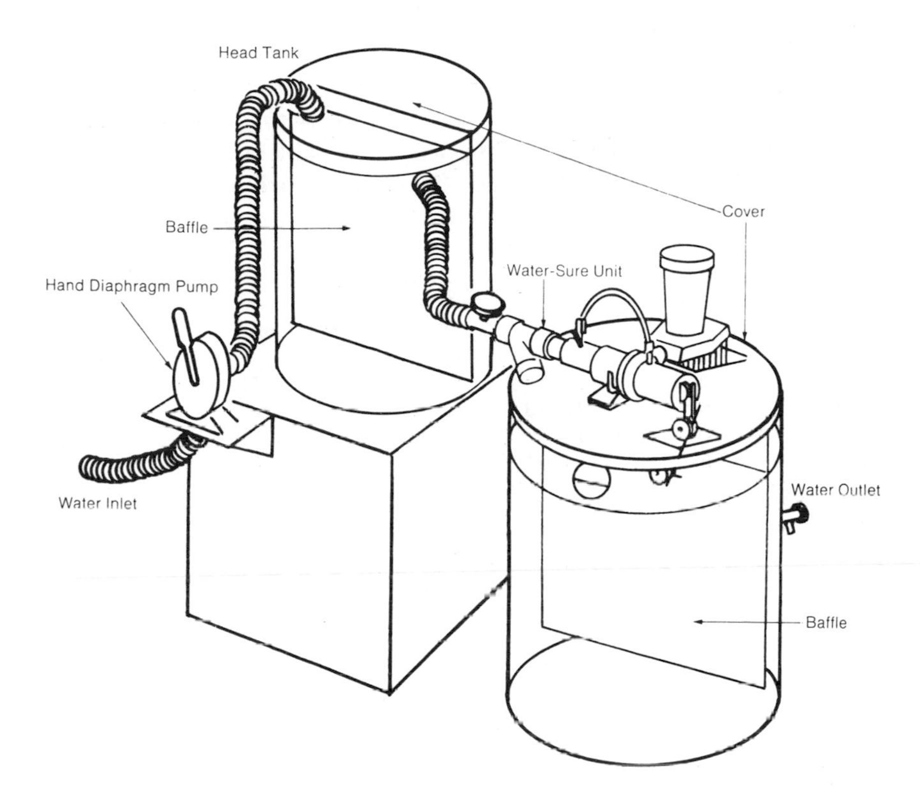

7. Diagrammatic sketch showing the installation of a split-flow erosion type hypochlorinator for field use in a situation lacking electrical power.

during which the water was flowing through the hypochlorinator was approximately four to five minutes.

Test conditions were established in an attempt to obtain a residual chlorine value of 0.6 mg/l after fifteen minutes of contact. Using "raw" water of uniform quality and chlorine demand, a residual value of 0.6 mg/l ± 0.1 mg/l was obtained approximately 95 percent of the time over a four-hour period (Table 1). Preliminary data suggest that even greater uniformity of residual values would be obtained at higher rates of chlorine dosing.

In order to obtain satisfactory performance with the type of split-flow erosion hypochlorinator used in these tests, two reservoirs for water are needed. One reservoir is for the raw or untreated water and should be of sufficient capacity to meet demand at peak flow conditions. The second reservoir is for the treated or chlorinated water and should have sufficient volume to insure an adequate chlorine contact time. Some public authorities believe ten minutes is necessary, while others recommend contact times of thirty or more minutes.

Table 1
Tabulation of Results of Residual Chlorine Determinations After Disinfection with a WaterSure, Type I, Split-Flow Type Hypochlorinator, Using High Test Calcium Hypochlorite Tablets and After a 15 Minute Chlorine Contact Period. Residual Chlorine Values Determined by the Orthotolidine Method.

Time	*Res. Chlorine (mg./1.)*	*Time*	*Res. Chlorine (mg./1.)*	*Time*	*Res. Chlorine (mg./1.)*
10:05	0.80[a]	10:58	0.70	11:50	0.70
10:06	0.70	10:59	0.60	11:51	0.50
10:07	0.70	11:00	0.55	11:52	0.55
10:08	0.65	11:01	0.55	11:54	0.55
10:18	0.65	11:11	0.60	1:30	0.65
10:19	0.65	11:12	0.60	1:31	0.55
10:20	0.65	11:13	0.60	1:32	0.45[a]
10:21	0.65	11:14	0.55	1:33	0.50
10:31	0.70	11:24	0.60	1:43	0.60
10:32	0.60	11:25	0.60	1:44	0.55
10:33	0.60	11:26	0.55	1:45	0.55
10:34	0.50	11:27	0.50	1:46	0.55
10:44	0.70	11:37	0.55	1:56	0.55
10:46	0.65	11:38	0.60	1:57	0.50
10:47	0.60	11:39	0.55	1:58	0.55
10:48	0.60	11:40	0.50	1:59	0.50

[a]Indicates observations that were more or less than 0.60 mg/1. ± 0.10 mg/1. of residual chlorine.

For public health reasons, the design of the chlorine contact chamber is very important. Since the efficiency and effectiveness of chlorine as a disinfecting agent is in part a function of time, the location of the inlet and the outlet to the receiving water tank or chlorine contact chamber should be such as to minimize short-circuiting. In the tests recently completed, a baffle was used to ensure adequate mixing and contact.

Typical Installations—Rural Water Supply

A typical farm installation is shown in Figure 8. This setup utilizes an elevated water tank from which the water flows by gravity to the various fixtures. The source of water is pictured as a surface supply.

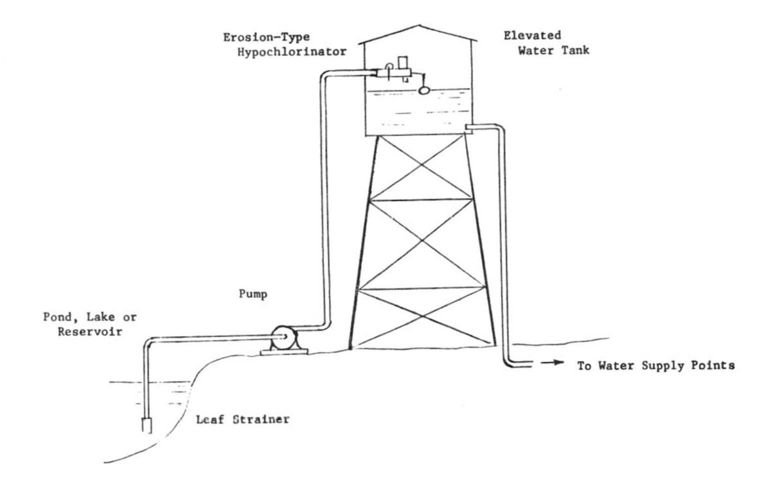

8. Schematic arrangement of an erosion-type hypochlorinator installed on a rural water supply system which uses an elevated tank.

A pump is activated when the water in the elevated tanks drops to a specified level and transfers the water from the source to the tank. The head pressure needed to operate the hypochlorinator is obtained by the pump. The elevated tank serves as the chlorine contact chamber.

Many variations of this basic design are possible. The source of energy to lift the water into the elevated tank may be from a windmill, hydraulic ram, or other available power source. In the hilly terrain of New England it may be possible to operate the entire system by gravity with the elevation of water at the source being several feet higher than the location of the hypochlorinator and the treated water reservoir.

To prevent recontamination of the water after disinfection, it is highly desirable to equip the treated water receiving tank with a cover. This protection device is demonstrated in Figures 7 and 8.

Typical Installations--Rural Wastewater Treatment

With the need to prevent water pollution at the source, the requirement to disinfect wastewater before discharge into the environment is greater than ever. Figure 9 shows an erosion-type hypochlorinator installed on a septic tank system. The disinfectant feeder is preceded by a siphon chamber which obviates the need for a ball float-con-

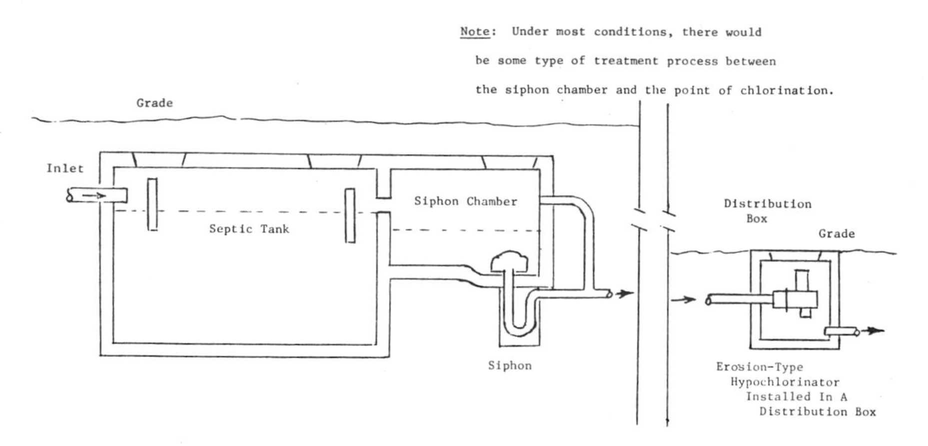

9. Cross-sectional sketch of a septic tank and a siphon chamber with a split-flow type erosion hypochlorinator installed in distribution box for final disinfection of the treated wastewater.

trolled quick-opening and closing valve. The intermittent action of disinfection is controlled by the siphon.

In order to obtain proper operation of a split-flow type erosion hypochlorinator, it is necessary that the wastewater to be treated be relatively free of coarse suspended material. All particles in the wastewater must be smaller than the diameter of the weep hole in the dissolving chamber. If there are large pieces of suspended material in the wastewater, there is a possibility that the weep hole may become clogged, thereby interfering with the draining of the dissolving chamber during the shut-off period.

A split-flow type erosion feeder is being planned for use on a spray irrigation system after the wastewater has been treated by a secondary aeration treatment process and oxidation lagoon. The pumps on the spray irrigation will be activated automatically by a time-clock mechanism. The split-flow type erosion hypochlorinator will be located on the discharge side of the pump. Preliminary tests are most promising.

Summary and Conclusion

The need for a simple, economical, efficient, effective, and reliable means to disinfect drinking water and wastewater in low density areas is great. Heretofore, practical means have not been available. The erosion-type hypochlorinator, particularly of the split-flow design, appears to have the potential to provide such means. This type of disinfectant feeder is a simple, low cost, dependable device. Operation and maintenance do not require special skills. No power source is required to provide continuous automatic operation with a high degree of reliability. The principal shortcoming, which may be overcome in most situations, is that the dissolving chamber cannot be satisfactorily operated under a hydraulic head.

References

Behrman, A. S. 1968. *Water is everybody's business*. Garden City, New York: Doubleday and Company.

Craun, G. F., and McCabe, L. J. 1973. Review of the causes of waterborne-disease outbreaks. *Journal of the American Water Works Association* 65: 74–84.

Crossland, J., and Brodine V. 1973. Drinking water. *Environment* 15: 11–16.

Environmental Health Letter 12: 2–3, December 1, 1973.

McCabe, L. J., Symons, J. M., Lee, R. D., Robeck, G. G. 1970. Survey of community water supply systems. *Journal of the American Water Works Association* 62: 670–687.

National Sanitation Foundation. 1973. *Standard no. 47 for flow through chemical feeding equipment.* Ann Arbor: The Foundation.

U.S. Public Health Service. 1962. *1962 Public Health Service Drinking Water Standards*. Washington, D.C.: Government Printing Office, PHS publication no. 956.

U.S. Government Printing Office. 1971. *Safe drinking water, hearings before the Subcommittee on Public Health and Environment of the Committee on Interstate and Foreign Commerce, House of Representatives, May 24–26, 1971.* Washington, D.C.: Government Printing Office.

The cooperation obtained from World Water Resources, Incorporated, New York, New York, in the conduct of these research studies is acknowledged with much gratitude.

17. Water Quality and Solid Waste Problems in Rural New Mexico and Some Solutions

John R. Wright

Introduction

The geography of New Mexico ranges from high Alpine forest to lower Sonoran Desert. The varied topography allows for six of the seven life zones found in North America. The southern part of the state is generally flat and arid with occasional mountain peaks. The Gila National Forest in the southwestern portion of the state is mountainous and provides water for the Mimbres Basin and Gila River. The northern part of the state is more mountainous. Elevations range from 13,150 foot Wheeler Peak in the Santa Fe National Forest near Taos to 2,850 feet as the Pecos River enters Texas south of Carlsbad.

The water resources of the state cannot provide sufficient irrigation. About three million acre feet of water from rainfall appear yearly as runoff in New Mexico streams, and an additional 2.5 million acre feet are received from other states. Downstream states are entitled to about 2.5 million acre feet as a result of interstate compacts and court decrees. The remaining three million acre feet will support about one million people, or the present population in an agricultural economy, and about six million if the waters were diverted to industrial use. Twenty-six million people could be supported at a rate of 100 gallons per capita per day for strictly domestic use. The economy of the state is basically agricultural, with tourism and the federal government contributing greatly to the state's income.

Northern New Mexico is predominantly composed of Spanish surname families and a Spanish agrarian culture. The eastern and particularly southeastern part of the state has an Anglo-American culture engaged mostly in cattle ranching and the oil industry. The Middle Rio Grande Basin has a Spanish culture and highly mechanized agriculture. Two of the state's largest irrigation systems lie in this area.

The area later called New Mexico was first colonized in 1598, and Santa Fe was established in 1610 as the territorial capital ten years before the pilgrims landed at Plymouth Rock. The state covers 121,666 square miles and has a population of 1,016,000. Thirty-six percent of the people or approximately 358,000 persons live in the Middle Rio Grande area. There are 21 municipalities over 5,500, only six of which are over 25,000. These communities, including Albuquerque, constitute only 59 percent of New Mexico's population. The rest of the people live in communities of under 5,500, and many communities range in size from ten to twenty families.

Problems

Environmental problems of New Mexico will be discussed in three categories: (1) Water Supply, (2) Sewage Disposal, and (3) Solid Waste.

Water Supply

New Mexico does not have an abundance of water. Surface water sources are very limited. The mountainous regions are spotted with acequias diverting the waters from the streams for irrigation purposes. Rock formations in these regions make water difficult to obtain. Groundwater aquifers include the alluvium of the stream bed, deep artesian aquifers, and many nonartesian underground aquifers ranging from a few hundred feet to over 3,600 feet deep.

In 1912, when the state entered the Union, the doctrine of prior appropriation was already well established. The constitution of the state recognizes and preserves the right to use water separate from the use of land—that is, it is based on the doctrine of prior appropriation, not on riparian rights.

Many regard the doctrine of prior appropriation as the most equitable way to distribute the water. The constitution states that water belongs to the state but may be diverted for beneficial use. All communities must obtain rights to water, as must an industry, a person irrigating a farm, or any other water user. The right to use the water does not include the right to pollute or damage the rights of an individual downstream. Rights are based upon the amount of water that is evaporated or totally consumed and not returned to the stream. Some of the problems of acquiring a water right are caused by regulations relating to changes in ownership of the right and the market for water not being well established in a particular area.

Development of a water supply system is frequently very difficult. Well systems in New Mexico range from a few feet deep, as in the alluvium of streams, to more than 3,000 feet deep. A water-supply well for a small community can easily cost thousands of dollars.

Distribution of water presents problems unique to New Mexico. When the houses within a small community are scattered, the cost of pipeline construction per connection can become exorbitant. In recent years, though, the increased use of plastic pipe has helped to reduce the cost of constructing water lines for rural communities. Another problem is the inability of small associations to operate and maintain the facilities properly after they are constructed.

In many areas of the state, water quality is very poor. Total dissolved concentrations over 2,000 mg/l are not uncommon in many communities. Extremely soft water, which is aggressive to the point of attacking metal piping and storage reservoirs, also causes operational problems. Hardness values range from 1.25 mg/l to 2,430 mg/l as calcium carbonate, and total dissolved solids range from 67 mg/l to 4,500 mg/l. The shallow groundwater is frequently polluted by water from the irrigation systems in the Middle Rio Grande and Pecos rivers. Special considerations must be given to system design to compensate for these water quality problems.

Sewage Disposal

Poorly designed septic tanks and sewage collection systems constructed in the early days on shoestring financing are causing problems today. Fifteen years ago the Department of Health began conducting privy eradication programs in many small communities throughout the state and encouraging persons to construct inside plumbing with septic systems. In the Middle Rio Grande Basin it has been estimated that there are 16,000 septic tanks along a thirty-mile stretch of stream. The sand and gravel soils do not purify the septic-tank effluent well and have thus allowed extensive pollution of water-supply wells.

Some of the early sewage systems designed for small communities utilized the lagoon system of liquid waste disposal. Early design criteria were similar to those of North Dakota, where use of the lagoon was first established. Conservative designs needed in North Dakota were overdesigned for New Mexico. The use of water was overestimated, and the evaporative powers of the sun in New Mexico were underestimated. Mosquito breeding and nonuse of significant capacity have resulted. Tesuque, New Mexico, an Indian Pueblo, had a sewage system built using lagoons in 1958, and today, fifteen years later, the bottom of the first cell is still not covered with water.

Finally, as with water supply systems, small sewage systems do not receive good operation or maintenance. For example, in Texico, New Mexico, the sewage lagoons produce large amounts of moss because weed growth is not controlled by the operating personnel.

Solid Waste Disposal

Methods of solid waste disposal in rural New Mexico are similar to those of the rest of the country. Recent packaging methods for beer, soft drinks, and packaged foods, however, cause a particular problem because the dry climate precludes destruction of the discarded materials by biological action, as highways, rest stops, and parks frequently evidence.

Ownership of land is one problem New Mexico faces with solid waste disposal. A large portion of New Mexico's land is federally owned and controlled either by the Bureau of Land Management, U.S. Forest Service, or the National Park Service. Recently the Environmental Protection Agency established a rule that any landfills constructed on federal land would have to be *sanitary* landfill systems (i.e. with daily cover). Superficially, the rule sounds reasonable; however, communities the size of Truchas (approximately 28 families) cannot possibly afford the land-moving equipment and manpower to cover a sanitary landfill daily. Catron County in southwestern New Mexico, for example, has only 2,198 people and encompasses an area approximately 71 percent of Vermont. There are five landfills in the county, and the population would create only about one cubic yard of waste per day per landfill. It seems apparent that a solid waste collection and management system as sophisticated as required by the federal regulations is impossible in this area of New Mexico. A city the size of Albuquerque can and does use the sanitary landfill method for collection and disposal of all solid waste materials, but small communities and rural areas should be allowed to use the modified landfill method with weekly or monthly cover. Some form of the landfill method, as distinguished from incineration, is used exclusively for solid waste disposal in New Mexico.

Solutions

Possible solutions will be discussed in the same categories used to outline the problems: (1) Water Supply, (2) Sewage Disposal, and (3) Solid Waste.

Water Supply

Water supply has been a particular problem throughout New Mexico because rural communities have relied upon the irrigation ditch for drinking water for over 300 years. Consequently, the infant mortality rate in 1940 was 2.1 times the national average. In 1946, Charles Caldwell, Director of Environmental Sanitation Services of the New Mexico Department of Health, conducted a survey which found that more than 190 communities with a population of over 200 had no water supply system and no source of safe water. He set out to write a bill which was later entitled the Sanitary Projects Act. The social and legislative process which brought the bill from concept to law was rather colorful. In the fall of 1946, Senator Filiberto Maestas came to Caldwell requesting help for his people because they did not have safe drinking water. When Caldwell showed Maestas his Sanitary Projects Act, Maestas agreed to sponsor the bill. Later, however, he told Caldwell that he hesitated to sponsor it because he was Spanish. Since many of the people that would benefit from the bill would be Spanish, he felt that an Anglo-American would have a better opportunity of carrying the bill through the legislature. Mary Pollard, then a Health Educator with the Department of Public Health, suggested that she throw a cocktail party to get the legislators drunk and coerce them into backing the bill. The party, held at Judge Kyker's home in Santa Fe, was a huge success, and the legislators who attended agreed to back the bill. With the able sponsorship of Senator Maestas, the bill was finally passed in the 1947 session of the state legislature.

The provisions of the bill are simple. The legislature appropriates approximately $100,000 a year for grants in aid to rural communities of ten or more families if they will form what is called a Mutual Domestic Water Consumers and Sewage Works Association. The state provides up to $12,000 per project for the cost of a well, pump, and pipe. The community of local citizens must provide the ditching, adobes for construction of a wellhouse, and the labor to install the piping. In the early days these arrangements worked very well, since people in a small community were used to working. By December 1972, 159 projects serving 6,570 families had been constructed.

The success of the program has been seen in a reduction of infant mortality. The reduction of water-borne disease cases such as typhoid has been relatively well documented since the passage of the act. A comparison of the five-year period of 1944 to 1948 to the five-year period of 1964 to 1968 has shown an 84 percent decline in cases of typhoid fever for the state as a whole. In four of the five counties with a large number of projects under the Sanitary Projects Act the decline was greater than 89 percent. Rio Arriba County has shown a 97½ percent decline, Taos County, 100 percent, Santa Fe County, 93 percent, and San Miguel County, 100 percent. However, no conclusions can be reached about the fifth one, Mora County, for it had no typhoid records during the 1940's. The state as a whole has shown a 70 percent decline in infant mortality during the same period. Of the counties that received aid under the Sanitary Projects Act, San Miguel County showed a 74 percent decline, Rio Arriba County, 70 percent, Taos County, 80 percent, and Santa Fe County, 79 percent. Safe drinking water and better overall sanitation have been responsible for these reductions in disease.

Rising construction costs and lack of interest among the local people have decreased the effectiveness of the Sanitary Projects Act in recent years. To make the act more workable, some legislators have recently proposed increasing the percentage of money contributed by the state for construction of the systems if the local community will pay for operation and maintenance by a professional manager who will take care of perhaps forty communities.

Many incorporated villages and towns also have water supply construction problems. In 1972 the President decided that the Farmers Home Administration would stop federal aid for the construction of water-supply systems. As a consequence, Farmers Home Administration, the Environmental Improvement Agency, and many community leaders approached the state legislature together and requested that they appropriate funds to pick up the Farmers Home appropriation. The legislature responded admirably and passed the Water Supply Construction Act. The act appropriated $1,500,000, which the Environment Improvement Agency has been administering since June of this year. The act provides a state grant of up to 40 percent for communities under 5,500 and one up to 25 percent for communities over 5,500. The agency has adopted guidelines for administration of the act, and as of September 1, 1973, all monies appropriated have been allocated to projects.

Public Law 89–121 is a federal act administered by the U.S. Public Health Service, which is similar to the Sanitary Projects Act and the Water Supply Construction Act of New Mexico but is for the Indian population. The Indian water supply systems are constructed to a much higher standard than the systems constructed under state funding. The U.S. Public Health Service proposes projects each year, and Congress appropriates funds for construction the following year. At this time, all pueblos and reservations have water-supply facilities. With the aid of the Sanitary Projects Act, Water Supply Construction Act, and Public Law 89–121, the majority of the population of New Mexico will soon enjoy safe drinking water.

The Environmental Improvement Agency is presently working with the Demonstration Water Project (DWP), a foundation located in Washington, D.C., to correct the operational problems of small systems. DWP has given a grant to HELP, a New Mexico foundation, which has established two corporations to provide support for system improvements and operation. The plan is to bring a number of the small associations under the arm of the support corporations for better operation and maintenance.

Sewage Disposal

Sewage disposal is a problem in New Mexico, as in any other part of the country. Historically, wastewater disposal regulation was the responsibility of the New Mexico Department of Public Health. In 1912 the territorial health officer was the only agent responsible for public health. In 1937, the territorial health officer's powers were transferred to the newly established Department of Public Health. The Department made a regulation which required a permit from the director of the department before anyone supplied water or disposed of sewage. The department was so poorly staffed, however, that permits were granted simply on the basis of plans and specifications.

In 1967, the Water Quality Control Commission was established by passage of the New Mexico Water Quality Act. The commission is made up of the heads of eight state agencies and one member at large. The agencies represented on the commission are considered constituent agencies. The commission has no staff but acts as the policy-making body and delegates administrative duties to various constituent agencies. Regulatory authority for water pollution control now rests with the Water Quality Control Commission. The Environmental Improvement Agency, however, performs the bulk of the administrative duties in water-pollution control.

The state has aid-to-construction programs as well as regulatory authority. For example, in 1957 the Sanitary Projects Act, previously discussed with reference to water-supply problems, was amended to allow the state to participate in the construction of sewage-disposal systems for unincorporated rural communities. Under the Sanitary Projects Act, eleven sewage disposal projects have been constructed serving approximately 550 families. These small communities, however, simply do not have the resources to operate and maintain their systems properly. Nine of the eleven projects are polluting the environment with poorly treated sewage. One project serving only fifty families has never been operated, and very poorly treated lagoon effluent is now running into the stream. Another project was constructed for thirty families, but only two have tied on because the community lacks the leadership to provide the extra piping and connections for other citizens. New EPA requirements placed on communities which apply for Federal Water Pollution Control Act construction funds have prevented the Sanitary Projects Act communities from receiving federal funds in New Mexico because they lack the authority to control the use of the systems by industries and private citizens.

The Environmental Improvement Agency believes that the sewage disposal systems under the Sanitary Projects Act, in contrast to the water supply systems, have been unsuccessful because of the higher costs of sewage system construction and operation and because of public apathy about sewage disposal. Also, the problems of red tape in dealing with both a state and federal bureaucracy tend to wear out the local citizens before a project is completed.

In order to take advantage of increased federal funding for municipal sewage treatment grants, New Mexico passed a one million dollar bond issue in 1970. In 1971 a 2.5 million dollar bond issue was passed to augment the funds. In 1973, six million dollars were appropriated from the general fund, making a total of 9½ million dollars that have been appropriated by the state since 1970. These funds are matched with funds from the Federal Water Pollution Control Act, and a total construction program of over 40 million dollars is now underway.

The state of New Mexico claims that 100 percent of all communities having a sewage collection system have secondary sewage treatment. In many communities, however, secondary sewage treatment does not meet the standards of the present definition established by federal regulation under the Federal Water Pollution Control Act Amendments of 1972. For example, some communities have trickling filters without secondary sedimentation tanks. Some communities have lagoon systems that produce relatively good effluent, but the algae content runs the BOD well above the 30 milligrams per liter limit of present regulations. State programs are now being implemented to solve these problems.

Solid Waste

Solid waste problems in New Mexico are significant. The wide dispersal of the population in many counties causes serious collection and disposal problems. The state's Environmental Improvement Agency has developed a comprehensive solid waste plan for New Mexico from its surveys to determine the amount of solid waste produced per capita in recreation areas and in municipal systems. It has been a catalyst in bringing together the National Forest Service, the

State Highway Department and various communities to develop regional landfill sites for some recreation areas. It has also worked to develop regional sites for more than one municipality and county use.

At the present time, the agency is conducting a training program for solid waste management and operating personnel. This program includes training in landfill construction for bulldozer personnel, safety training for collection personnel, and management training for county and city managers and engineers who must plan and develop landfill sites and collection routes.

In 1972, a bill was introduced in the state legislature to provide 50 percent state support to communities under 5,500 to purchase equipment and develop proper landfill sites. The bill, however, failed to pass the legislature. It is the intent of the Environmental Improvement Agency to work with the New Mexico Municipal League during the next session of the legislature for legislation that will establish a grant-in-aid program for construction of solid waste handling facilities.

It is interesting to note that the 1973 session of the legislature provided six million dollars for sewage disposal construction and 1.5 million dollars for water-supply construction but turned down the solid-waste proposal. Again, public interest in safe water supply was keen and helped sell the water bill. The sewage disposal program was already well established. But the legislature did not care to begin funding a new program for solid-waste disposal, especially without broad public support.

18.

Thirty-Five Years of Use of a Natural Sand Bed for Polishing a Secondary-Treated Effluent

Donald B. Aulenbach, Nicholas L. Clesceri, James Tofflemire, and James J. Ferris

Introduction

When the sewage treatment plant for Lake George Village, New York, was constructed in the late 1930's, laws were already in effect preventing the discharge of any sewage, raw or treated, into the waters of Lake George or into any streams which emptied into this beautiful recreational lake. Therefore, extra steps were taken to provide for "complete" treatment of the sewage. The basic treatment plant is a common design with circular Imhoff tanks for primary sedimentation and sludge digestion, trickling filters, and secondary sedimentation of the trickling filter effluent. The unique portion of the plant, other than its dual pumping system lifting the sewage approximately 200 feet from the collection point by the lake to the treatment plant, is the discharge of the effluent from the secondary sedimentation tanks directly onto natural sand beds without chlorination. Morrell Vrooman, Jr. (1940) has determined that the sand beds are "more than 25 feet in depth" and that the "final effluent becomes groundwater, which, in all probability, seeps eventually to some water course as a highly purified liquid which cannot be identified as a sewage effluent."

When the original treatment plant was designed and constructed, the population estimates for the area varied from approximately 1,500 persons in winter to about 5,000 at the peak of the summer season. In order to allow for this approximately threefold change in population, the treatment system was built essentially in triplicate, using one third of the system for the wintertime flows and the entire plant for summertime flows. The present population being served by the treatment system is an estimated 2,100 persons in winter and 12,300 in summer (Aulenbach and Clesceri, 1973). Initially, there were six sand beds with a total area of approximately 72,000 square feet (6,690 square meters). Currently, there are fourteen sand beds where the original six were located and an additional seven sand beds at a higher elevation above the primary settling tanks. The total area of these combined beds is 6.4 acres (2.6 hectares). A general layout of the plant is shown in Figure 1.

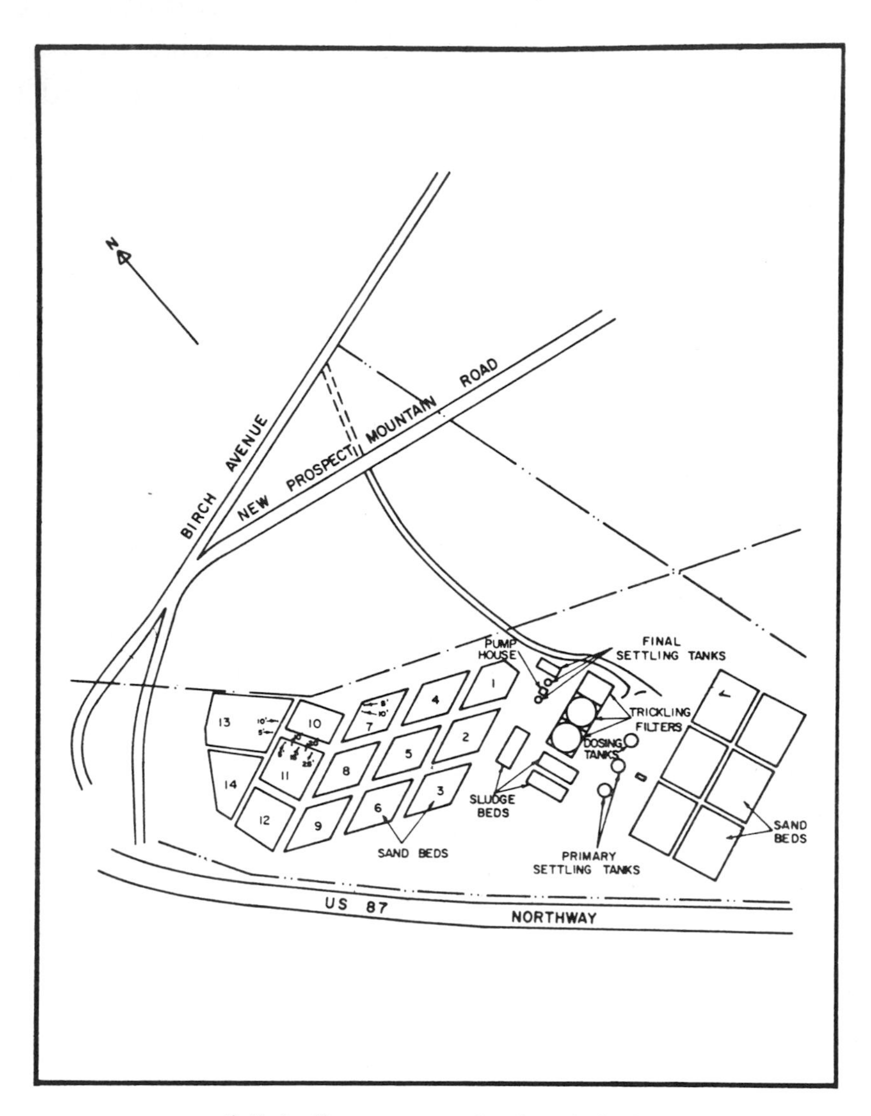

1. Lake George sewage treatment plant.

Previous Studies

When the original plant was built, the primary concerns in sewage treatment were the disposal of the liquid effluent and the sanitary quality as measured by the coliform count. Twenty-five feet of soil was considered adequate to remove the coliform bacteria or pathogens. However, no tests have ever been performed to show positively the removal of coliforms by passing effluent through sand beds. In recent years, engineers have realized that sanitary quality is not the only parameter of treatment efficiency. Inorganic nutrients may be unaffected by treatment but may influence secondary biological growth in the receiving water. Nitrogen and phosphorus compounds are frequently limiting nutrients in a lake, and their input should be minimized. In order to determine the potential discharge of these and other substances into Lake George, studies were performed in the sand beds during the late 1960's by students at Rensselaer Polytechnic Institute. Wells were installed in beds 7, 11, and 13 at depths of 5, 10, 15, 20, and 25 feet. Bed 7 was an extremely slow bed and took up to a week to dry, while the water percolated away in a day or two from Bed 11. Bed 13 was chosen because it had had only limited use. (The control valve for this bed was accessible only through a manhole, while valves for the other beds were operated from above the ground.)

It was found that no sample could be obtained from Bed 7 and that samples could be obtained from the wells at the 5 and 10 foot depths only in beds 11 and 13. It was assumed that samples obtained at the 5 and 10 foot depths indicated a continuous water column to these depths, but that by the time the water percolated 15 feet down into beds 11 and 13, the water became dispersed and could not be pumped directly.

The results of these studies (Aulenbach, Glavin, Romero-Rojas, 1970) are summarized in Table 1. Essentially all of the coliforms are removed in the first five feet of the bed depth. The influent BOD, which was about 40 mg/l on top of the bed, was reduced to less than 8 mg/l at the 5 foot depth to less than 2 mg/l at the 10 foot depth. This indicates very satisfactory BOD removal. There is no significant reduction in the removal of chlorides with depth. The overall reduction with 10 feet of depth in bed 11 was 5 percent, whereas in bed 13 it was 10 percent. Essentially all of the organic nitrogen was removed within the first 10 feet of the filter bed. The ammonia nitrogen, on the other hand, was removed only to an extent of about 80 percent in the first 10 feet of the sand beds. Since the removal appears to be nearly linear, it may be estimated that nearly complete ammonia removal could be accomplished with an additional 5 feet of depth of sand or 15 feet total. Although the results of the nitrate determination varied considerably, the concentration did appear to increase slightly with depth. There was no quantitative balance between the reduction of ammonia and organic nitrogen and the increase in nitrate nitrogen. It may be concluded, however, that there is an overall oxidation process involved in converting organic and ammonia nitrogen to nitrate.

Table 1
Changes in Water Quality Parameters in Vertical Transport through Sand Beds

	Bed 11			Bed 13		
		Depth, ft.			*Depth, ft.*	
Median concentrations	*Applied Effluent*	*5*	*10*	*Applied Effluent*	*5*	*10*
Coliforms, #/100 ml	970.0	46.0	8	1260	20.0	1
BOD, mg/1	38.5	7.5	1.45	46.0	6.8	1.2
Chloride, mg/1	37.5	36.5	35.5	39.0	35.0	35.0
Organic nitrogen, mg N/l	4.25	0.75	0	5.6	0.28	0
Ammonia, mg N/1	9.7	7.8	1.8	13.2	6.6	2.8
Nitrate, mg N/1	2.4	2.4	8.7	3.3	4.5	8.0
Polyphosphate, mg PO_4/1	3.0	0.8	0.4	1.8	1.2	1.4
Orthophosphate, mg PO_4/1	22.0	27.9	24.8	22.8	6.4	7.6
Total phosphate, mg PO_4/1	25.4	33.4	27.5	25.8	7.2	10.0

The phosphorus concentrations alone showed a significant difference between beds 11 and 13. There seems to be a marked reduction in polyphosphates in the first 5 feet of bed 11 but little further reduction between 5 and 10 feet in bed 11. Bed 13 showed no significant change in polyphosphate concentration. The orthophosphate showed no reduction with depth in bed 11 and even a slight increase at the 5 foot depth. In bed 13, which was little used, there was a significant reduction in the first 5 feet but no further reduction in the second 5 feet of bed depth. The total phosphate results are similar to those of the orthophosphate. It was concluded that, for the depths studied, bed 11, which had received considerable use, had reached a saturation point for adsorption of phosphorus, whereas bed 13, which had received little use, still had an active phosphorus adsorption capacity.

The original plan for the previously described studies was to install a series of wells in and around the treatment plant in order to determine the direction of flow of the treated effluent, which was applied to the percolation bed. However, a large number of wells was deemed infeasible at that time. In another attempt to determine the flow of the effluent through the ground, resistivity studies were performed in the area around the sewage treatment plant (Fink, 1974). The studies were made using a Wenner array with fifty-foot electrode spacing. The results showed (Figure 2) that the path of least resistance flowed first in a slightly eastward direction and then turned northward toward Lake George.

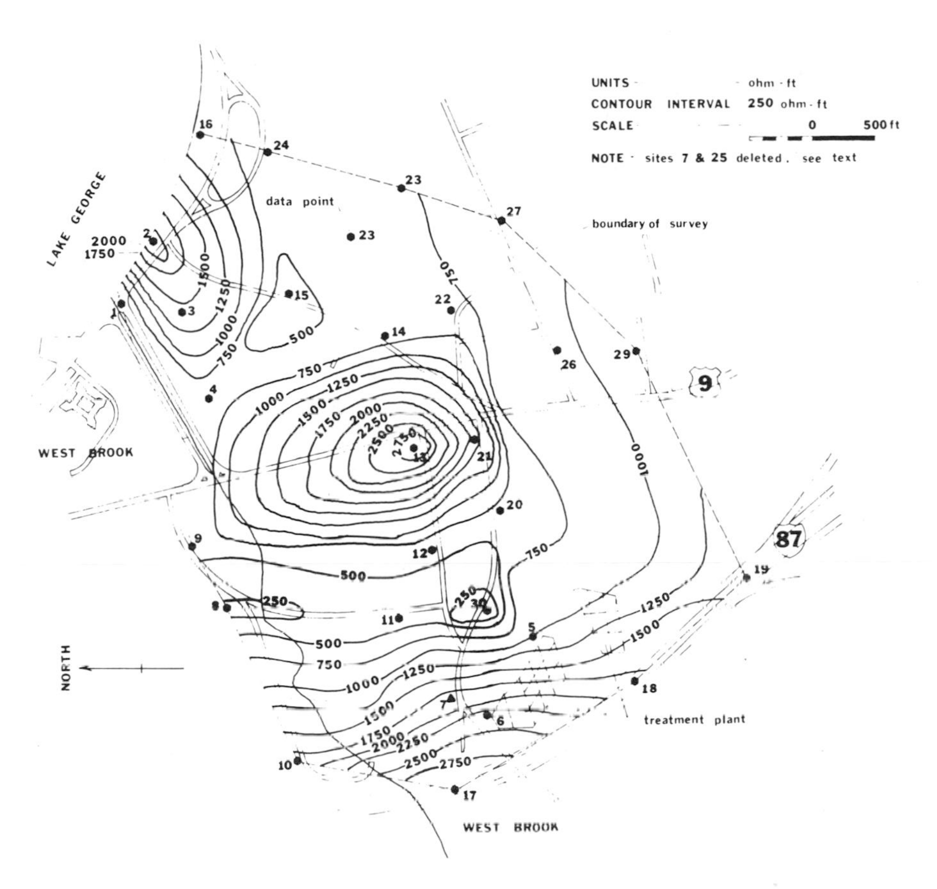

2. Iso-resistivity map with interpretation.

Present Studies

With recommendations from a consulting firm, Rist-Frost Associates, funds provided by the Village of Lake George, and manpower from the New York State Department of Environmental Conservation, wells were located in and around the sewage treatment facility, as shown in Figure 3. Well 1 is in the approximate center of the lower seepage bed no. 4. Well 2 is located approximately halfway down the steep portion of the slope just south of West Brook. Well 3 actually consists of three wells, A, B, and C, and just recently well 3D has been installed. The three wells obtain water from three different depths. Well 3A is very near to the surface, 3B is approximately 11 feet deep, and 3C is 21 feet deep. Wells 3A–C are located adjacent to a significant seepage of water from the base of the hill close to West Brook. Well 4,

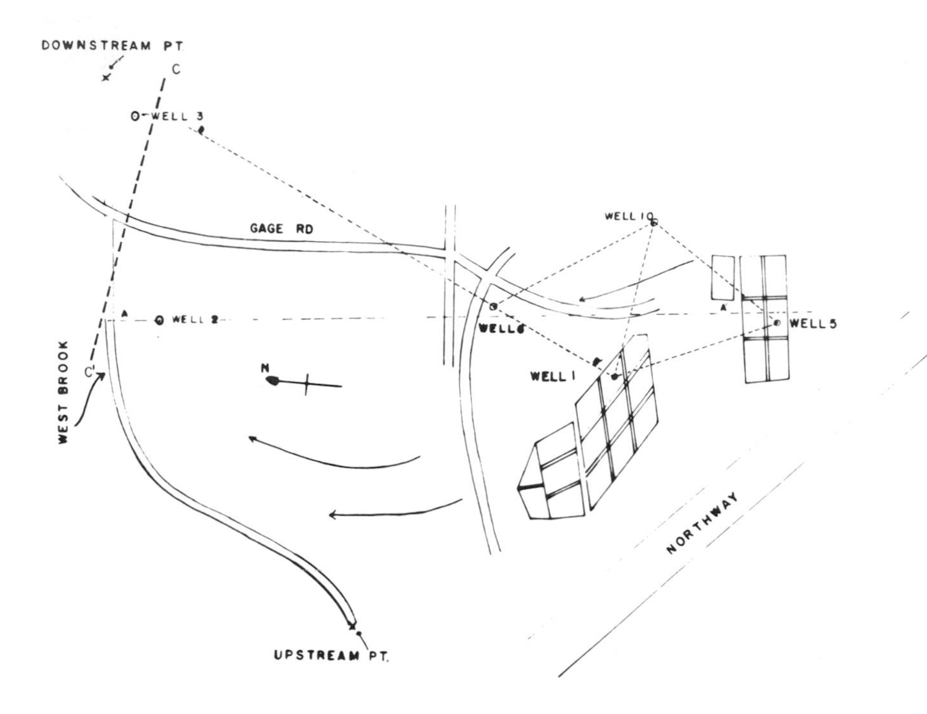

3. Lake George sewage disposal beds.

intended to be installed upstream from all potential sources of groundwater contamination, was never installed, because of rocks above the groundwater. Well 5 is located near the center of upper sand bed no. 16. Well 6 is located approximately 500 feet downstream from the percolation beds. There are no wells numbered 7, 8, and 9. Well 10 is located in the field slightly east of the sewage treatment plant and was intended to represent uncontaminated groundwater. The samples designated West Brook upstream were taken where Sewell Road crosses West Brook, and the samples designated West Brook downstream were taken sufficiently far downstream from Wells 3A–C to ensure adequate mixing of the drainage that enters West Brook near 3A–C. Specific data for the wells are summarized in Figure 4.

Results

Samples were obtained from the various locations in the area during the spring and summer of 1973 and were analyzed for the following characteristics:

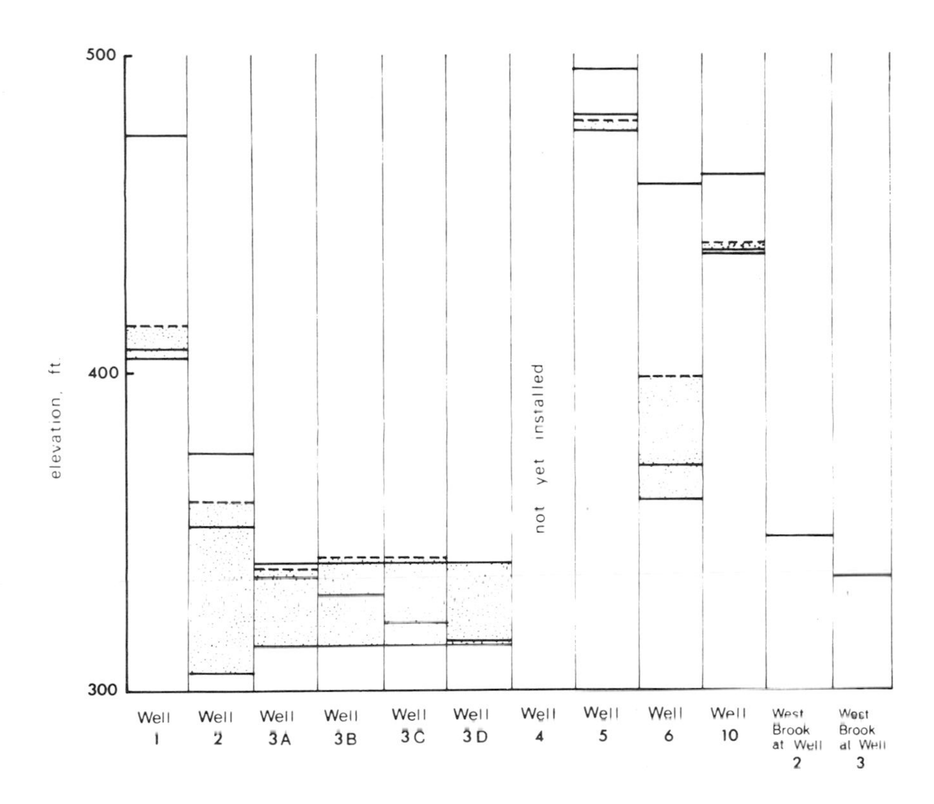

4. Well data.

Groundwater condition. In general, the temperature of the groundwater decreased with depth from the sand beds. Wells 1 had a lower temperature when the bed was dry, which indicated the direct effect of the sewage effluent on the bed. Wells 3A–C decreased in temperature with depth in the summer and increased with depth in the winter. It may be noted that West Brook is a cold-water stream, never exceeding 15°C during the period of this study.

The presence of dissolved oxygen (DO) is considered desirable for the efficient treatment of sewage effluent discharged into the ground. In general, the DO increased with distance from the sand beds. Well

2 always had a high DO. Well 3C always had a lower DO than 3A. West Brook was nearly always saturated with DO at the upstream location, but on several days a measurable decrease in DO was noted at the downstream location, which was attributed to the discharge of groundwater into the brook.

Similarities to treated sewage effluent. The water seeping from the ground at the base of the sand hill enters West Brook at two major locations. The two seepage streams add some 10 to 15 percent to the flow of West Brook. This inflow is reflected in the dissolved solids content of West Brook. The dissolved solids varied between 49 and 75 mg/l at the upstream location and from 70 to 120 mg/l at the downstream location. In all cases there was an increase, which varied between 20 and 50 mg/l. The dissolved solids in West Brook increased from April through August, when the population and thus the flow at the sewage treatment plant also increased. In general, the wells in the sand beds (wells 1 and 5) exhibited noticeable fluctuation in dissolved solids with time and in particular with flooding of the bed. On the other hand, the wells not located in sand beds had a relatively constant dissolved solids content throughout the period of the sampling.

The high alkalinity in the wells is reflected in an increase in alkalinity from the upstream to the downstream sampling locations within West Brook. The lowest alkalinities were found in well 10 consistently and in well 5 when it was dry. This latter would indicate that the alkalinity is being flushed away by fresh groundwater after the treated sewage has percolated through upper bed 16. In each case, higher alkalinities were found in well 3A, with noticeably lower values in well 3B and slightly lower values in well 3C. This obviously represents the depth profile and indicates that the high alkalinity liquid is coming in near the surface and that more dilution, because of lower alkalinity groundwater, occurs at deeper locations.

The chloride concentration of the sewage treatment plant effluent varied between 30 and 35 mg/l during this sampling period. The chloride concentration may generally be used as a tracer of groundwater because there is little soil adsorption of chloride. In general, wells 2 and 6 had chloride concentrations similar to that of the sewage effluent. Well 5 had a concentration similar to the sewage effluent during the period that the bed was flooded, but when the bed was dry, the concentration was only 8 mg/l, probably representing the chloride content of the natural groundwater in the area. Wells 3A and

3B, however, were consistently higher in chloride content than the sewage effluent discharged onto the sand beds, and the chloride content of well 3B was consistently higher than that of 3A and 3C. In one instance, the chloride content of the seepage in the small drainage ditch behind the highway department garage reached a maximum of 153 mg/l. Since this concentration is much greater than that of the sewage effluent, it raises some doubt as to the validity of using chloride as a tracer in this particular case. The highway department stores salt on their property for use on the roads during the winter.

Nutrients. Having reasonably well established that the water recovered from wells 2, 3, and 6 and the seepage emanating from the ground near West Brook is the same water that was applied to the sand beds at the sewage treatment plant, the transport of the nutrients nitrogen and phosphorus through the soil may be observed. At Well 6, approximately 500 feet downstream from the percolation beds, the phosphorus content was less than one percent of that found in the sewage effluent applied to the beds. Similar low results were found in wells 2 and 3, with generally higher phosphorus content being found nearer the surface at well 3A than at the deeper wells 3B and 3C. With the exception of the one high value of 92 mg/l of soluble phosphorus on April 17 at the seepage area above Gage Road, all the results showed that the soluble phosphorus content in the area of West Brook was less than 50 mg/l. In West Brook on two occasions, the soluble phosphorus content appeared to decrease from the upstream to the downstream location, whereas on three occasions there was a slight increase.

The nitrogen results are not quite so conclusive, because some data, particularly the nitrate content in the sewage effluent, was unobtainable. No exact balance of the nitrogen content was made, because there was a large fluctuation in the nitrogen content of the sewage influent and the effluent discharged onto the sand beds. Obviously, daily fluctuations in the nitrogen content would not be reflected in the content in the wells on the same day. In all instances, there was a significant increase in the nitrogen content of West Brook between the samples obtained above and below the entrance of the groundwater seepage.

One set of samples (July 25, 1973) from the influent, effluent West Brook (upstream and downstream) and several wells was examined for coliphage as a preliminary indication of the possible presence of other such infectious agents. Infectious coliphage was found in all samples tested.

Discussion

There has been considerable concern over disposal of domestic wastes from the rural areas and small towns. If the discharge of wastes may affect a local recreational lake, concern is particularly justified. Whereas information is fairly readily available as to the effect of wastewaters that are collected and treated and discharged into a stream entering a lake or into a lake directly, there is insufficient information on the effectiveness of discharging wastewaters, either treated or untreated, directly into the soil for purification. It has been fairly well established that under properly operating aerobic conditions, soils are capable of removing pathogenic bacteria and organic matter in the applied wastewater. However, much less information is available relating to the removal of nutrients, primarily nitrogen and phosphorus, by the soil. In general, sand has been considered less capable of removing phosphorus than finer claylike soil, but the total capacity of any soil to remove nutrients over a long period of time has not been sufficiently evaluated.

The treatment system at Lake George Village offers an opportunity to study both the treatment efficiency of sandy soils and the effect of relatively long years of use. For approximately thirty-five years, the village treatment plant has discharged biologically treated effluent onto natural delta sands. Studies have confirmed the ability of the sand to remove coliforms, BOD, and organic and ammonia nitrogen, with some apparent oxidation of the reduced nitrogen compounds to nitrate. The sand showed little ability to reduce the chloride concentration. Phosphate removal was variable, apparently because of the ion exchange capacity, at least within the first 10 feet of bed depth.

By studying the water seeping out of the base of the sand hill approximately 2,000 feet (600 m) from the treatment plant and wells at various locations in the area, the characteristics of the purified wastewaters discharged into the ground have been determined. The presence of dissolved oxygen, although low at some locations, confirmed in all of the well samples that the soil has an oxidizing capability. This is considered desirable, for it aids in oxidizing any organic matter remaining in the wastewaters as well as removing pathogenic bacteria and oxidizing reduced nitrogen compounds to nitrates. The lowest dissolved oxygen recorded (1.7 mg/l) was in well 5 during the period of time when this bed was flooded. At all times, a small residual DO was found in the treatment plant effluent as discharged onto the sand percolation beds. Thus in this case, the soil system receiving the effluent from the Lake George Village sewage treatment plant appears adequate to provide what is normally considered desirable treatment or polishing of the treated wastes discharged into the sand.

The dissolved solids, alkalinity, and chloride may normally be used to identify the location of the sewage effluent in the ground with reasonable accuracy. Well 10, the control, had consistently low dissolved solids, alkalinity, and chloride as compared to other wells. The low concentrations of these substances in Well 5 when the bed was dry indicated that the high solids content during the sewage application was flushed away after application. These factors appear to indicate that the wastewater applied to the percolation beds does reach wells 2 and 6, even though the point of well 6 is approximately 27 feet below the surface of the groundwater.

Wells 3A–C present problems of interpretation. Fairly consistently, Well 3B, the medium depth sample, indicated higher dissolved solids and chloride than the other two depths, while Well 3A, the shallow location, had consistently higher alkalinity than the deeper samples. Moreover, all three parameters were higher in wells 3A–C than the equivalent parameters in the effluent discharged to the sand beds. This indicates that dissolved solids are being added to the groundwater somewhere between the treatment plant and the locations of wells 3A–C from some other source, possibly from the leaching of road salt stored at the highway department garage on the hill above wells 3A–C. This would not, however, explain the increased alkalinity in this area. The extreme concentrations of chloride found in the area of wells 3A–C make it impossible to trace the water discharged onto the sand beds at the treatment plant with chloride. Thus additional tracer studies are necessary in order to confirm positively the flow of the treatment plant effluent to the area of wells 3A–C.

Within West Brook, the dissolved solids, alkalinity, and chloride were consistently increased between the upstream and downstream sampling locations. Although no stream flow measurements were made of the small tributary streams entering West Brook in the area, a rough estimate was that the flow in West Brook was approximately ten times that of the flows in the small tributaries. A rough calculation was made based upon only the dissolved solids content, and a dilution factor of 1:9 agreed fairly well with the increase in the dissolved solids in this general area.

Additional studies will have to be made in order to make a more realistic nitrogen balance in this system. In any event, it may be concluded that significant amounts of nitrogen, particularly nitrate, do reach West Brook and that the nitrate concentration in West Brook increases significantly as it passes through this area.

Probably the nutrient of most concern is phosphorus. The effluent from the sewage treatment plant has in the order of 4 mg/l (4000 μg/l) of total soluble phosphorus. The phosphorus content of the wells located in the sand beds indicated a reduction of about 90 percent of the phosphorus at the point of securing of the samples. In well 1, this represents a vertical passage through approximately 67 feet of sand and in well 5, through approximately 16 feet of sand. However, caution must be used in interpreting these results. It is possible that this low phosphorus content represents some dilution by the groundwater. Approximately 500 feet (150 m) from the lower sand beds, less than one percent of the original phosphorus applied to the sand-beds was recovered. Tracer studies must be made to confirm the direct passage of water from the percolation beds to the point of sampling. The effect of the phosphate in West Brook appears to be negligible.

Conclusions

It appears that the sewage effluent applied to the sand beds at the Lake George Village sewage treatment plant enters the groundwater, flows in a generally northerly direction, and emerges in some form or another from the ground near West Brook, following which it flows into West Brook and ultimately into Lake George. The sandy soil of this area appears to be adequate for the conventional treatment of the applied effluent. Reflecting the quality of the sewage effluent applied to the sand beds, the water emerging from the ground near West Brook is much higher in dissolved solids, alkalinity, and chloride than the natural groundwater in the area. These parameters are increased in West Brook as the stream flows past this area. Although insufficient data are available to make a positive statement concerning the nitrogen contributions from the sewage effluent, it does appear that the reduced nitrogen is completely oxidized to nitrate prior to its emergence from the ground and increases the nitrate content of the water of West Brook significantly.

It appears that the total phosphorus content of the applied sewage effluent has been reduced by more than 99 percent in its passage through approximately 2,000 feet of sand from the treatment plant to West Brook. There is little increase in the phosphate content of West Brook from the effluent.

References

Aulenbach, D. B., and Clesceri, N. L. 1973. Sources and sinks of nitrogen and phosphorus: water quality management of Lake George (N.Y.). In *Water 1972*, vol. 69, AIChE Symposium Series no. 129, ed. G. F. Bennett, pp. 253–262. New York: American Institute of Chemical Engineers.

Aulenbach, D. B., Glavin, T., Romero-Rojas, J. A. 1970. Effectiveness of a deep natural sand filter for finishing of a secondary treatment plant effluent. Paper presented at the New York Water Pollution Control Association, 29 January 1970, New York, New York.

Fink, W. B., Jr. 1974. Ground water contamination due to treated sewage spreading—a case study. Master's thesis, Rensselaer Polytechnic Institute.

Hershey, A. D., Kalmonson, G., Bronfenbrenner, J. 1943. Quantitative relationships in the phage-antiphage reaction: unity and homogeneity of the reactants. *Journal of immunology* 46: 281–299.

Vrooman, Morrell, Jr. 1940. Complete sewage disposal for a small community. *Water works and sewerage* 87: 130–133.

19. On-Site Household Wastewater Treatment Alternatives—Laboratory and Field Studies

Richard J. Otis, Neil J. Hutzler, and William C. Boyle

The management of wastewater discharged from the singlefamily dwelling, motel, restaurant, or other establishment which is not connected to central sewerage is a major problem throughout the country. The septic tank-soil absorption system has been relied upon almost entirely to treat and ultimately dispose of the wastewater in unsewered areas, but it is often ineffective in preventing public health hazards and nuisances. In the past, septic tank system failures could be ignored because rural population densities were not high. Today, however, with the rapid growth of high density suburban and summer home communities, serious public health problems have arisen because of poorly functioning waste disposal systems.

The failure of the system to treat and dispose of wastewater safely does not seem to be due to inherent shortcomings of the septic tank-soil absorption field, but more to its misapplication and misuse. Patterson, Minear, and Nedved (1971) in a review of the literature dealing with septic tanks found the principal causes of malfunctions to be poor site conditions and improperly constructed or maintained systems. Specifically, the most common causes noted were shallow bedrock, impervious soil, high groundwater, overloaded soil, improper installation, and poor maintenance.

Assuming that better supervision and control can correct the problems of improper installation and maintenance, the fact does remain that there are many areas simply unsuitable for on-site treatment and disposal by the septic tank system. The U.S. Soil Conservation Service estimates that less than 32 percent of all the land area in the United States is suitable for the installation of septic tank-soil absorption field systems (Wenk, 1971). Either there is insufficient soil in other areas to provide adequate purification of the waste before it reaches the groundwater or the soil is not permeable enough to absorb the liquid. Since development is continuing in areas where central sewerage is unavailable and site conditions are poor for septic tank systems, alternative systems need to be developed to prevent public health hazards from occurring.

Currently the only alternatives to the septic tank system are (1) the prevention of all development in unsuitable areas, (2) holding tanks, or (3) central sewerage. None of these is satisfactory. Anything short of public acquisition of the land areas that are considered unsuitable seems ineffective in preventing development.* Holding tanks are currently discouraged whenever possible by many public health officials because of the regular pumping that is required and the lack of adequate disposal fields. This leaves community collection and treatment as the favored alternative. Developers, however, are not eager to provide sewers because of the added responsibility placed on them for maintaining a sewerage system and the larger investment required with a slower return. Thus future economic development will be inhibited in many areas, and existing hazards will go uncorrected, unless effective alternative systems are found. While not all lands should be opened to development, it is felt that restrictions to development should be based on rational zoning policies rather than on the suitability of the soil.

Objectives of Study

A household waste disposal system can be broken down into three component parts. These are (1) the wastewater source, (2) the treatment module, and (3) the point of final disposal. For each of the three components, various methods or devices can be employed to provide a safe and effective system designed to function within the particular site limitations. The treatment module will be concentrated on here. (For methods and devices that might be employed to reduce the volume and strength of the waste at the source see the General Dynamics Electric Boat Division Study, 1973.) Alternative soil disposal systems are discussed by Bouma in "Use of Soil for Disposal and Treatment of Septic Tank Effluent," this text (pp. 89–93).

Any treatment and disposal system fails if it allows harmful pollutants to accumulate to dangerous levels in the receiving environment. Though the septic tank provides limited treatment in the form of settleable solids removal, this is sufficient for a conventional system if the proper site conditions prevail. Where site conditions are not optimal, however, greater amounts of pollutants must be removed before the liquid is released to the environment. One method of accomplishing this is to provide more effective treatment.

Currently there are numerous alternatives available for the treatment of household wastewaters. The selection of any one of these

*Jon Kusler, attorney-at-law, 1971: personal communication.

alternatives and the design of the entire system around it will depend upon several criteria: (1) the effluent quality produced, (2) the variability in effluent quality, (3) servicing and maintenance requirements, and (4) total annual costs.

The design of an on-site wastewater treatment and disposal system becomes critical where site conditions are poor because the treatment system must have great control over the effluent quality discharged. Variability in performance may become one of the more important criteria by which to judge the system. If the effluent produced by one unit within the system is highly variable and unpredictable, it is necessary to design subsequent units able to handle the fluctuations. This adds to the cost of the system. It may be cheaper to select a unit that reliably produces a poorer average effluent quality with little variation than one which may perform much better but is susceptible to major upsets. If a pattern for the variability could be predicted, servicing at the proper interval might reduce substantially these changes in quality, but this also would increase the cost of the system.

It is the objective of this phase of the Small Scale Waste Management Project in progress at the University of Wisconsin to evaluate various treatment processes and designs on the basis of the four criteria outlined above. Several treatment units have been selected from the variety on the market today for evaluation in both field and laboratory environments. An attempt was made to include as wide a range of treatment processes as possible. The mechanically less complex and less expensive units were favored because they were considered more suitable to single household use.

Test Facilities

In the spring of 1972, both laboratory and field studies of full-scale household treatment units were initiated. Treatment units under investigation include (1) single-compartmented septic tanks, (2) multiple-compartmented septic tanks, (3) continuous flow-extended aeration units receiving either raw or septic tank wastes, (4) batch-extended aeration units receiving raw wastes only, (5) rotating biological disk units, (6) a chemical precipitation and coagulation unit, (7) intermittent sand filters loaded with either septic tank effluent or aerobically treated waste, and (8) dry-feed chlorinators following sand filtration. Only the septic tanks and aerobic treatment units will be discussed here.

Laboratory

The laboratory facility has been fully operational since January 1973. It was built to explore questions which could not be answered from monitoring treatment systems in the field. Waste flows from individual households are highly variable and depend on the residents' habits; therefore, it is difficult to compare the relative treatment efficiencies of various systems using data from field studies. There is also a higher risk factor in the field. Any system installed must offer a good chance of success to avoid early failure and costly replacement to the homeowner. Monitoring the systems, too, is a problem with many scattered field sites. The laboratory, however, is able to make up for shortcomings in the field work by providing controlled conditions for more accurate results and greater testing flexibility.

To evaluate and compare treatment units in the laboratory realistically it is necessary to simulate household wastes. Wastes from single households are discharged intermittently, and each discharge varies widely in its hydraulic and organic characteristics. This variability can have a marked effect on the treatment efficiency of small units. Major household wastewaters were identified and characterized for simulation in the laboratory (Ligman, Hutzler, Boyle, 1974). Waste feeders were developed to simulate clothes washing, dishwashing, dish rinsing, garbage grinding, bathing, showering, and toilet use. Each feeder is controlled by an electronic system and can be automatically activated. Thus the various household wastes can be fed to each unit in a variety of sequences and frequencies. The system allows each unit to be fed identical waste and in identical patterns.

Five treatment processes are under study in the laboratory. Representative, commercially available treatment units were installed for testing. A septic tank, a continuous flow aerobic unit, a batch aerobic unit, a rotating biological disk unit, and a chemical coagulation and precipitation unit are included. A three-chambered septic tank which feeds the chemical treatment unit is also monitored. Sand filters are being added (Figure 1). Each unit is fed wastes in a sequence that is typical of household water use patterns, as shown in Figure 2, which yields a total daily flow of 170 gallons.

The single-chambered septic tank is a 610 gallon rectangular tank similar to those found in many home installations. The three compartment tank has a liquid volume of 260 gallons. The continuous flow aerobic unit is a 1200 gallon tank consisting of a trash trap which acts as a small septic tank, an aeration chamber where air is injected with a mechanical aerator, and a settling chamber with gravity sludge return. The batch aerobic unit is one tank, which acts as both an

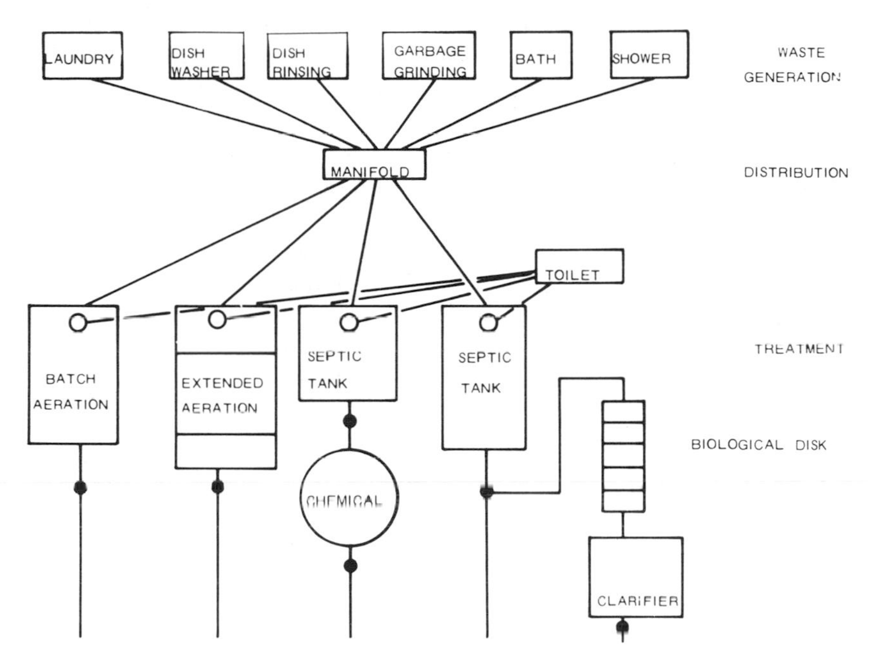

1. Laboratory flow sheet.

SCHEDULE OF SIMULATED EVENTS

EVENT	FLOW	VOLUME	DAILY VOLUME
T - TOILET USE	15 GPM	3 GAL	45 GAL
B - BATH	20	25	25
S - SHOWER	2.5	25	25
D - DISHWASHER	2	14	28
G - GARBAGE	3	2	4
K - KITCHEN	3	1.5	3
L - LAUNDRY	16	40	40

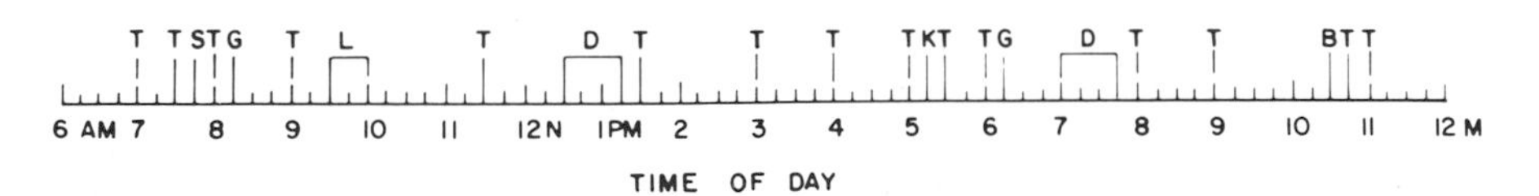

2. Schedule of simulated events.

aeration chamber and a settling chamber. It has a treatment capacity of 450 gallons. A centrifugal blower injects air into the tank for 18 hours, then shuts off in the early morning to allow the tank solids to settle. A submerged centrifugal pump discharges the supernatant from the tank after a five-hour settling period, and the daily cycle begins again. The rotating biological disk unit consists of four stages of plastic disks with a treatment capacity of 300 gallons per day. Solids sloughing off the disks are settled out in a clarifier following the disk module.

The effluents from these units are being continuously sampled. Daily flow composite samples are collected and further composited for analysis two or three times per week.

Field Installations

Field installations of conventional and alternative treatment systems began in the fall of 1971. The last were completed in the summer of 1973. All installations were made at private homes to take advantage of different soil conditions and life styles. Table 1 gives a description of the twelve systems currently included in the project. The homeowner is responsible for obtaining a contractor, but supervision is provided by the project. The cost of the system is borne by the owner except for the purchase and installation of experimental treatment units or equipment, which is paid by the project. No guarantee is made that the system will be successful. By-passes around all experimental equipment are provided if equipment failure could result in damage to the homeowner's property. Each system is designed such that the experimental equipment could be removed, leaving a functioning septic tank system. For participation in the project, each homeowner receives a substantial payment to allow access to the system by project personnel for monitoring and maintenance.

Each installation is visited regularly, usually monthly or bi-monthly for sampling. Twenty-four hour flow-composited samples are taken automatically from the effluents of each unit within the system. These samples are stored in ice during sampling and transported to Madison for analysis. Field measurements are made of temperature and pH of all sampling points and dissolved oxygen (DO) and chlorine residual where applicable. Grab samples are taken from the aeration chambers of the aerobic units for solids analysis. A thirty-minute settling test is also run in the field on the mixed liquor to determine the sludge volume index. Water and electric meters are installed on each system and are read before and after the sampling period.

Table 1
Description of Field Installations

A. Systems in Slowly Permeable Soils (percolation rate $>$ 60 min/inch)

1. Meyer, Clark County (5 persons)
   1000 gal. Septic Tank → Mound
   data since August 1972

2. Brevak, Ashland County (3 persons; $\sim$ 47 gal/cap/day)
   1000 gal. Septic Tank → Mound
   data since May 1972

3. Huhn, Ashland County (3 persons; $\sim$ 42 gal/cap/day)
   1000 gal. Septic Tank → Mound
   data since May 1972

4. Ashland Experimental Farm, Bayfield County (7 persons; $\sim$ 45 gal/cap/day)
   1200 gal. Septic Tank → Mound
   data since August 1972

5. Peterlik, Chippewa County (5 persons)
   1000 gal. Septic Tank → Mound
   no data

6. Ashland Experimental Farm. Bayfield County (4 persons)
   1000 gal. Septic Tank → 500 gal. Septic Tank → two 8′ × 8′ Sand Filters → Chlorinator → 150 gal. Contact Chamber
   data since October 1972

7. Electric Research Farm, Dane County (4 persons; $\sim$ 45 gal/cap/day)
   600 gal. Aeration Tank → 180 gal. Clarifier → two 4′ × 4′ Sand Filters → Chlorinator → 150 gal. Contact Chamber
   data since August 1972

8. Michal, Chippewa County (8 persons)
   1200 gal. Septic Tank → Rotating Disk Unit → Inverted Pipe Soil Field
   no data

9. Kessler, Outagamie County (5 persons)
   Batch Aeration → Pressure Distribution Soil Field
   data since August 1972

B. Systems in Shallow Bedrock

10. Fredrickson, Door County (6 persons; $\sim$ 36 gal/cap/day)
    936 gal. Septic Tank → 550 gal. Aeration → 200 gal. Clarifier → Mound
    data since June 1972

11. Peninsular Experimental Farm, Door County (2 persons; $\sim$ 87 gal/cap/day)
    1000 gal. Septic Tank → Mound
    data since November 1972

C. Systems in Permeable Soils (percolation rate $<$ 45 min/inch)

12. Marshke, Price County (4 persons)
    Septic Tank → Conventional Soil Field
    data since October 1972

Any servicing or repairs necessary at each site are noted and performed by project personnel or private firms. The homeowner is not asked to do anything differently from what he would normally do with a conventional system. If he should notice a malfunction, he is asked to notify the project immediately by phone.

Analysis

All samples are stored at 4°C and analyzed within 18 hours of collection. Constituents analyzed include five-day biochemical oxygen demand (BOD_5) filtered and unfiltered, chemical oxygen demand (COD) filtered and unfiltered, total solids, total volatile solids, total suspended solids (TSS), volatile suspended solids (VSS), organic, ammonia, and nitrite-nitrate nitrogen, total and dissolved phosphorus, total coliforms, fecal coliforms, fecal streptococcus, and total bacteria. All analyses are run according to *Standard Methods.*

Evaluation

It is most realistic to look at the performance of the treatment units in terms of the probability that a given effluent quality will be reached. All data from the analyses have been plotted on logarithmic-probability paper. In nearly every case the data have seemed to fit a log-normal distribution. These plots are used to determine the effluent quality, taken as the median value of which each unit is capable.

To obtain a numerical measure of variability, the standard deviation of the mean of the logarithms of the data (grouped as a whole in the case of the field data) is used. This gives a number which, independent from the median value, is useful in determining the variability of the effluent quality relative to other units. Those units with smaller standard deviations associated with them are considered to be more stable because their effluent is of a more consistant quality. This is further used to compute the 95 percent confidence intervals of the means.

Servicing and maintenance requirements are best evaluated in terms of total annual costs for comparisons between units. Experience with malfunctions of mechanical components, however, is insufficient to assign an annual cost to repairs, so they will be discussed separately.

The computation of annual costs for each unit includes (1) purchase price, (2) installation cost, (3) operational costs, and (4) maintenance costs. In this computation, a maximum unit lifetime of 25 years is used and an 8 percent interest rate assumed.

Results

(1) Effluent Quality and Variation

Laboratory

Data presented here were collected for a three-month period beginning June 1, 1973. During this time the amount of feed materials used daily were tabulated and then composited, using their particular waste strengths to obtain daily wastewater characteristics. The average concentrations of COD, BOD, and TSS are listed in Table 2.

Table 2
Average Laboratory Wastewater Characteristics

Month	*COD (mg/1)*	*BOD (mg/1)*	*TSS (mg/1)*
June	520	260	320
July	485	240	291
August	400	200	195
Average	470	235	275

The variations of the BOD and suspended solids of the various effluents are shown in Figures 3 and 4 respectively. The daily variations of the influent are also shown. The logarithmic probability plots for this data are made in Figures 5 through 8. The median, mean, standard deviation of the mean of the logarithms, and the 95 percent confidence interval of the means for various data, are presented in Table 3.

Septic Tanks

The multichambered septic tank (3 chambers) appears to give slightly better treatment than the one-chambered septic tank. The differences in TSS removals become significant at the 95 percent confidence level, but the differences in BOD_5 removals are not signficant at this level.

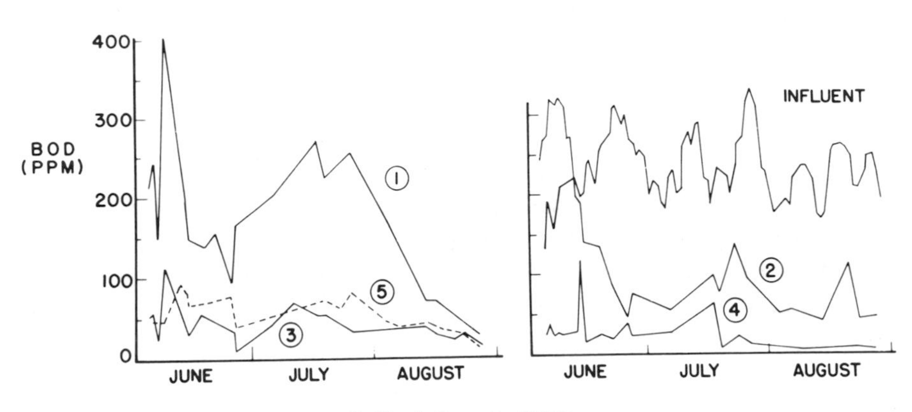

3. Variations in BOD.

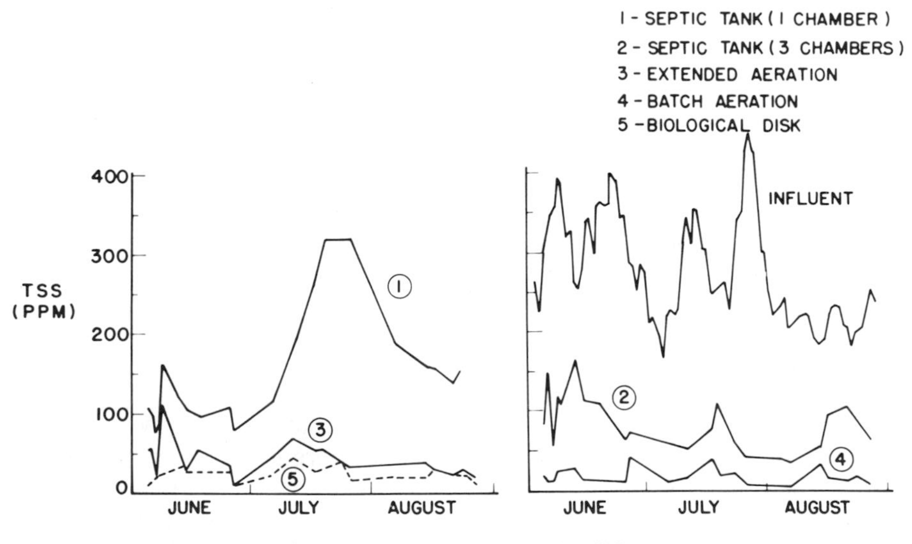

4. Variations in suspended solids.

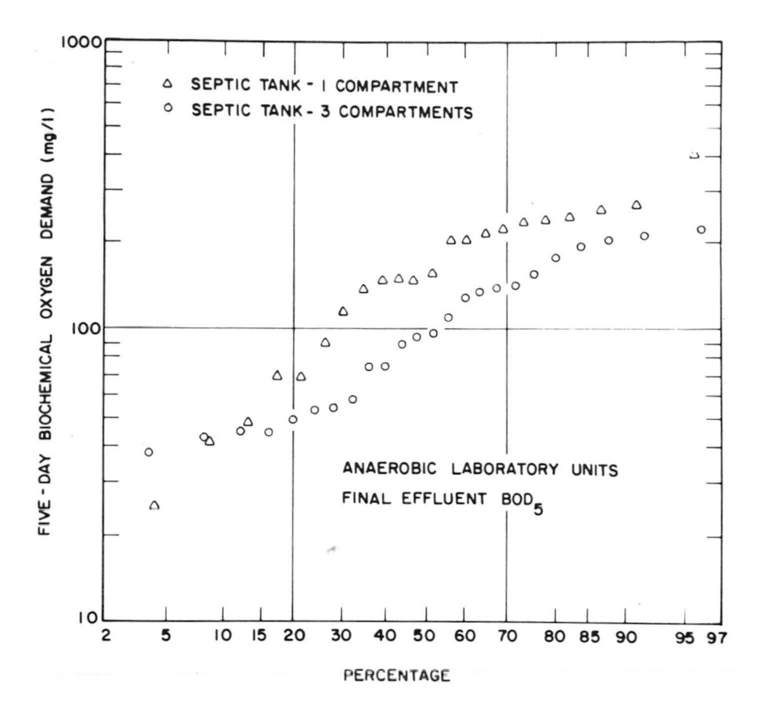

Figure 5.

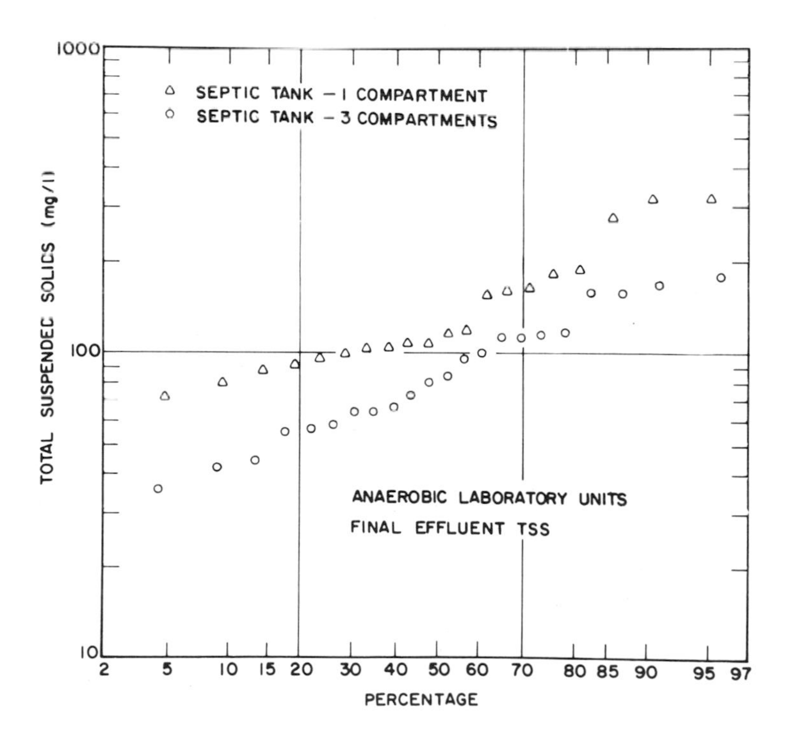

Figure 6.

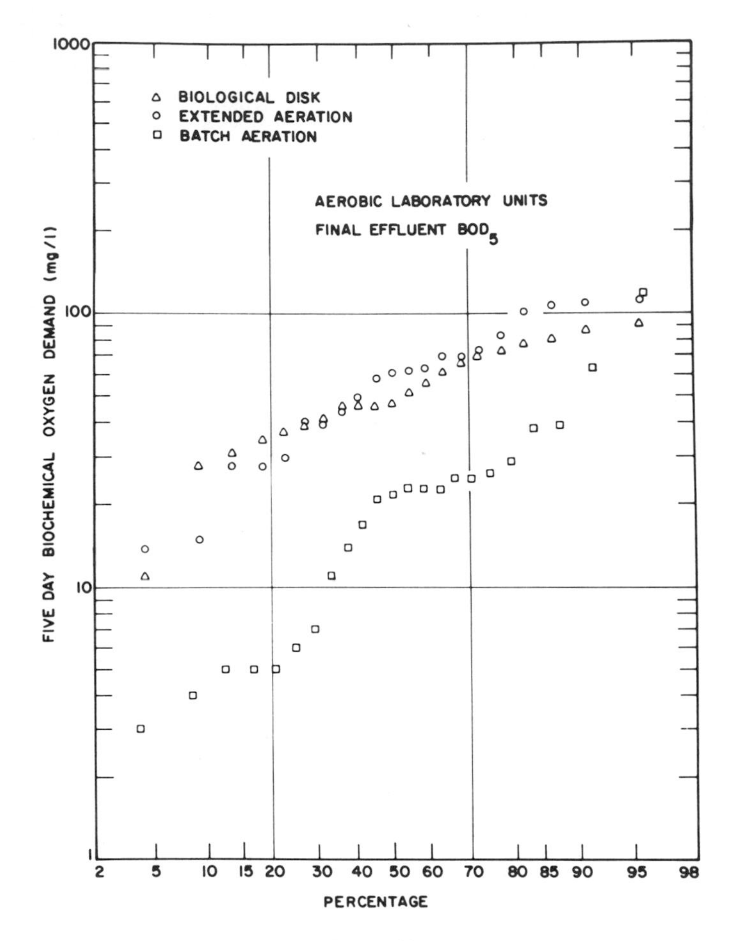

Figure 7.

be slightly better suspended solids removal with the multichambered tank.

During the test period, the single-compartmented septic tank discharged larger amounts of solids on two occasions. Apparently gas formation in the anaerobically decomposing sludge caused it to float to the surface. Hydraulic surges pushed these solids out of the system. This same phenomenon was noted in the trash trap of the continuous flow aerobic unit where floating black solids appeared and disappeared periodically.

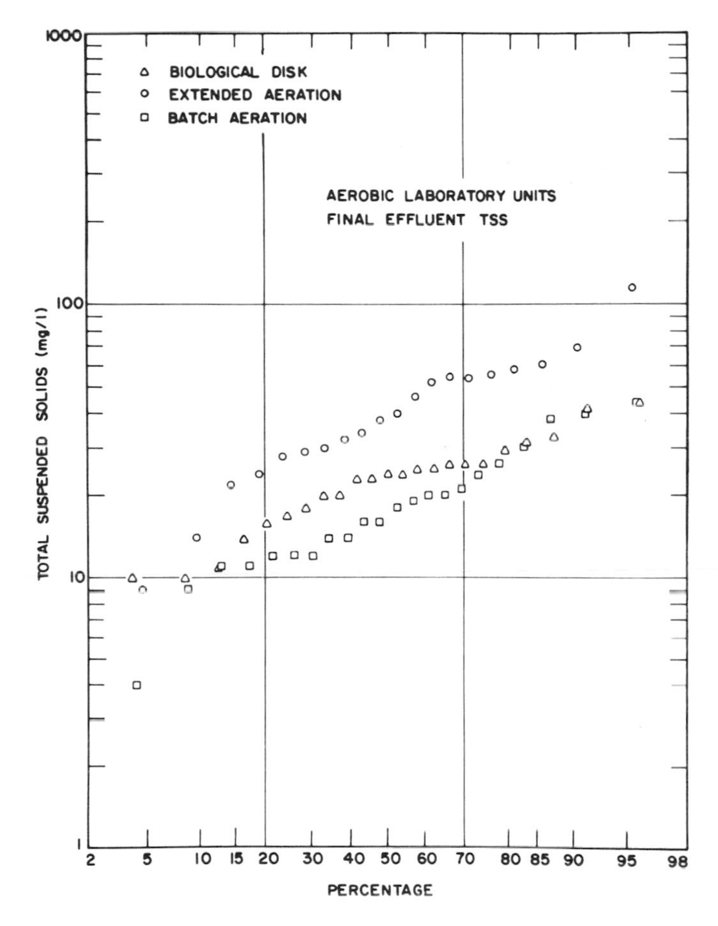

Figure 8.

Aerobic Units

The data indicate that during this particular run the aerobic units, as a group, performed significantly better than the septic tanks. The variability in effluent quality as indicated by the standard deviation of the logarithms for both groups is generally the same except in the batch aeration process, where the batch aerobic unit had a higher BOD_5 and TSS removal but also exhibited higher variability. The variability experienced for the laboratory aerobic units is generally less than that

Table 3
Summary of Effluent Quality Data from Laboratory Units

Characteristic (All Concentrations Are in mg/1)	*Septic Tank (Single-Chamber)*	*Septic Tank (Multi-Chamber)*	*Extended Aeration*	*Batch Aeration*	*Biological Disk*
COD					
Median	310	230	65	30	70
Mean	335	220	65	35	65
Std. Dev of Log[a]	.15	.14	.17	.28	.15
95% Confidence Interval	290–385	195–250	55–75	25–45	55–70
# of Data Points	24	26	25	24	25
BOD					
Median	155	95	60	20	45
Mean	140	95	50	15	50
Std. Dev. of Logs	.31	.25	.26	.40	.21
95% Confidence Interval	100–195	75–120	40–70	10–25	40–60
# of Data Points	22	24	21	23	21
SUSPENDED SOLIDS					
Median	140	95	40	17	25
Mean	135	85	35	17	20
Std. Dev of Logs	.19	.20	.25	.24	.17
95% Confidence Interval	110–165	70–100	30–50	13–22	18–26
# of Data Points	20	22	20	22	23

[a]Standard Deviation of the mean of logarithms.

in the field units. This can be explained by the higher degree of supervision in the laboratory and the fact that the concentration of solids under aeration never reached as high levels as did units in the field. The average suspended solids concentration for the batch aerobic unit was 4,800 mg/l, while the extended aeration concentration was only 140 mg/l. No problems with bulking have been experienced in either set of units.

Results for the rotating biological disk are dependent upon clarifier design. Poor design prevents proper sludge removal, and both denitrification and anaerobic decomposition of sludge will degrade effluent quality as a result. The solids sloughing off the disk are dense and ordinarily settle rapidly.

The continuous flow aerobic unit also demonstrated clarification problems. Sludge built up on the surface of the clarifier regularly.

Attempts to break up the blanket were generally unsuccessful. Grease also appeared on the surface.

Field

Field samples have been collected since May of 1972 from Systems 2, 3, 4, 6, 7, 9, 10, and 11 (Table 2). The results of the BOD_5, TSS, and fecal coliform analyses plotted on logarithmic probability paper appear in Figures 9 through 14. The median ranges, means, and 95 percent confidence intervals of the means are summarized in Table 4 for BOD_5, TSS, forms of nitrogen, and fecal coliforms. The standard deviation of the logarithms for BOD_5 and TSS are also included in Table 4.

Septic Tanks

Septic tanks can be characterized as producing an effluent very high in BOD_5 and TSS. The survival rate of fecal coliforms is also high. Median values of the field units vary between approximately 90 mg/l and 200 mg/l for BOD_5 and 35 mg/l and 95 mg/l for TSS. Grouped as a whole, the means were computed to be 125 mg/l and 60 mg/l respectively (assuming a log-normal distribution). This compares to averages of 130 mg/l BOD_5 and 40 mg/l TSS reported by Preul (1964) from six septic tanks. The standard deviation of the logarithms is 0.26 and 0.31 for BOD_5 and TSS respectively.

Multiple-compartmented septic tanks have not yet been shown to provide significantly better treatment over the single compartmented tank. It may be that compartmentalizing becomes more beneficial as the sludge accumulates to greater depths.

Aerobic Units

Aerobic units are capable of much higher degrees of treatment than septic tanks, but periodic upsets cause great variability in effluent quality. From three field units (Systems 7, 9, and 10) the median range found for BOD_5 is approximately 25 mg/l to 55 mg/l (Figure 9). The mean for all units combined is 30 mg/l, which compares favorably with that claimed by manufacturers and is significantly lower than that from septic tanks. However, the standard deviation of

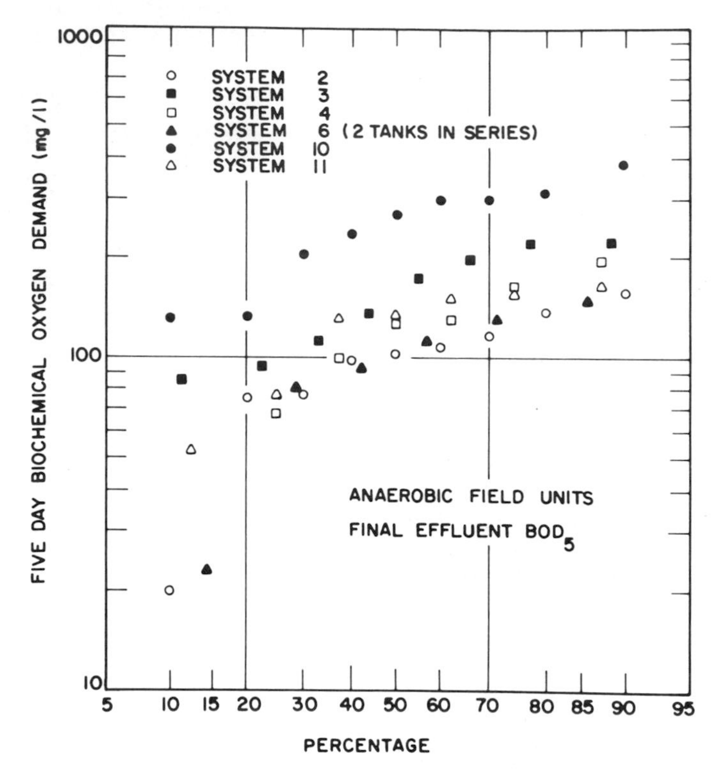

Figure 9.

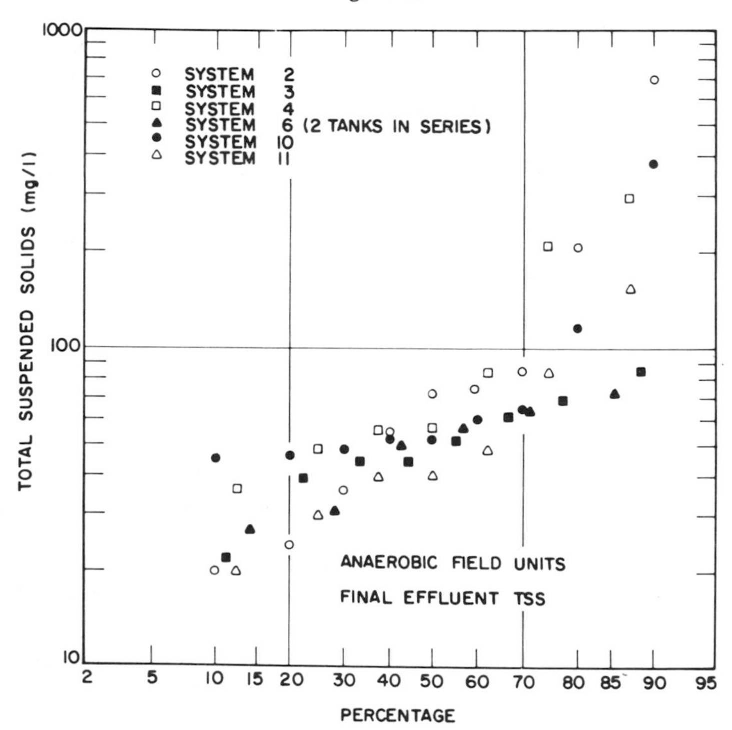

Figure 10.

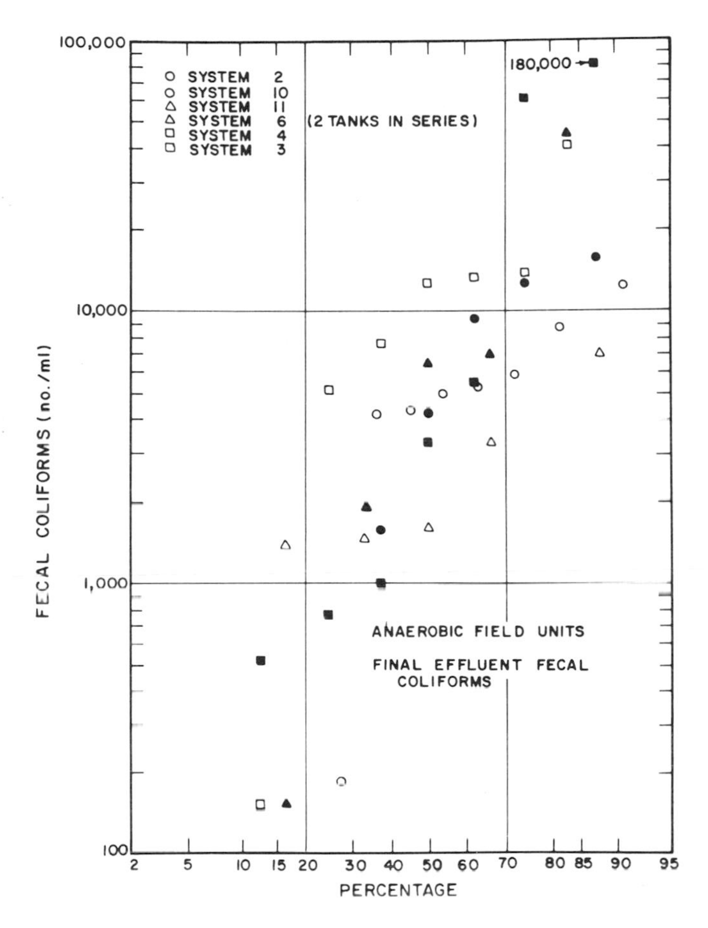

Figure 11.

the logarithms is 0.53 as compared to 0.26 for septic tanks. This indicates a much greater variability in the performance of the aerobic units.

Much of this variability can be explained by the TSS data. The removals of TSS are no better than in septic tanks. A median range of from approximately 35 mg/l to 60 mg/l is found (Figure 10) with a grouped mean of 65 mg/l. This compares to 60 mg/l found for the septic tanks. Also the standard deviation of the logarithms is 0.51, which is much greater than the 0.31 found for septic tanks because of occasional washouts of biological solids. The mixed liquor concentrations in the units build up to a point where a discharge will take place. Bulking also causes discharge of solids. Bulking occurred in System 7

Table 4
Summary of Effluent Qualities from Septic Tank and Aerobic Unit Field Installations

	Septic Tank Effluent (6 Septic Tanks Sampled)				*Extended Aeration Effluent (3 Aerobic Units Sampled)*			
	Ungrouped	*Grouped Data*			*Ungrouped*	*Grouped Data*		
Waste Fraction	*Median Range*	*Mean*	*Standard Deviation of the Loga-rithms*	*95% Confidence Interval*	*Median Range*	*Mean*	*Standard Deviation of the Loga-rithms*	*95% Confidence Interval*
BOD_5 (mg/1)	90–290 (46)[a]	125	0.26	105–150	25–55 (27)[a]	30	0.48	20–50
TSS (mg/1)	35–95 (46)[a]	60	0.32	50–75	35–60 (28)[a]	65	0.51	45–105
Organic-N (mg-N/1)	3–25 (28)[a]	—	—	—	4–14 (18)[a]	—	—	—
Ammonium-N (mg-N/1)	17–56 (38)[a]	—	—	—	0–2 (25)[a]	—	—	—
NO_2 & NO_3-N (mg-N/1)	0.15–0.8 (38)[a]	0–2	6–12	—	20–40 (25)[a]	2–8	40–60	—
Fecal Coliforms (no/ml)	1200–14000 (41)[a]	2645	0.96	1315–5320	90–600 (17)[a]	190	0.62	90–395

[a]Number of samples analyzed.

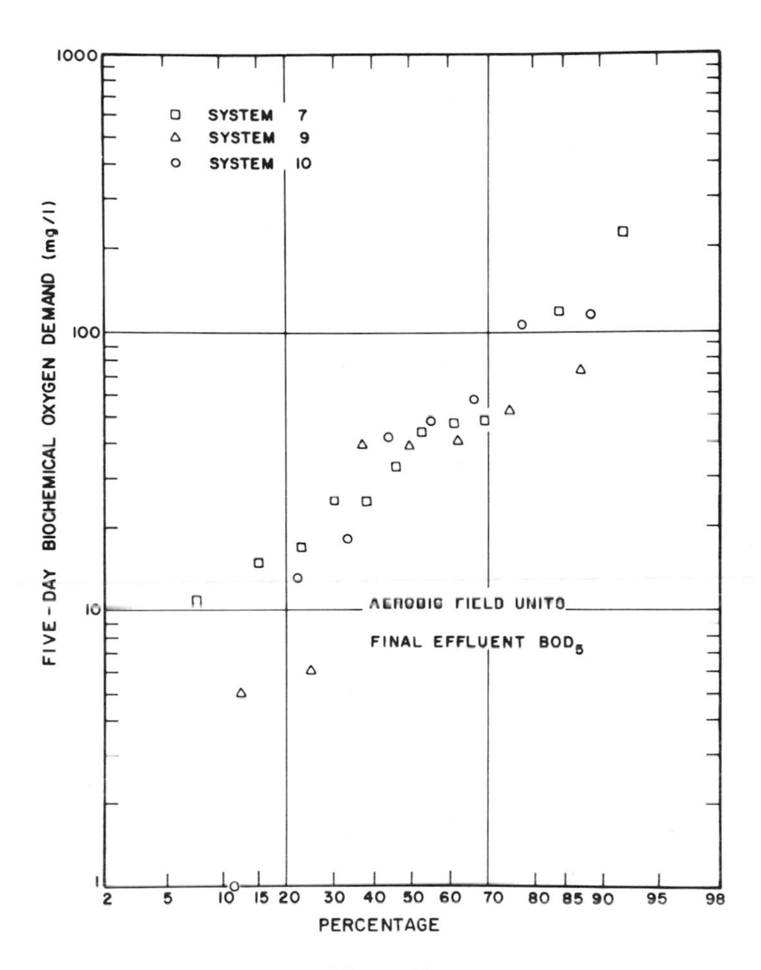

Figure 12.

from January to June. Filimentous bacteria proliferated, and the resultant sludge would not settle. No cause was found for this. Daily chlorination for a week reduced the bulking, but it quickly returned.

Analyses showed a marked decrease of fecal coliform bacteria through the extended aeration units compared to the septic tanks. This can be explained by the competition among the microorganisms within the aeration chamber.

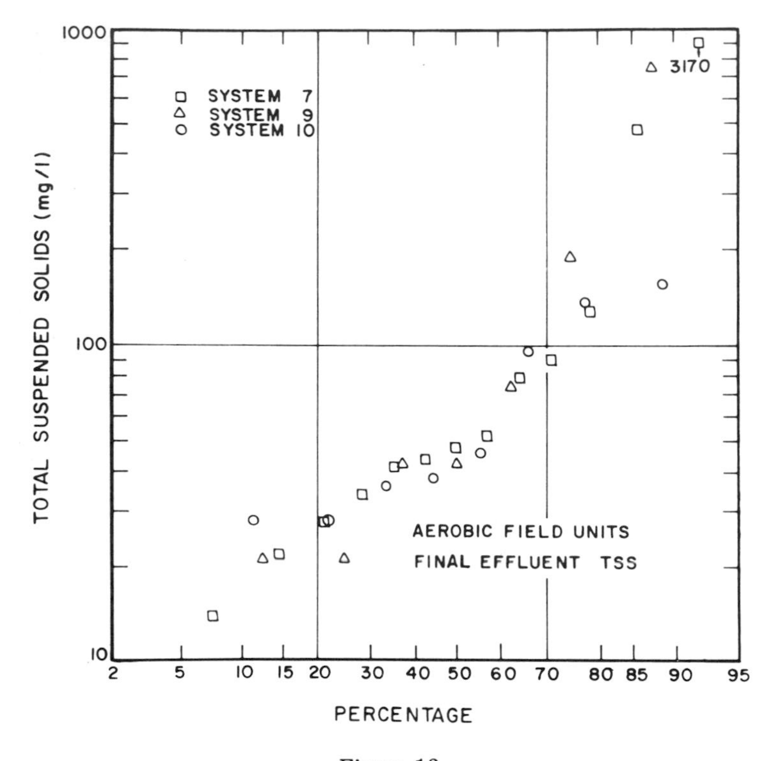

Figure 13.

(2) Operation and Maintenance

Septic Tanks

Maintenance of the septic tanks in terms of sludge removal has not been necessary in any system. The oldest system monitored is only sixteen months old. No cost data can be reported.

Aerobic Units

In addition to bulking and periodic solids discharges, several other operational problems have occurred in the field. Many solids in System 10 were lost to treatment because of floating sludge on the final clarifier. A sludge blanket of over one foot thick developed and rose above the effluent weir before the tank was pumped. Later, the use of toilet bowl deoderizers caused a loss of biologically active solids in

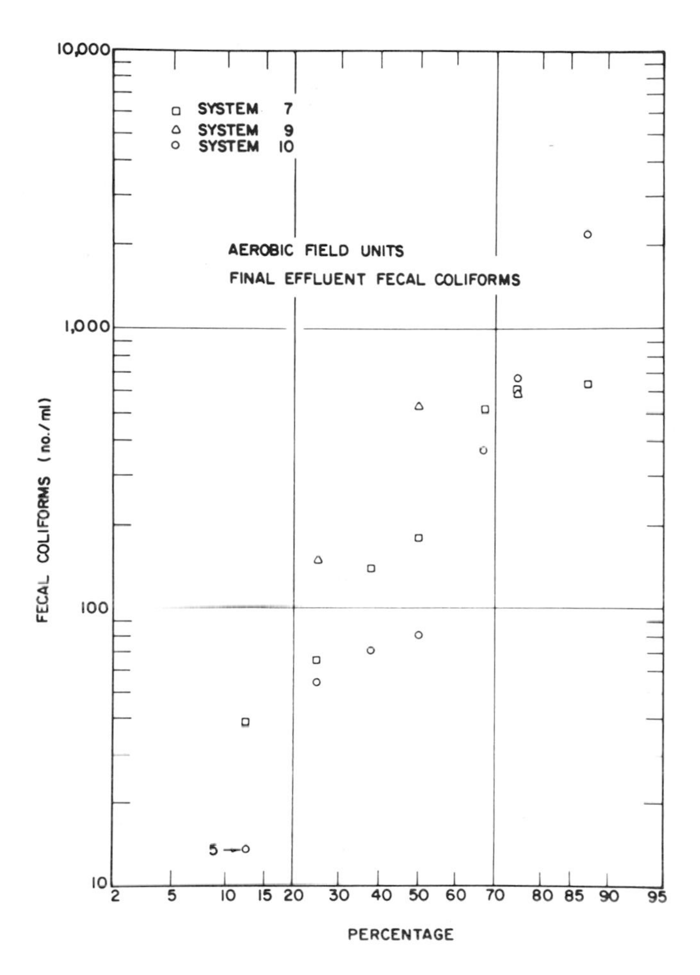

Figure 14.

the aeration chamber for a period of seven months. During this time, the mixed liquor suspended solids concentrations dropped from 3000 mg/l to 500 mg/l. Grease and garbage grindings are a problem in System 7. They appear floating on the final clarifier. This particular unit does have a skimmer to remove floating material, but it becomes clogged by the grease which coats the pipe.

It becomes obvious that regular and competent maintenance is necessary with the aerobic units. This includes pumping to remove excess solids. One unit operated 11 months and another 13 months before this became necessary. Other maintenance included pump and aeration equipment servicing once every six months according to manufacturers' recommendations.

Very few mechanical malfunctions have occurred to date. Clogging of mechanical components by debris in the sewage to the point of malfunction seems to be the greatest problem. In System 9 the intake to the pump in the batch unit has clogged twice with toilet paper. The skimmer in the continuous flow unit at System 7 has plugged several times with grease.

(3) Costs

Septic Tanks

Costs of septic tanks vary with the materials used and the site conditions. As of June 1973 in Wisconsin, the delivered cost of precast septic tanks can be calculated using the formula

$$C = \$125 + \$0.15\,(V - 500)$$

where C is the delivered cost and V is the capacity in gallons. For capacities above 1800 gallons, cast-in-place tanks are used. Costs for steel and iron tanks were found to be similar. Fiberglass tanks are not available in Wisconsin.

Installation costs vary with the contractor and the site conditions. Generally, however, installation costs including the plumbing from the house to the tank are about $250–$300.

Operation and maintenance costs have not yet been determined. The oldest system installed is only sixteen months old. Sludge accumulations have not been great enough to require pumping. To compute total annual costs over a 25-year lifetime, it will be assumed, therefore, that pumping should be performed once every three years as recommended by the *Manual of Septic Tank Practice* (1969) at a cost of $30 per pumping.

The total annual cost for installation and operation of a 1,000-gallon septic tank (less soil absorption field) amortized at 8 percent over a 25-year system lifetime is calculated as follows:

1,000 gallon pre-cast septic tank	$200
Installation	$275
Operation	$ 0
Maintenance (1 pumping/3 years @ $30/pumping)	$ 10/year
Total Annual Cost (1973 dollars)	$ 54

Aerobic Units

The cost of aerobic units is high because of the poor market. Costs could drop if the units gained widespread acceptance. Currently, costs range from $750 to over $2,000. Most units market for about $900 to $1,000. Cost of operation can also vary widely among the units available today. Power consumption for most units ranges from about 3 kwhrs per day to 7 kwhrs per day depending on the on-off cycle set for the aerator. About 4 kwhrs per day is the average value experienced. At 2.5¢ per kwhr this is approximately $3.00 per month. Other regular costs include pumping, which should be performed every six to ten months. Since the unit is easily accessible and only a portion of the liquid removed, the costs will be assumed to be cheaper than that for septic tanks. A rate of $20 per pumping will be used.

The total annual cost is computed as follows:

Equipment Costs	
Tank, control panel, etc. (25 yr. life)	$750
Aeration equipment (10 yr. life)	$150
Installation costs	$350
Operational Costs (power)	$ 36
Maintenance Costs (includes pumping every 10 months at $20/pumping)	$ 40
Total Annual Cost	$203

Discussion

Aerobic units were found to provide significantly higher BOD_5 removals in comparison with septic tanks for units run in both the field and laboratory settings. Total suspended solids removals, however, were not significantly different between the two types of treatment in the field though significant differences were found in the laboratory.

The variability in the quality of effluents produced is similar for all laboratory units (Table 3). In the field, however, the septic tanks were found to be more stable (Table 4). Nearly all fluctuations in quality can be attributed to periodic discharge of solids in the effluent from each of the units, but the field-extended aeration units are particularly susceptible to upset. The buildup of excess solids, hydraulic surges from influent flows, floating sludge and debris on the final clarifier, and bulking sludge cause most of the increases in TSS effluent concentrations. The laboratory studies show that, for extended aeration, batch treatment, which provides completely quiescent conditions during settling, performs better than the overflow clarifiers studied. The rotating biological disk unit in the laboratory shows superior TSS removal, however, because the solids sloughed from the disks are more dense and settle rapidly.

Multiple-compartmented septic tanks do not show improved BOD removals over single-compartment tanks, but the differences in TSS removals are significant at the 95 percent level.

Regular maintenance is shown to be required. The lower variability in effluent quality from units in the laboratory can be attributed largely to the constant attention they receive. Mixed-liquor suspended-solids concentrations have not been allowed to build up to excessive levels, and floating sludge is kept at a minimum. Experience in the field shows that the extended aeration units should be inspected and serviced if necessary at least every two months, and excess solids removed every eight to ten months.

The computation of total annual costs show that aerobic units become competitive with septic tanks only if surface discharge is allowed. Data from the literature (Laak, 1970) on the comparative effects of anaerobically and aerobically treated wastes on rates of clogging in tight soils do not indicate that substitution of the septic tank with an aerobic process unit is justified where a soil absorption field is used for disposal.

Recommendations

Results of this study indicate that increased attention must be given TSS removal if a consistently high effluent quality is to be attained. Methods of controlling hydraulic surges through the units, simple procedures for removal of excess solids, and methods preventing biological upsets must be developed. Emphasis must also be given to disinfection and nutrient removal if surface discharge is ever to be allowed. It is highly recommended that all units marketed be tested for performance and reliability by a recognized laboratory before approvals are granted for use.

References

Electric Boat Division, General Dynamics Corporation. 1973. Flow reduction and water recycle. EPA report.

Laak, R. 1970. Influences of domestic wastewater pretreatment on soil clogging. *Journal of Water Pollution Control Federation* 42: 1495–1500.

Ligman, K., Hutzler, N. H., Boyle, W. C. 1974. Household wastewater characterization. *Journal of the Environmental Engineering Division,* ASCE 100: 201–213.

Patterson, J. W., Minear, R. A., Nedved, T. K. 1971. *Septic tanks and the environment.* National Technical Information Service publication no. PB-204519.

Preul, H. C. 1964. Travel of Nitrogen Compounds in Soils. Ph.D. dissertation. University of Minnesota.

U.S. Public Health Service. 1969. *Manual of septic tank practice.* U.S.P.H.S. publication no. 526.

Wenk, V. D. 1971. Water pollution: domestic wastes. *A technology assessment methodology* 6. Washington, D.C.: Mitre Corporation in cooperation with Office of Science and Technology publication no. PB-202778-06.

V
Non-Point Pollution

20. Non-Point Pollution Sources and Control

Raymond C. Loehr

Introduction

Interest in potential pollution sources other than municipal and industrial sources has increased as the nation has expanded its water-pollution concerns and recognized that water, air, and land resources are not isolated components. The emphasis of national water-pollution control policy is now on the amount of wastes that can be kept out of surface waters rather than on the amount of wastes that can be assimilated by the waters.

Certain potential pollution sources such as precipitation, drainage from urban areas, runoff from forested and pasture lands, return irrigation flows, decaying vegetation, and wastes from wild animals generally have been assumed to be small compared to point sources such as municipal and industrial waste discharges. Greater information on these non-point sources has challenged this assumption, however. It is appropriate to compare the magnitude of certain non-point sources and to evaluate which may be amenable to control. As these sources are identified and evaluated, feasible remedial measures and available technology can be applied.

This paper will (a) summarize available information on certain non-point sources entering surface waters and (b) comment on the feasibility of controlling these non-point sources. The non-point sources to be compared are precipitation, runoff from forested land, discharges from agricultural lands, animal feedlot runoff, urban land runoff. This paper is limited to a comparison of organic and inorganic constituents, excluding pesticides, and their relative importance to water-pollution control efforts.

The impact of non-point sources depends on intensity and yield factors as well as on their seasonal distribution. Since overland flow and stream flow are the major transport agents for non-point sources, knowledge of changes in concentration and flow is essential for calculations of yield rates. Comparison of non-point sources based solely on concentration units is difficult because of the flow-dependent, intermittent nature of these sources. A better method of comparison is the use of area yield rates, such as quantity of characteristic per unit of drainage area per unit of rainfall or runoff. Rarely are adequate data available to develop this type of yield rate. In this paper the data are presented in terms of both concentration (mg/l) and potential area yield rate (kg/ha/yr).

These rates should be considered as potential rather than actual surface-water loading rates. Some studies have obtained data on the characteristics of non-point sources by difference, using available point-source data. Other studies have obtained data by direct measurement. In most cases, yearly or monthly average flow data were used to determine the yield rates. It would be inappropriate and inaccurate to multiply the indicated yield rates by the area of a particular land-use category in a given watershed and to use the result as the amount actually added to a specific body of water. The area yield rates in this paper are presented for comparative purposes only.

Sources and Characteristics

Precipitation

Tables 1 and 2 summarize reported data. Area yield rates are not constant and are a function of rainfall, which is variable in time and space. Data are presented as reported in the literature and are assumed to be typical of bulk precipitation.

Fuel burning, automobiles, manufacturing operations, forest and other fires, volcanic eruptions, and wind erosion contribute to the constituents of precipitation. Constituents placed in the atmosphere at one location will return to the earth at other locations. From air-movement patterns one can predict where the atmospheric constituents will be deposited by precipitation.

Sulfate concentrations in precipitation have increased probably because of the sulfur dioxide and other sulfur products added by man-made combustion (Fisher et al., 1968; Likens, 1973; Johnson et al., 1972). The sulfur in precipitation collected close to industrial locations was usually greater than that collected from other locations (Hoeft et al., 1972). In addition, sulfur concentrations were lower in the summer and increased in the winter. In the study by Hoeft et al., the amount of ammonia and organic nitrogen in precipitation was higher in the spring and lower during the winter. It also was higher in areas adjacent to barnyards than in areas removed from the barnyards. Average Wisconsin sulfur and total nitrogen contributions due to

Table 1
Precipitation Characteristics (mg/1)[a]

Constituent	References							
	54[c]	54	8	46	31	11	49	25
Nitrogen								
NH_4-N				0.06	0.16	0.17–1.5	1.1	
NO_3-N			0.14	0.31	0.33	0.56	1.15	
Inorganic N[b]	0.7	0.9						
Total N	1.27	1.17						0.73
Phosphorus								
Total PO_4-P							0.02	0.04
Hydrolyzable PO_4-P	0.24	0.08						
Suspended Solids	13	11.7						
COD	16	9						
Major Ions								
Ca			0.65		0.21			
C1			0.57					
Na			0.56		0.13			
K			0.11		0.05			
Mg			0.14		0.04			
SO_4			2.18		3.2			
HCO_3					0			
Year	1963–64	1963–64	—	—	1965–68		4 yrs. data	
Environment	Urban	Rural	—	Northern Europe	Forest		Ohio	

[a]data are primarily yearly averages.
[b]inorganic N = NH_4, NO_2, and NO_3-N.
[c]Reference numbers in the tables refer to the References on pp. 296–299.

Table 2
Reported Precipitation Characteristics: Average Nitrogen and Phosphorus kg/ha/yr[a]

Location	*N*		*P*	*Reference*
	Total	*NO_3 + NH_4-N*		
World mean	6.2 (0.8–7.0)[c]	—	—	23 (pre 1954)
World mean	8.7 (1.8–22.2)	—	—	23 (1905)
Europe and U.S.	—	0.8–2.1	—	23 (pre 1952)
Temperate zone	—	6.8	—	23 (1938)
Humid temperature zone	5.6	—	—	23 (1960)
New York	10.0	—	—	23 (pre 1948)
United Kingdom				
upland	8.2	—	0.27	38
northern	8.7–19	—	0.2–1.0	38
Netherlands				
rural	8.5	—	—	28
industrial	16–100	—	—	28
Canada				
Hamilton, Ont.	—	6.0	—	34 (1948)
Ottawa	7.7 (4.8–12.9)	—	—	34 (1924–25)
Ceylon	—	12.9	—	34 (1941)
Western Australia	0.5–3.2	—	—	46
Scotland	8.2	—	0.45–0.7	9 (1962–63)
Ithaca, N.Y.	—	7.4	0.05	29
Aurora, N.Y.	—	7.6	0.06	29
Geneva, N.Y.	—	8.3	0.05	29
Hubbard Brook, N.H.	—	5.8	0.10	19
Cincinnati, Ohio[b]	9.6	5.2	0.6	53

[a]1.12#/acre = 1.0 kg/ha.
[b]average U.S. rainfall of 30"/year assumed.
[c]data in parentheses indicate range of data.

precipitation were estimated as 30 kg S/ha/yr and 20 kg N/ha/yr respectively.

Precipitation is a variable and intermittent source of potential contaminants to surface waters. Once the contaminants are in precipitation, they are uncontrollable. Man, however, can exert some control over the contaminants that are released to the atmosphere through management practices that minimize particulates and gases from combustion processes, particulates from soil disturbance by rainfall, and volatiles from industrial operations.

Rural Land

General. In recent decades, agricultural production has become more efficient as a result of mechanization, increased use of agricultural chemicals, application of modern business methods to farm management, and the application of research results to production. During the same period, the potential water quality problems associated with runoff from crop and pasture lands and animal feedlots and with leachate from fertilized and manured fields have increased. These problems are not new to agriculture but have become more noticeable because of the specialization and intensification of agriculture.

Constituents contained in the runoff from rural land originate in rainfall, wastes from wildlife, decay from leaf and plant residue, applied nutrients, herbicides and pesticides, nutrients and organic matter initially in the soil, and wastes from pastured animals. To separate the natural from the controllable non-point sources requires that water volumes be accurately measured, storm characteristics described, representative sampling initiated, sufficient data obtained, and potential control measures identified.

Organic and inorganic constituents are released at varying rates from all soils and exposed geological formations. The natural weathering of rocks and minerals and the oxidation and leaching of organic matter contribute organic and inorganic matter in runoff, even in the absence of man's activities.

Vegetation on the land is a major control over the rate of runoff, subsequent erosion, and nutrient loss. Surface runoff from rural lands will have a higher load of clay and organic matter, a higher load of adsorbed phosphorus, and a lower load of soluble salts than will tile drainage water. In an area drained by tile, surface waters may be expected to receive both adsorbed and soluble phosphorus in times of high rainfall and runoff. The variation in crop land runoff can be wide. Data from a three-year study (Weibel et al., 1966) demonstrated that the runoff from a cultivated wheat field contained 5–2070 mg/l of suspended solids, 30–160 mg/l of COD, 0.5–23 mg/l of BOD, 2.2–12.7 mg/l of total N, and 0.2–3.3 mg/l of total P.

Erosion can be a major contributor of organics and nutrients to surface waters in rural areas. Soil erosion is a selective process, in the sense that the fine particles are more vulnerable to erosion than are the coarser soil fractions. Eroded material may have three to five times as much organic nitrogen content as the original soil.

Exposed plant residues and animal wastes will undergo decomposition and leaching. Thus the exposure of the residues prior to runoff will affect the soluble material in the runoff. Snowmelt runoff occurs in the spring, frequently on frozen ground, and can carry a higher contaminant load than runoff that has an opportunity to infiltrate into the ground.

Crop fertilizers have been blamed as major contributors to nutrients in surface waters. Generally only 5–10 percent of the fertilizer phosphorus added to a soil is taken up by the following crop. The remaining applied phosphorus is converted to insoluble forms, may be solubilized later, and may become a source of available phosphorus for crops in subsequent years. Rarely does phosphorus from commercial fertilizers or spread manures occur in groundwater. Phosphorus added to soils as a fertilizer or released in the decomposition of organic matter will be rapidly converted to iron and aluminum phosphates in acid soils and calcium phosphates in alkaline soils. The rate at which the phosphorus is converted to the insoluble form is regulated by the time taken for the phosphorus to come in contact with the soil. This is affected by the ground cover, rainfall rate, slope of the ground, and permeability of the soil. With proper erosion control measures, little phosphorus should reach surface water.

The constituents of base-water flows from agricultural land represent a small fraction of the potential organic and inorganic loss from crop lands. In a rural area in which farms occupied 90 percent of the land, the average base-flow nitrogen contribution was 1.25 kg/ha/yr or about 3 percent of the nitrogen applied, and the average base-flow phosphorus contribution was 0.11 kg/ha/yr or about 2 percent of the applied phosphorus (Minshall et al., 1969). Surface runoff and interflow during runoff can contribute greater amounts of nutrients to surface waters than can base flows.

Nitrogen in fertilizers is oxidized to nitrate, which is soluble in water and can move with surface runoff or soil water. Efforts have been made to find a relationship between applied fertilizer nitrogen and the concentration of nitrogen in stream and drainage waters. But a direct relationship is not likely because of a large preexisting reservoir of nitrogen in the soil and the biological transformation that the reservoir can undergo.

The efficiency of utilization of applied nitrogen in terms of increase in crop uptake per unit of nitrogen applied is always less than one. On the average, no more than 50 percent of the applied nitrogen in fertilizer is recovered by crops. Grass may recover 80 percent or more of the applied nitrogen. The recovery of the nitrogen varies with the season and the age of the crop. Denitrification in soils is perhaps 10–20 percent of the total mineral nitrogen in the soil. Ammonia volatilization can be 5–10 percent of any ammonia applied as a fertilizer.

Some nutrients leave crop lands. It does not necessarily follow that the same amount reaches surface waters and lakes. Intermediate vegetation and other environmental factors may remove most of the material in suspension and part of the material in solution.

The time distribution of nitrates in a number of rivers suggests that the nitrate concentration and load is the greatest when the flow is high, i.e., when rainfall has the greatest opportunity to leach and transport the soluble nitrates from the soil. A study of nitrate concentrations in English rivers (Tomlinson, 1970) was unable to correlate nitrogen fertilizer usage in adjacent areas with the nitrate concentrations.

One of the requirements for satisfactory crop production is the availability of nutrients when the crop is growing. Intensive farming operations are designed to satisfy this requirement either by the addition of fertilizers or by the incorporation of readily decomposable organic material such as manures in the soil. It is inevitable that there is a risk of nitrate loss during runoff and leaching. It is difficult to conceive of any approach that will eliminate that risk. It is possible, however, to decrease the risk by proper application of inorganic fertilizers at periods when the crops are growing coupled with the use of slow-release organic fertilizers.

Constituents in runoff from different types of agriculture should not be lumped together, since there are diverse agricultural production situations each with its own potential pollution problems and solutions.

Forest land. Forested areas represent areas that have not been grossly contaminated directly by man's activities. The runoff from these areas can be one of the better indicators of constituents that result from natural conditions. The increased world demand for wood fiber has accelerated cultural production practices in managed forested areas. Practices such as forest fertilization, block cutting of mature trees, and other forest management practices will alter the characteristics of runoff from forested areas. Table 3 summarizes data on the characteristics of forest land runoff. With few exceptions, the data represent unmanaged forest situations.

The effect of forest fertilization on surface water quality has been reviewed (Groman, 1972). The leaching of nitrates into ground water

Table 3.
Drainage from Forested Areas—Average Values and Ranges

Location	Nitrogen			Phosphorus		Calcium	
	TKN	*NH_4-N*	*NO_3-N*	*Total P*	*Soluble P*	*Ca^{++}*	*Reference*
Concentration—mg/1							
Yakima River	0.08 (0–0.22)		0.2 (0.05–0.5)	0.07 (0.02–0.14)	0.009 (0–0.023)	—	47 (12 mo. data)
Tieton River	0.068 (0–0.13)		0.126 (0.03–0.18)	0.115 (0.03–0.20)	0.008 (0–0.23)	—	47 (7 mo. data)
Cedar River	—		0.065 (0.018–0.15)	0.022 (0.015–0.085)	0.004 (0.002–0.007)	—	47 (12 mo. data)
Minnesota							
Watershed 2	0.22	0.04	0.50	0.03	—	54	8 (Aug.–Nov. 1966)
" 3	0.77	0.08	0.91	0.06	—	64	8 " "
" 4	0.72	0.06	0.42	0.05	—	38	8 " "
" 6	0.41	0.03	0.34	0.02	—	17	8 " "
Ohio	0.46	—	1.3	0.015	—	—	49
Area Yield Rate—							
Yakima River	3.3 (TKN + NO_3-N)			0.83	—	—	47
Tieton River	1.5 (TKN + NO_3-N)			0.86	—	—	47
Cedar River	—			0.36	—	—	47
Minnesota (g/ha/day during the period noted)							
Watershed 2	6.3	0.4	7.3	0.3	—	646	8 (Aug.–Nov. 1966)
" 3	6.2	0.5	6.3	0.5	—	646	8 "
" 4	2.9	0.3	0.7	0.2	—	113	8 "
" 6	5.9	0.4	3.2	0.4	—	259	8 "

Patterson Creek	1.8	9.0 (NO_2 + NO_3-N)		0.7	—	—	22
Ohio	1.4	—		0.05	—	—	49 (4 yr. ave.)
Hubbard Brook							
New Hampshire							
—undisturbed	—	0.39[a]	8.3[a]	0.028[a]	—	—	19
		0.11[c]		0.037[b]	—	—	19
—deforested	—	0.70[a]	652[a]	0.061[a]	—	—	30
		0.37[c]		0.561[b]	—	—	30

[a]dissolved and fine suspended.

[b]particulate matter (sediment) and bed load.

[c]total N, particulate matter (sediment) and bed load.

following fertilization was insignificant. Major loss of fertilizers to surface waters occurred by inadvertent application to water courses and compacted areas such as roads and trails. Nutrients from forest fertilization do not appear to be a problem under current conditions and practices.

Range land. Where rainfall is sparse or the land not very fertile, intensive crop production is not possible and few animals can be supported per acre of land. Some domestic animals are permitted to range over the land in search of available food. Runoff from range land, with its low animal and human density and low fertilized acreage, contains background or "natural" constituents. Runoff from range land that had low intensity agriculture and no evidence of chemical fertilizers contained 0.65 kg NO_3-N, 0.76 kg total P, and 0.024 kg soluble P per hectare per year (Campbell and Webber, 1970).

Crop land. As used in this paper, crop land includes nonforested land that is or could be under agricultural production. Examples include orchards, grassland, idle farm land, and land used for production of grains and vegetables. The runoff from crop land has different characteristics than range land runoff since the former is under the direct management of man.

In a complex watershed, the analysis of the nutrient loss from crop lands is made difficult by interactions among many uncontrollable variables. The variations in the data of Table 4 reflect the effects of these interactions.

(a) Subsurface drainage. The soil of many farms has poor drainage. Subsurface tile drains can be installed to permit better water movement and crop yields. This tile drainage will contain soluble constituents from the soil and materials added to the soils. The drainage can enter surface streams at many places and is essentially a non-point source. Examples of the characteristics of such drainage are presented in Table 5.

(b) Irrigation return flows. Where rainfall is insufficient, imported or impounded water can be used for crop production. Adequate water is required to flush excess minerals from the soil as well as to meet the crop and evaporative requirements. The surface and subsurface irrigation drains discharge to surface waters at diverse points.

The term irrigation return flow is all inclusive and indicates that portion of the applied water which is not lost to evapotranspiration or to a deep aquifier and which finds its way back to the stream. The factors affecting the quality of irrigation return flows include evaporation, transpiration, ion exchange, erosion, filtration and aeration characteristics, the chemical composition of the soil, types of crops, fertilizer use, amount and rate of precipitation, and the quality and quantity of the irrigation water.

Animal Production

(a) Pasture and lands used for manure disposal. These lands are used either for grazing animals or for disposal of animal wastes. Where animals have direct access to streams, animal urine and feces can be discharged directly to these waters. Table 6 indicates some of the contribution of these sources to surface waters.

The data obtained by Howells et al. (1971) illustrated the relative pollutional effects of manure disposal and pastured animals. Their major conclusions included the following:

1. Pollution indices for land drainage from manure spreading paralleled stream hydrographs with extended dragout on cessation of surface runoff.
2. The extent of water pollution from farm animal production units is more dependent on waste management methods than on the volume of the waste involved.
3. Direct access of animal waste to surface waters should be prevented.
4. Points of animal concentrations should be located away from streams and away from hillsides leading directly to streams.
5. Vegetation should be provided between areas of animal concentration and drainage paths or surface waters to intercept any contaminant.

It is possible to control the non-point pollution of surface waters by animal wastes. Animal manures should not be disposed of where rainfall or snow melt will result in their direct discharge to watercourses. The manure should be incorporated with crop or disposal land shortly after being spread there.

(b) Manure storage. Where manure cannot be disposed on land in all seasons, such as in the winter, it may be stacked or stored until conditions permit land disposal and/or integration with crop production. During storage, seepage from the manure can occur and be a source of pollution. Although the volume of seepage is small, the quantity of contaminants is not insignificant. The stored manure seepage can be controlled by retention ponds and subsequent distribution on crop land in a nonpollutional manner. Even though manure seepage can occur in a large number of locations throughout the country and, as

Table 4
Nitrogen and Phosphorus in Runoff from Rural and Crop Land: kg/ha/yr

Location	Constituent						
	TKN	*NO_3-N*	*Total N*	*Soluble P*	*Total P*	*Remarks*	*Reference*
Rural and Crop Land							
Catoctin Creek (Potomac)	2.8	23.5[a]	—	—	1.8	80% farm 20% forest	22
England	—	—	4	—	—	unused land	28
	—	—	8	—	—	grassland	
	—	—	13	—	—	clay soils	
Germany/Switzerland	—	—	3	—	0.9	unused land	28
	—	—	8	—	0.4	grassland	
	—	—	33	—	0.1	arable land	
North Carolina	—	0.06	0.8	—	0.033	unfertilized (Apr.-Oct.)	26
(kg/ha)	—	0.07	0.9	—	0.045	fertilized (Oct.-Apr.)	
Ohio	1.2	3.4	—	—	0.056	farmland (1967-69)	49
Canadarago Lake, New York	—	—	7.13	0.056	0.187	agriculture plus one small town	18
Ohio[b]	—	43	—	—	6.2	tillage:corn	42
	—	53	—	—	3.0	no tillage:corn	
England							
—Great Ouse River	—	—	11.7	—	0.06	drainage from rural lands	38
—Other rivers	—	—	0.6-22.5	—	0.6-2.3		38
Wisconsin							
—continuous corn	—	—	14	—	—	two year averages, loam soil	56
—corn	—	—	5.5	—			
—oats	—	—	6.0	—	—		
—hay	—	—	3.5	—	—		

—fallow	—	—	63	—	—		
Missouri (kg/ha)							
—fallow	—	0.9	—	—	—	data from two rains totaling 4.5″	44
—corn, oats	—	0.33	—	—	—		
—rotation	—	0.45	—	—	—		
—continuous corns	—	0.10	—	—	—		
Wisconsin							41
drainage area tributary to						runoff from agricultural areas with no domestic or industrial waste contribution	
—Lake Monona	—	—	6.6	—	—		
—Lake Waubesa	—	—	7.5	—	1.3		
—Lake Kegonsa	—	—	9.2	—	0.45		13
Arkansas	—	—	3.6	—	2.3	watershed 80% agriculture	52
Illinois	—	8.1	—	—	0.06	no significant municipal or industrial waste—80% crop land—rich organic soil	
Wisconsin	—	—	1.2	—	0.11	stream base flow, 90% farm land	36

[a] $NO_2 + NO_3$-N.

[b] Data indicate the effect of an 11″ rain and resultant flooding.

Table 5
Characteristics of Effluent from Drains under Cropland

Location	mg/1	kh/ha/100 mm Drainage Water	mg/1	kh/ha/100 mm Drainage Water	Reference
Netherlands					
—marine clay soil	0.06 (0–0.55)	0.06	17 (2–53)	16.6	17 (Sept. 1970–March 1971)
—river clay soil	0.035 (0.01–0.06)	0.04	13 (1–23)	13	
—sandy soil	0.02 (0.01–0.022)	0.02	25 (14–43)	25	
—old cutover soil[a]	0.022 (0.01–0.035)	0.02	14 (5–18)	14	
—young cutover soil	0.7 (0.035–1.1)	0.72	9.5 (2–17)	9.5	
	No Fertilizer—kg/ha/yr		Fertilizer added—kg/ha/yr		
	N	P	N	P	
United States					10 (Average of 1961–67 300# fertilizer per acre added to all crops, additional 100# per acre added to corn)
corn	5.0	0.12	13.5	0.21	
oats—alfalfa	3.8	0.12	5.1	0.12	
alfalfa—first year	4.3	0.12	3.5	0.13	
alfalfa—second year	4.2	0.07	7.7	0.20	
continuous corn	5.9	0.23	12.5	0.26	
bluegrass sod	0.3	0.01	0.6	0.11	

	Country		
	England	*Germany/Switzerland*	
	N	*N*	*P*
River discharages from agricultural soils (kg/ha/yr) (3)			
Item			
land			
unused	4	3	0.9
grassland	8	8	0.4
arable	—	33	0.1
clay	13	—	—

[a]cutover soils are soils that have had the top organic peat layer removed for fuel; the remainder is an organic sandy soil.

Table 6
Characteristics of Runoff from Pasture Land and Land used for Manure Disposal

	Constituent							
	BOD	*COD*	*TOC*	*NO_3-N*	*Total N*	*Total P*	*Remarks*	*Reference*
North Carolina (g/ha/day)	30	—	63	—	7.6	2.8	—mixed grains and orchard, swine waste spread	21
	17	—	124	1.8	10.4	0.6	—pasture, corn, orchard, swine waste spread	
	35	—	97	1.7	7.7	3.5	—250 hog drylot, row crops, wood, grassland	
	94	750	150	4.0	38	25	—pasture for 50–100 dairy cows, plus corn	
	3450	10,700	3750	46.5	400	130	—pasture for 160 cows on 15 acres	
	26	320	72	8	10	1.8	—poultry waste spread on 5 acres three times a year	
	1850	4350	2250	14	435	—	—22 tons of poultry waste spread on 4 acres once	
	46	720	150	—	5.8	1.8	—35 beef cows on 15 acres of pasture	
Wisconsin (kg/ha/yr)								
—manure not spread	—	—	—	—	4.4 (3.6–5.5)	1.3 (1.2–1.5)	—three year average—1967–69, dairy cattle manure spread at the rate of 15 ton/acre	57

—manure applied[a]						
—winter	—	—	—	—	12.7	2.9
					(3.0–27)	(1.0–5.8)
—spring	—	—	—	—	3.8	0.8
					(3.0–5.2)	(0.7–1.0)

[a]High values due to a thaw and a 0.75″ rain immediately after spreading manure in winter; manure spread in the spring was incorporated into the soil after spreading.

such, approximates a non-point source, it is a controllable source of pollution.

(c) Animal feedlots. Animal feedlots are confined feeding operations in lots or pens which are not normally used for raising crops and in which no vegetation intended for animal food is grown. Runoff from unenclosed feedlots can be a water pollution problem. Feedlot wastes reach a stream during periods of runoff. Uncovered livestock wastes pollute surface waters through the drainage of pollutants from a broad surface area. Pollution caused by runoff is reduced when the animals are housed.

The highest concentrations in feedlot runoff occur in the initial runoff and decrease to a lower, more uniform level as runoff continues. This lower level results from the gradual dissolving of materials in the surface layer of manure on the lots. The important variables that affect the quality of the feedlot runoff include rainfall intensity and duration, antecedent water content of the manure pack, type of feedlot surface, and temperature. Typical characteristics of cattle feedlot runoff are presented in Tables 7 and 8. Only a small proportion of the wastes on a feedlot, perhaps 2 to 10 percent, are washed away in runoff. The concentration of contaminants in winter runoff from feedlots can be considerably higher than the concentration in summer runoff.

Urban Runoff

Street litter, gas-combustion products, ice-control chemicals, rubber and metals lost from vehicles, decaying vegetation, domestic pet wastes, fallout of industrial and residential combustion products, and chemicals applied to lawns and parks can be sources of contaminants in urban runoff.

A portion of the urban runoff can drain to sewerage systems while the remainder may reach surface waters by natural drainage channels without receiving any treatment. The discharge to streams or sewage treatment plants results in an intermittent load that receptors may be unable to handle.

The characteristics and control of urban runoff have been studied in detail by a number of investigators, and the fundamental factors delineated. Characteristics of urban runoff as recorded in a number of studies are presented in Table 9. Besides the conventional water pollution parameters, constituents such as chlorinated hydrocarbon and organic phosphate compounds, a number of heavy metals and

Table 7
Major Ions in Cattle Feedlot Runoff (43) (mg/1)

Ion	*Mean*	*Range*
Na	840	40–2750
K	2520	50–8250
Ca	790	75–3460
Mg	490	30–2350
Zn	110	1–415
Cu	7.6	0.6–28
Fe	765	24–4170
Mn	27	0.5–146

polychlorinated biphenyls have been found in urban runoff (Sartor et al., 1972). Runoff from residential streets contained the highest concentrations of total phosphorus, runoff from arterial streets contained the highest concentrations of soluble phosphorus, and runoff from arterial highways contained the highest concentration of nitrogen (Sylvester, 1961).

A study in the Roanoke River Basin has provided direct comparative information on the characteristics of urban and rural runoff (Grizzard and Jennelle, 1972) (Table 10). The rural area consisted of forested and pasture land. Only a small fraction of the watershed was used for crop production. The urban area consisted primarily of cities and suburbs. Only 10–15 percent of the urban drainage area was agricultural land. Urban runoff had higher yields of all constituents. The effect of such runoff on surface waters will depend upon the relative amounts of each type of area in a watershed.

Comparison of Sources

Non-point sources are the result of complex interactions in and on the soil, making definitive comparisons difficult. Analytical methods and differences in sampling methods and parameters cause additional problems when comparing data from various studies. In this attempt to compare the noted sources, the order of magnitude of their characteristics and the differences between sources are more significant than the values themselves.

Because of the interest in nutrients, Figures 1 and 2 were prepared to compare the total nitrogen and total phosphorus contributions of the non-point sources discussed in this paper. All of

Table 8
Beef Cattle Feedlot Runoff Characteristics

	Constituent			
Location	*Total Solids*	*Volatile Solids*	*COD*	*BOD*
Concentration—mg/1				
Nebraska				
—snowmelt runoff	—	—	41,000 (14,100-77,000)[b]	—
—rainfall runoff	—	—	3,100 (1300-8200)	—
Texas				
—dirt lots	—	—	9,500 (2900-28,000)	1,460 (1010-2200)
—concrete lots	—	—	21,500 (8400-32,800)	8,000 (3300-12,700)
Colorado	—	100-7000	—	300-6000
Kansas	10,000-25,000	—	4000-40,000	1000-11,000
Area Yield Rate—kg/ha/yr				
Nebraska				
—snowmelt runoff				
—100 ft^2/head—1969	60,000	31,000	—	—
—1970	1,000	770	—	—
—200 ft^2/head—1969	14,000	7,200	—	—
—1970	640	450	—	—
—rainfall runoff				
—100 ft^2/head—1969	7,500	3,400	—	—
—1970	27,000	16,000	—	—
—200 ft^2/head—1969	7,300	3,400	—	—
—1970	18,000	10,000	—	—
South Dakota[a]	10,500 (1600-23,200)	4,900 (800-10,000)	7,200 (720-16,000)	1,560 (135-3800)

[a]runoff from beef feeding, dairy confinement, and lamb finishing operations.
[b]data in parentheses indicate the reported range of data.

the comparisons are presented in terms of the items in the precipitation or runoff. These should not be interpreted as characteristics necessarily reaching surface waters. As indicated earlier, many of these characteristics can be altered during transport across additional land and through creeks and streams.

Comparing the data in terms of concentration units (Figure 1), precipitation, forest-land runoff, and surface-irrigation return flow have comparable values. The wide range in crop-land phosphorus is the effect of erosion. When erosion is controlled, the total phosphorus of crop-land runoff should be near the lower end of the range. Subsurface-irrigation return-flow nitrogen concentration is greater than that from surface-irrigation return flows, undoubtedly because of the leaching of nitrogen compounds from the soil. Crop land tile drainage

TKN	NH_4-N	NO_3-N	Total N	Total P	Alk	Reference
—	780 (6-2020)	17 (0-280)	2,100 (190-6530)	290 (5-920)	—	35
—	140 (2-1240)	10 (0-220)	920 (11-8590)	360 (4-5200)	—	35
128 (9-280)	56 (2-85)	23 (0-103)	—	—	730 (100-1400)	55
550 (70-1070)	350 (33-775)	220 (0-880)	—	—	1,250 (85-2400)	55
—	—	—	—	—	—	37
200-450	—	—	—	—	—	32
—	—	—	1,600	620	—	35
—	—	—	100	10	—	
—	—	—	450	200	—	35
—	—	—	360	60	—	
—	—	—	330	22	—	35
—	—	—	900	130	—	
—	—	—	220	33	—	35
—	—	—	350	70	—	
510 (46-3110)	—	—	—	130 (9-470)	—	33

has nitrogen and phosphorus concentrations comparable to subsurface irrigation return flows.

The total phosphorus concentration of crop land runoff, irrigation return flows, crop-land tile drainage, and urban-land drainage span comparable ranges. Because of the difficulties of comparing different studies, it is impossible to state that the concentrations of total phosphorus in the five non-point sources discussed here are likely to be grossly different. There is, however, a wide range of values for all non-point sources.

Animal feedlot runoff and manure seepage have characteristics that are many orders of magnitude greater than those of the other non-point sources.

The water pollution effect on surface water can be evaluated better

Table 9
Urban Land Drainage and Stormwater Overflow Characteristics[a]

				Constituent
Location	*Total Solids*	*Suspended Solids*	*COD*	*BOD*
Concentration—mg/1				
Cincinnati	—	227 (5-1200)	111 (20-610)	17 (1-173)
Seattle	—	—	—	—
Tulsa	545 (200-2240)	367 (84-2050)	85 (42-138)	12 (8-18)
East Bay Sanitary District	1400	1400	—	87 (3-7700)
Los Angeles County	2910	—	—	160
Washington, D.C.	—	2100 (26-36,200)	—	126 (6-625)
Detroit	(310-910)	(102-210)	—	(96-234)
Madison, Wisc.	—	(20-340)	—	—
Durham, N.C.	2730 (274-13,800)	—	179 (40-600)	15 (2-232)
Russia				
—rainwater runoff	—	(450-5000)	—	(12-145)
—street washing water	—	(31-14,500)	—	(6-220)
—melting snow	—	(570-4950)	—	(5-105)
Moscow	(1000-3500)	—	—	(186-285)
Leningrad	14,500	—	—	36
Stockholm	(30-8000)	—	—	(17-80)
Area Yield Rate: kg/ha/yr				
Cincinnati				
—storm water runoff	—	640	310	47
—combined sewer overflow	—	280	—	—
Tulsa	1400 (550-5700)	—	220 (67-530)	30 (13-54)
Detroit (kg/ha)	—	220	—	100
Ann Arbor (kg/ha)	—	1230	—	35
Rock Creek (Potomac)	—	—	—	—
Durham, N.C. (1968-70)	14,200	—	93	75
Stockholm				
—highway runoff	2040	1185	—	100
—terrace house runoff	470	173	—	14
—residential block runoff	930	620	—	43

[a]average characteristics, range of values shown in brackets.
[b]organic N.
[c]$NO_2 + NO_3$-N 1.12#/acre = 1.0 kg/ha.

TKN	*NO_3-N*	*Total N*	*Soluble P*	*Total P*	*C1*	*Reference*
—	—	3.1 (0.3-75)	—	1.1 (0-7.3)	—	54 (1962-64)
2.0	0.53	—	0.08	0.21	—	47
0.9[b] (0.4-1.5)	—	—	0.38 (0.18-1.2)	—	12 (2-46)	2
—	—	—	—	—	5100	1
—	—	—	—	—	200	1
—	—	—	—	—	42 (11-160)	1
—	—	—	—	—	—	1
—	(0.4-2.0)	—	(0.2-1.8)	(0.5-4.0)	—	27
—	—	—	—	0.6 (0.15-2.50)	13 (3-390)	3 (1968-70)
—	—	—	—	—	(6-32)	39
—	—	—	—	—	(11-17)	39
—	—	—	—	—	(6-58)	39
—	—	—	—	—	—	1
—	—	—	—	—	—	1
—	—	—	—	—	—	1
—	—	8.8	—	1.1	—	53
—	—	6.8	—	5.6	—	53
2.1 (1.2-4.0)	—	—	1.0 (0.4-3.0)	—	—	2
8.8	0.2		2.1	4.0	—	4 (June Aug., 1965)
1.2	0.9	—	0.3	1.0	—	4
2.9	12[c]	—	—	1.6	—	22 (60% urban, 30% farm)
—	—	—	—	3.0		3
—	—	5.1	—	0.2		45
—	—	1.4	—	0.04		45
—	—	3.5	—	0.16		45

Table 10
Area Yields from Land Runoff Sources (56) (kg/ha/yr)

Item	*Urban Runoff*	*Rural Runoff*
Total organic carbon	345	144
Total PO_4	150	7.8
TKN	7.5	1.9
Nitrate N	13.3	5.0
Sodium	235	42.5
Potassium	133	21.7
Calcium	960	625
Magnesium	290	290

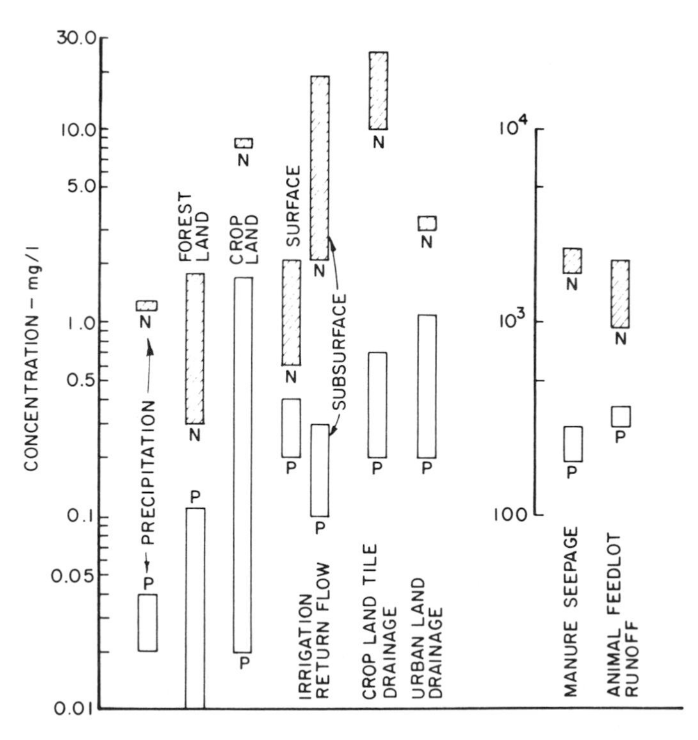

1. Comparison of non-point sources: range of total nitrogen and total phosphorus concentrations—(mg/1).

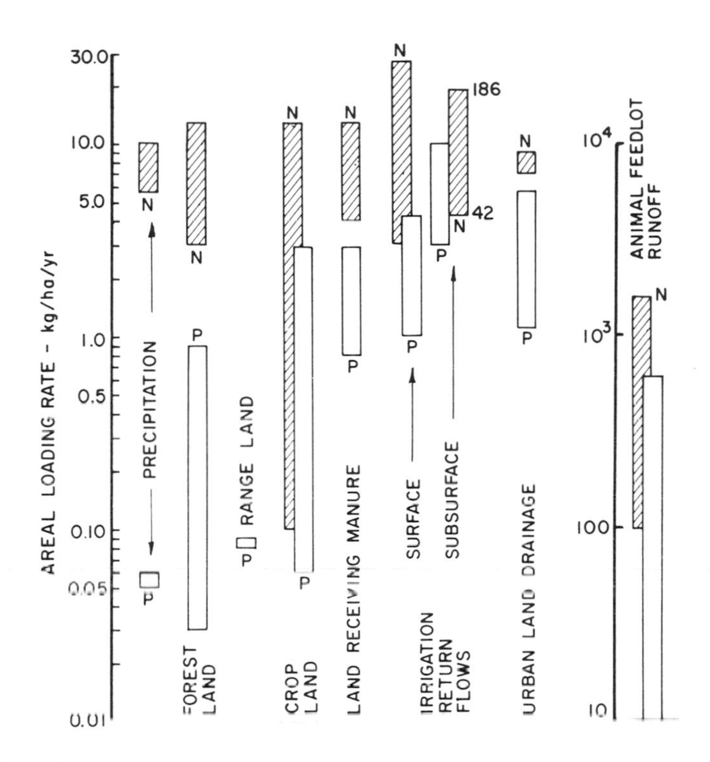

2. Contribution of total nitrogen and total phosphorus: area yield ranges from non-point sources (kg/ha/yr).

when the data are in terms of quantity of characteristic per unit of drainage area (kg/ha or kg/ha/yr), since non-point source contributions are flow dependent. Figure 2 illustrates the range of reported data and permits a reasonable comparison of the relative contribution from these non-point sources. The yields of total nitrogen from precipitation, forest land, crop land, land-receiving manure, surface-irrigation return flows, and urban-land drainage span a considerable range. The yields of total phosphorus from precipitation and range land are comparable, and the yields of total phosphorus from land-receiving manure, surface-irrigation return flows, and urban-land drainage are comparable. The yields of total phosphorus from forest land and crop land span a wide but comparable range, again undoubtedly because of the effect of erosion.

The yields of nitrogen and phosphorus in animal feedlot runoff also are orders of magnitude greater than that of the other non-point sources.

Although these comparisons cover wide ranges and may be considered gross, they do permit an assessment of the non-point sources that are controllable, especially in the light of available technology. Actual decisions on the control of non-point sources should be made on the basis of the relative importance of the respective sources in specific locations and on what is technically and logically controllable.

The term uncontrollable is used in this paper to describe both sources with characteristics which man cannot change significantly in type or magnitude and also sources which do not need to be controlled at this time.

The following is an assessment of non-point sources and the degrees of control required or feasible for them. Uncontrollable non-point sources include precipitation, forest land runoff, and range-land runoff. Precipitation is a natural phenomenon. The only way that man can control the characteristics of precipitation is to exert some control over the types and amount of material exhausted to the atmosphere. Air quality and gaseous emission standards are an attempt to exert this control. They are part of a total environmental protection effort and not solely a water pollution control matter.

Control of forest-land runoff and range-land runoff does not appear to be needed at this time, since the concentrations and yields of constituents are comparable to that of precipitation. These two non-point sources, forest runoff and range-land runoff, generally can be considered as background sources unless current practices or available data change drastically. Erosion control should be practiced in forests and on range land as with all other lands to prevent silt and associated material from entering surface waters.

Non-point sources that may require control include crop-land runoff, runoff from land receiving manure, crop-land tile drainage, and irrigation-return flow. The decision to control these non-point sources in a specific location should depend upon an evaluation of whether their constituents adversely affect the quality of a particular receiving water. It would be an inappropriate use of resources to attempt control of any of these sources where they do not have an adverse effect or where such control will not produce a positive effect on water quality. Routine control of crop-land runoff and runoff from land receiving manure is technically difficult and may not be needed in all cases. If manure is disposed of on land under good crop and land management conditions, contaminants from these lands should be comparable to those of crop land.

Crop-land tile drainage and irrigation-return flows are difficult to intercept and control. The nitrogen content of this drainage is the major item of concern. Where the drainage will have a significant effect on water quality, control of these sources will be necessary.

Non-point sources that should be controlled include urban land runoff, manure seepage, and feedlot runoff. Technology is available for the control of urban-land runoff. Collection and treatment of urban runoff prior to release to surface waters is becoming increasingly feasible. Manure seepage and feedlot runoff represent significant contaminant sources. Unsophisticated control procedures such as runoff retention ponds are available for the control of these sources and should be included in water-pollution control policies.

Direct discharge of animal urine and manure can occur when animals have access to surface waters. Where such animals cause pollution problems, it may be necessary to restrict their access to streams as part of local or regional water pollution control policy.

Non-point sources can also be compared to other sources, using a variety of parameters to assess their relative pollution contribution. Such comparisons indicate that the relative contribution in a watershed will be related to the type and magnitude of the cultural activities (i.e., activities controlled by man) in the watershed. Domestic and industrial wastewaters are major contributors of phosphorus in many lake or river basins.

Summary

Based upon the reported range of characteristics in non-point sources and upon available technology for their control, the non-point sources were identified as follows:

(a) uncontrollable or not needing control: precipitation, unmanaged forest-land runoff, range-land runoff;
(b) possibly needing control: crop-land runoff, runoff from land receiving manure, crop-land tile drainage, irrigation-return flows;
(c) requiring control: urban-land runoff, manure seepage, feedlot runoff.

Many non-point sources will be difficult to control if control is defined as the ability to establish and enforce effluent standards for a given runoff event. However, if the control is approached by utilizing appropriate management practices, many of the sources are controllable. Proper land and crop management practices, such as erosion control, timing and integration of manure disposal practices on the land, and timing of fertilizer application to coincide closely with crop need offer opportunities to minimize runoff of potential pollutants from agricultural lands.

References

1. American Public Works Association. 1969. *Water pollution aspects of urban runoff.* Federal Water Pollution Control Administration, United States Deprtment of the Interior final report, contract WA 66-23.
2. Avco Economic Systems Corporation. 1970. *Storm water pollution from urban land activity.* Federal Water Pollution Control Administration, United States Department of the Interior final report, contract 14-12-187.
3. Bryan, E. H. 1970. *Quality of stormwater drainage from urban land areas in North Carolina.* Raleigh: Water Resources Research Institute of North Carolina. Report 37.
4. Burm, R. J., Krawezyk, D. F., Harlow, G. L. 1968. Chemical and physical comparison of combined and separate sewer discharge. *Journal of Water Pollution Control Federation* 40: 112-16.
5. California Department of Natural Resources. 1971. *Nutrients from tile drainage systems.* EPA Report 13030 ELY.
6. Campbell, F. R., and Webber, L. R. 1970. Contribution of range land runoff to lake eutrophication. Paper read at Fifth International Water Pollution Research Conference, July-August 1970, San Francisco, California
7. Carter, D. L., Bondurant, J. A., Robbins, C. W. 1969. Water soluble NO_3-nitrogen, PO_4-phosphorus, and total salt balances on a large irrigation tract. *Soil Science Society of America proceedings* 35: 331-35.
8. Cooper, C. F. 1969 Nutrient output from managed forests. In *Eutrophication: causes, consequences, correctives,* pp. 404-445. Washington, D.C.: National Academy of Sciences.
9. Crisp, D. T. 1966. Input and output of minerals for an area of pennine moorland: the importance of precipitation, drainage, peat erosion and animals. *Journal of applied ecology* 3: 327-48.
10. Cywin, A., and Ward, D. 1971. *Agricultural pollution of the Great Lakes basin.* EPA report 13020.
11. Feth, J. H. 1966. Nitrogen compounds in natural water. a review. *Water resources research* 2: 41-58.
12. Fisher, D. W., Gambell, A. W., Likens, G. E., Bormann, F. H. 1968. Atmospheric contributions to water quality of streams in the Hubbard Brook Experimental Forest, New Hampshire. *Water resources research* 4: 1115-26.
13. Gearheart, R. A. 1969. Agricultural contribution to the eutrophication process in Beaver Reservoir. Paper 69-708 read at the 1969 winter meeting, American Society of Agricultural Engineers, 9-12 December, Chicago, Illinois.
14. Graham, T. R. 1967. Pollution prevention in Northern Ireland. *Effluent and water treatment journal* 7: 87-89.
15. Grizzard, T. J., and Jennelle, E. M.

1972. Will wastewater treatment stop eutrophication of impoundments? Paper read at 27th Purdue Industrial Waste Conference, May 1972, Lafayette, Indiana.

16. Groman, W. A. 1972. Forest fertilization: a state of the art review and description of environmental effects. Environmental Protection Technology Series, EPA-R2-72-016. Corvallis, Oregon: National Environmental Research Center, EPA.
17. Henkens, H. 1972. Fertilizer and the quality of surface waters. *Stikstof* 15: 28–40.
18. Hetling, L. J., and Sykes, R. M. 1973. Sources of nutrients in Canadarago Lake. *Journal of Water Pollution Control Federation* 45: 145–56.
19. Hobbie, J. E., and Likens, G. E. The output of phosphorus, dissolved organic carbon, and fine particulate carbon from Hubbard Brook Watershed. *Limnology and Oceanography* 18: 734–42.
20. Hoeft, R. G., Keeney, D. R., Walsh, L. M. 1972. Nitrogen and sulfur in precipation and sulfur dioxide in the atmosphere in Wisconsin. *Journal of environmental quality* 1: 203–208.
21. Howells, D. H., Kriz, G. J., Robbins, J. W. D. 1971. Role of animal wastes in agricultural land runoff. Environmental Protection Agency final project report, project 13020 DGX.
22. Jaworski, N. A., and Hetling, L. J. 1970. Relative contribution of nutrients to the Potomac River basin from various sources. *Proceedings of the Agricultural Waste Management Conference*, pp. 134–146. Ithaca: Cornell University.
23. Johnson, J. D., and Straub, C. P. 1971. Development of a mathematical model to predict the roll of surface runoff and groundwater flow in overfertilization of surface waters. Water Resources Research Center, University of Minnesota bulletin no. 35.
24. Johnson, N. M., Reynolds, R. C., Likens, G. E. 1972. Atmospheric sulfur: its effect on the chemical weathering of New England. *Science* 177: 514–516.
25. Joyner, B. F. 1971. Appraisal of chemical and biological condition of Lake Okeechobee. U.S. Geological Survey open file report 71006.
26. Kilmer, V. S., and Joyce, R. T. Fertilizer use in relation to water quality with special reference to the southeastern United States. Paper read at 22nd Annual Fertilizer Conference of the Pacific Northwest, July 1971, Bozeman, Montana.
27. Kluesener, J. W., and Lee, G. F. Nutrient loading from a separate storm sewer in Madison, Wisconsin. Paper read at the 45th Water Pollution Control Federation Conference, 8–12 October, 1972, Atlanta, Georgia.
28. Kolenbrander, G. J. 1972. The eutrophication of surface waters by agriculture and the urban population. *Stikstof* 15: 56–67.
29. Likens, G. E. 1973. The chemistry of precipitation in the central Finger Lakes region. Ithaca: Cornell University Water Resources and Marine Sciences Center.
30. Likens, G. E., and Bormann, F. H. Nutrient cycling in ecosystems. 1971. In *Ecosystem: structure and function*, ed. J. Wiens, pp. 26–67. Corvallis, Oregon: Oregon State University Press.
31. Likens, G. E., Bormann, F. H., Johnson, N. M., Fisher, D. W., Pierce, R. S. 1970. Effects of forest cutting and herbicide cutting and

herbicide treatment on nutrient budgets in the Hubbard Brook Watershed. *Ecological monographs* 40: 23–47.

32. Loehr, R. C. 1970. Drainage and pollution from beef cattle feedlots. *Journal of Sanitary Engineering Division, American Society of Civil Engineers* 96: 1295–1309.
33. Madden, J. M., and Dornbush, J. N. 1971. Measurement of runoff and runoff carried waste from commercial feedlots. In *Livestock waste management and pollution abatement*, pp. 44–47. St. Joseph, Michigan: American Society of Agricultural Engineers publication no. PROC-271.
34. Matheson, D. H. 1951. Inorganic nitrogen in precipitation and atmospheric sediments. *Canada journal of technology* 19: 406–412.
35. McCalla, T. M., Ellis, J. R., Gilbertson, C. B., Woods, W. R. 1972. Chemical studies of solids, runoff, soil profile, and groundwater from beef cattle feedlots at Mead, Nebraska. In *Proceedings of the Agricultural Waste Management Conference*, pp. 211–223. Ithaca: Cornell University.
36. Minshall, N., Nichols, M. S., Witzel, S. A. 1969. Plant nutrients in base flows of streams. *Water resources research* 5: 706–713.
37. Norton, T. E., and Hansen, R. W. 1969. Cattle feedlot water quality hydrology. In *Proceedings of Agricultural Waste Management Conference*, pp. 203–216. Ithaca: Cornell University.
38. Owens, M. 1970. Nutrient balances in rivers. *Water treatment and examination* 19: 239–252.
39. Pravoshinsky, N. A., and Gatello, P. D. 1968. Calculation of water pollution by surface runoff. *Water Research* 2: 24–26.
40. Sartor, J. D., Boyd, G. B., Agardy, F. J. 1972. Water pollution aspects of street surface contaminants. Paper read at the 45th Annual Water Pollution Control Federation Conference, 8–12 October 1972, Atlanta, Georgia.
41. Sawyer, C. N. 1947. Fertilization of lakes by agricultural and urban drainage. *Journal of New England Water Works Association* 61: 109–127.
42. Schwab, G. O., Taylor, G. S., Waldron, A. C. 1969. Pollutants in drainage water. Paper 69-710 read at the 1969 winter meeting of the American Society of Agricultural Engineers, 9–12 December, Chicago, Illinois.
43. Shannon, E. E., and Berzonik, P. L. 1972. Relationships between lake trophic state and nitrogen and phosphorus loading rates. *Environmental science and technology* 6: 719–725.
44. Smith, G. E. 1967. Fertilizer nutrients as contaminants in water supplies. *Agriculture and the quality of the environment*, ed. N. C. Brady, pp. 173–186. Washington, D.C.: American Association for the Advancement of Science publication no. 85.
45. Soderlund, G., and Lehtinen, H. 1972. Comparison of discharges from urban storm water runoff, mixed storm overflow, and treated sewage. Paper read at the Sixth International Water Pollution Research Conference, June 1972, Jerusalem, Israel.
46. Stewart, W. D. P. 1968. Nitrogen input into aquatic ecosystems. In *Algae, man, and the environment*, ed. D. F. Jackson, pp. 53–72. Syracuse: Syracuse University Press.
47. Sylvester, R. O. 1961. Nutrient content of drainage water from forested, urban, and agricultural areas. In *Algae and metropolitan wastes*, pp. 80–87. U.S. Dept. Health, Education, and Welfare publication no. SEC-TR-W61-3.

48. Sylvester, R. O., and Seabloom, R. W. 1962. A study on the character and significance of irrigation return flows in the Yakima River basin. Washington, D.C.: Public Health Service, U.S. Dept. Health, Education, and Welfare.
49. Taylor, A. W., Edwards, W. M., Simpson, E. C. 1971. Nutrients in streams draining woodland and farmland near Coschocton, Ohio. *Water resources research* 7: 81–89.
50. Timmons, D. R., Holt, R. F., Latterell, J. J. 1970. Leaching of crop residues as a source of nutrients in surface runoff water. *Water resources research* 6: 1367–75.
51. Tomlinson, T. W. 1970. Trends in nitrate concentrations in English rivers in relation to fertilizer use. *Water treatment and examination* 19: 277–293.
52. Wang, W. L., and Evans, R. L. 1970. Nutrients and quality in impounded water. *Journal of American Water Works Association* 62: 510–514.
53. Weibel, S. R. 1969. Urban drainage as a factor in eutrophication. In *Eutrophication: causes, consequences, correctives*, pp. 384–403. Washington, D.C.: National Academy of Sciences.
54. Weibel, S. R., Weidner, R. B., Cohen, J. M., Christianson, A. G. 1966. Pesticides and other contaminants from rainfall and runoff as observed in Ohio. *Journal of American Water Works Association* 58: 1075–84.
55. Wells, D. M., Grub, W., Albin, R. C. Meenaghan, G. F., Coleman, E. 1970. Control of water pollution from Southwestern cattle feedlots. Paper read at the Fifth International Water Pollution Conference, July–August 1970, San Francisco, California.
56. Keeney, D. R., and Walsh, L. M. 1971. Source and fate of "available" nitrogen in rural ecosystems. In *Farm animal wastes: Nitrates, and phosphates in rural Wisconsin ecosystems*, ed. T. J. Brevik and M. T. Beattey. Wisconsin, extension service.
57. Powell, R., and Dunsmore, J. Phosphorus in the rural ecosystem: Runoff from agricultural land. In Brevik and Beattey, above, no. 56.

21. Water Quality Improvement Through Control of Road Surface Runoff

James D. Sartor
and Gail B. Boyd

Introduction

During the past few years, it has become increasingly obvious that runoff from storms is by no means "rainwater" in terms of quality. Rather, storm runoff typically contains substantial quantities of impurities, so much so that it is a more serious source of pollutants than municipal sewage in many areas. Numerous studies have been and are being conducted to help define this problem with the amounts of polluting substances involved, their sources, their practical significance, and their possible means of control.

Runoff can contribute to a variety of problems, including direct pollution of receiving waters, overloading of treatment facilities, and impairment of sewer and catch basin functions. These problems are caused in part by hydraulic overloading, but also by the various pollutants contained within the runoff.

Previous studies by the American Public Works Association and AVCO Corporation (1969, 1970) provide much valuable information on the total problem of water pollution resulting from urban runoff. They both point out the shock pollution loads which storm runoff from urban areas can place on receiving waters. Among the sources of pollution in urban and rural runoff water are debris and contaminants from streets, contaminants from open land areas, publicly used chemicals, air-deposited substances, ice control chemicals, and dirt and contaminants washed from vehicles. The APWA report suggests various means of reducing the pollution problem created by runoff and emphasizes the need for more definitive investigations as to the source, cause, and extent of the pollutants; the interrelationships and significance of the variables; and the development of standard procedures, methods and / or techniques for measuring the street surface contaminants. Among the concepts proposed for limiting storm water pollution was the improvement of street cleaning methods and operations.

At the outset of this study, Woodward-Envicon, Inc. conducted a comprehensive search of the literature for existing data regarding the sources, quantities, and pollutional properties of street surface contaminants and refuse, which revealed the following:

(a) A considerable amount of data and information exists on pollutional loads associated with storm water and combined storm and sewer systems.
(b) The data available on storm-water pollutional loads are not directly relatable to the materials contributed by street surface contaminants.
(c) Information is lacking on relationships between street surface contaminants, their pollutional characteristics, and the manner in which they are transported during storm runoff periods.

Description of Study

The broad objectives of this study were to investigate and define the water pollution impact of storm water discharge and to develop alternative approaches suitable for reducing pollution from this source. The study focused on three principal areas:

(a) determining the amounts and types of materials which commonly collect on the surfaces of streets and roads;
(b) determining the effectiveness of conventional public works practices in preventing these materials from polluting receiving waters; and
(c) evaluating the significance of this source of water pollution relative to other sources.

The overall problem of controlling runoff pollution is complex indeed. The general approach to this problem involves dividing the overall problem into discrete segments which can be studied first separately and then in relationship to one another. This project is but a part of that overall approach; this segment concerns the street surface. As the overall problem becomes understood, effective control measures can be developed and implemented.

The major efforts of the eighteen-month study centered around three elements:

(a) collecting contaminant materials from street surfaces in cities and towns all over the country;
(b) analyzing those materials to determine their physical, chemical, and biological properties and, insofar as these pertain to source

identification, evaluating pollutional potential and / or possible means of control; and

(c) observing and evaluating various street-cleaning practices in several cities throughout the country.

The required information was developed through five means:

(a) field measurements and sample collection;
(b) physical, chemical, and biological analyses of samples;
(c) experimental studies;
(d) literature reviews; and
(e) surveys by questionnaire and interview.

In this study "street surface contaminants" are defined as being those materials found on street surfaces which are capable of being washed off during common rainstorms. Street surfaces are defined as being the paved traffic lanes, any parking lanes, and the gutter, i.e., the area typically bounded by curbs. In urban areas the total contribution of contaminants comes from a much larger area than just this street surface. For instance, there are substantial contributions from sidewalks, planter strips, yards, driveways, parking lots, roofs of buildings, etc. Thus the quality of the water entering the storm sewer inlet is only partly a function of the contaminants washed from the street per se. In rural areas, where paved and guttered streets occupy an even smaller fraction of the total runoff area, the discrepancy becomes more pronounced. Nonetheless, it is probable that the findings of this study will be of value to those interested in the control of rural environmental problems.

The study included investigations of twelve U.S. cities and towns having populations ranging from 13,200 persons in Bucyrus, Illinois, to 895,000 persons in Baltimore, Maryland. Street surface contaminant samplings in these communities represent approximately 25,000 curb miles of streets.

Investigations of the characteristics and quantities of street surface contaminants were reported in terms of a variety of parameters. A series of screening studies were conducted to determine the physical, chemical, and bacteriological properties of representative composite samples. The array of tests run routinely throughout the study was based on the results of these screening studies. Conventional sanitary engineering parameters included total and volatile solids, biochemical oxygen demand, chemical oxygen demand, Kjeldahl nitrogen, soluble nitrates, and phosphates. Included also were less common parameters, such as chromium, copper, zinc, nickel, mercury, lead, cadmium, chlorinated hydrocarbons, dieldrin, DDD, DDT, methoxychlor, endrin, methylparathion, lindane, and polychlorinated biphenyls. Additionally, studies were conducted on the presence and general abundance of both total and fecal coliform bacteria. Physical properties of the contaminant materials were also studied in detail, because these determine the effectiveness of any control approach.

Results and Discussion

Degree of Contamination of Runoff

Runoff from street and road surfaces is generally highly contaminated. In fact, it is similar to raw sanitary sewage in many respects. To place the research findings in perspective for the practicing engineer, some comparative computations based on a hypothetical but fairly typical United States city have been made. The runoff from the first hour of a moderate-to-heavy storm with brief peaks to at least ½ inch per hour would contribute considerably more pollutional load than would the same city's sanitary sewage during the same period of time (see Table 1).

The study determined that streets are cleaned intentionally or by rainfall about once every five days on the average. There is little question that street surface contaminants warrant serious consideration as a source of receiving water pollution, particularly in cases when such discharges of contaminants coincide with times of low stream flow or poor dispersion.

Composition and Properties of Contaminant Materials

The major constituent of street surface contaminants was consistently found to be inorganic, mineral-like matter, similar to common sand and silt. This inorganic material, most of which is probably blown, washed, or tracked in from surrounding land areas, does not constitute a serious water pollutant by itself. Along with this material, however, is organic matter, a small fraction of the total on the basis of mass. At a given location, both the organic and inorganic fractions increase in loading intensity (measured in pounds per curb mile) with increasing time since the last cleaning. Data indicate that the organic fraction tends to accumulate at a faster rate than the inorganic fraction, but within the time frame of interest here (a few days to a few weeks) the organic fraction is still much smaller than the inorganic.

Table 1
Calculated Quantities of Pollutants which would enter Receiving Waters (hypothetical city)[a]

	Street Surface Runoff (1b/hr)	*Raw Sanitary Sewage (1b/hr)*	*Secondary Plant Effluent (1b/hr)*
Settleable plus suspended solids	560,000	1,300	130
BOD_5	5,600	1,100	110
COD	13,000	1,200	120
Kjeldahl nitrogen	880	210	20
Phosphates	440	50	2.5
Total coliform bacteria (org/hr)	4000×10^{10}	$460{,}000 \times 10^{10}$	4.6×10^{10}

Runoff contaminant loadings are based upon the first hour of a moderate-to-heavy storm.

[a]The hypothetical city used for this comparison had the following characteristics:

Population: 100,000 persons
Total land area: 14,000 acres
Distribution of developed land residential: 75%, commercial 5%, industrial 20%
Streets (tributary to receiving waters): 400 curb miles
Sanitary sewage: 12×10^6 gal/day.

The quantity and character of contaminants found on street surfaces are summarized in Table 2. The tabulated values are for all cities tested. They are weighted averages in which data for larger cities are allowed to bias the reported loading intensities.

Significant amounts of heavy metals, particularly zinc and lead, were detected in the contaminant materials collected from street surfaces. Depending upon their specific chemical form, heavy metal compounds can be highly detrimental to biological systems. The samples collected in this study have been analyzed only so far as to indicate the total quantities of each metal present, not their specific chemical form. The Environmental Protection Agency has subsequently developed more definitive information from the samples collected in this study.

Substantial quantities of organic pesticides and related compounds were found in the street surface contaminants. On the order of 0.001 lb per curb mile total was found for the cities tested, although the data showed considerable variation from site to site. The chlorinated hydrocarbons, p, p-DDD and p, p-DDT, were found rather consis-

Table 2
Quantity and Character of Contaminants

Measured Constituents	*Weighted Mean for All Samples (lb/curb mile)*
Total solids	1400
Oxygen demand	
BOD_5	13.5
COD	95
Volatile solids	100
Algal nutrients	
Phosphates	1.1
Nitrates	0.094
Kjeldahl nitrogen	2.2
Bacteriological	
Total coliforms (org/curb mile)	99×10^9
Fecal coliforms (org/curb mile)	5.6×10^9
Heavy metals	
Zinc	0.65
Copper	0.20
Lead	0.57
Nickel	0.05
Mercury	0.073
Chromium	0.11
Pesticides	
p, p-DDD	67×10^{-6}
p, p-DDT	61×10^{-6}
Dieldrin	24×10^{-6}
Polychlorinated biphenyls	1100×10^{-6}

Note: "org" refers to the number of coliform organisms observed.

tently, as were polychlorinated biphenyl compounds. Although these have repeatedly been associated with adverse environmental effects in recent controversies, the actual significance of the findings cannot yet be stated, since the environmental consequences of such materials have not yet been established.

Perhaps one of the most important findings of this study is that such a great portion of the overall pollutional potential is associated with the fine solids fraction of the street surface contaminants. Furthermore, these fines account for only a minor portion of the total loading on street surfaces. As shown in Table 3, the very fine, siltlike material ($<$43 microns) accounts for only 5.9 percent of the total solids but about one fourth of the oxygen demand and perhaps one third to one half of the algal nutrients. It also accounts for over one half of the heavy metals and nearly three fourths of the total pesti-

Table 3
Fraction of Total Constituent Associated with each Particle Size Range (% by weight)

	< 43μ	43μ → 246μ	> 246μ
Total Solids	5.9	37.5	56.5
BOD_5	24.3	32.5	43.2
COD	22.7	57.4	19.9
Volatile solids	25.6	34.0	40.4
Phosphates	56.2	36.0	7.8
Nitrates	31.9	45.1	23.0
Kjeldahl nitrogen	18.7	39.8	41.5
Heavy metals (all)	51.2		48.7
Pesticides (all)	73		27
Polychlorinated biphenyls	34		66

cides. This concentration of pollutants in a small amount of very fine matter is of particular importance, considering that conventional street-sweeping operations are rather ineffective in removing fines (sweepers were observed to leave behind 85 percent of the material finer than 43 microns and 52 percent of the material finer than 246 microns).

The oxygen-demand potential of the contaminants was studied by several means. It was found that the presence of toxic materials in street surface contaminants seriously interfered with BOD measurements. Such materials (particularly heavy metals) were found to be present in many samples at levels far in excess of those known to cause substantial interference. The chemical oxygen (COD) test was found to be a better basis for expressing oxygen demand potential.

Quantity and Distribution of Contaminants

The quantity of contaminant material existing on street surfaces was found to vary widely with many factors. However, loading intensities averaged on the order of 1400 lb/curb mile of street for the cities tested. The total solids loading intensities for the various land-use areas tested are given in Table 4.

The quantity of contaminant material existing at a given test site was found to depend upon the length of time that had elapsed since the site was last cleaned, either intentionally by sweeping or flushing, or by rainfall. The field sampling program focused on collecting ma-

Table 4
Loading Intensities vs. Land Use[a]

Land Use	*Numerical Mean*	*Weighted Mean*
Residential		1,200
low/old/single	850	
low/old/multi	890	
med/new/single	430	
med/old/single	1,200	
med/old/multi	1,400	
Industrial		2,800
light	2,600	
medium	890	
heavy	3,500	
Commercial		290
central business district	290	
shopping center	290	
Overall		1,400

[a]Units expressed as lb/curb mile.

terials from street surfaces at a single point in time; there were no repeated samplings of a given site to develop information on how contaminants accumulate with time. However, information was collected for each site to define the elapsed time since the last substantial rain storm and / or cleaning. Computer analyses of such data revealed correlations between antecedent cleaning time and loading intensity. In general, industrial land-use areas tend to accumulate contaminants faster than commercial or residential areas. (Accumulation patterns as calculated here are shown in Figure 1.) The principal factors affecting the loading intensity at any given site include the following: surrounding land use, the elapsed time since streets were last cleaned, local traffic volume and character, street surface type and condition, public works practices, season of the year, etc.

Contaminant loading intensities were found to vary with respect to land-use patterns in the surrounding locale. In general, industrial areas have substantially heavier than average loadings. All twenty industrial test sites taken together have an average loading of some 2800 lb/curb mile, twice the mean for cities on the whole. This is probably because industrial areas tend to be swept less often and because generation rates of dust and dirt tend to be high on account of "fallout," spillage from vehicles, unpaved dirt areas, poor street conditions, etc. Heavy industrial areas showed the heaviest loadings of all industrial areas and

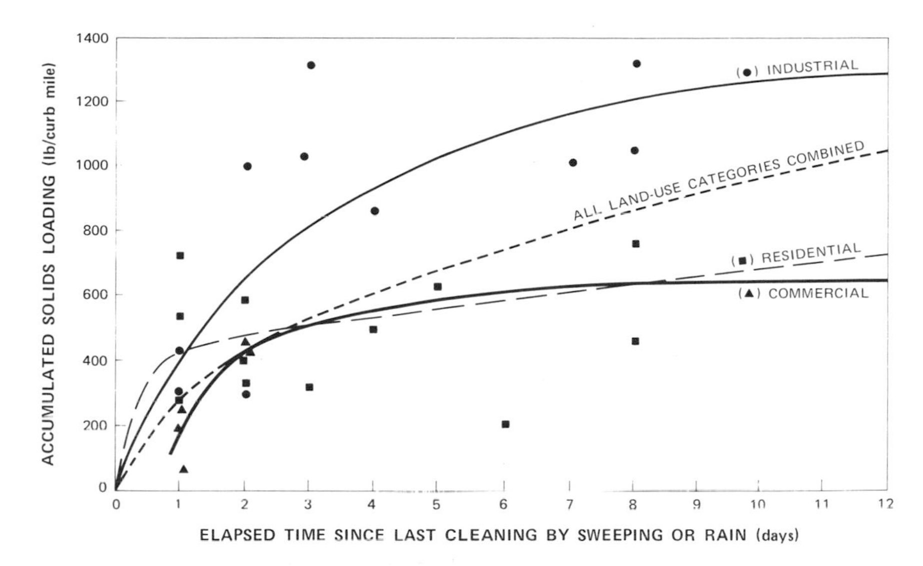

1. Accumulation patterns for land-use categories.

medium industrial areas the lightest. The loadings varied so widely between individual sites that it would be speculative to state why any one type of industrial area is dirtier than another.

Commercial areas have substantially lighter loading intensities than the mean for cities on the whole (290 lb/curb mile average vs 1400). This is probably because streets in these areas are swept so often—typically several times weekly and daily in prime areas.

Residential areas were found to have an average loading intensity comparable to the average for all land uses of all cities taken together: 1200 lb/curb mile. Here again, the loadings varied widely from site to site, and offering a conclusive explanation for these variations would be speculative. The data imply some tendency, however, for newer, more affluent neighborhoods to be cleaner, possibly because they are better maintained by residents and/or are further from sources of contamination.

Tests showed that street surface contaminants are distributed quite non-uniformly across the streets. The solids loading intensity across a typical street is given in Table 5. Typically, 78 percent of the material was found within 6 inches of the curb and over 95 percent within the first 40 inches, presumably because the direct impact and air currents from traffic push the material to the curb and gutter.

Street sweeping makes the gutter much cleaner, but also moves much of the material out of the gutter and redistributes it on areas

Table 5
Typical Solids Loading Intensity

Street Location (Distance from Curb) (in.)	*Solids Loading Intensity (% of total)*
0–6	78
6–12	10
12–40	9
40–96	1
96 to centerline	2

which were somewhat cleaner prior to sweeping (see Figure 2). The present design of gutter brooms is such that they tend to redistribute the dust and dirt fraction (<2,000 microns) over the surface of the street and indeed are not particularly efficient in moving the dust and dirt fraction out of the gutter.

Contaminant Transport

Rainfall washes particulate contaminants to the point of entry to the sewer system in a manner that is regular and predictable on the basis of a few, easily measured parameters descriptive of the site and the rainstorm: rainfall intensity, street surface characteristics, and particle size. Computer-assisted analysis of data from a special series of field experiments revealed that the wash-off phenomenon can be expressed mathematically by a simple exponential equation:

$$N_c = N_o\,(1 - e^{-krt})$$

where N_c is the weight of material of a given particle size washed off a street initially having a loading of N_o after t minutes of rainfall at an intensity of r inches per hour. The proportionality constant k (units of hours per inch minute) depends upon street surface characteristics but was found to be almost independent of particle size, at least within the 10 to 1,000 micron range of sizes of particular interest here.

Street surface characteristics were found to have an effect on the contaminant loadings observed at a given site. For example, asphalt streets had loadings about 80 percent heavier than all concrete streets. Streets paved partially with asphalt and partially with concrete were intermediate (loadings were about 65 percent heavier than for all concrete streets). The condition of street pavement is also important. Streets in fair to poor condition had loadings about 2½ times higher than streets in good to excellent condition.

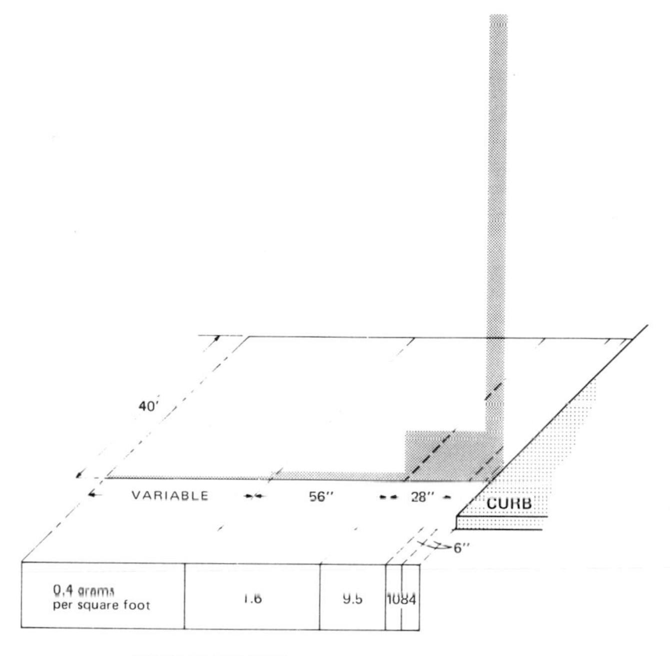

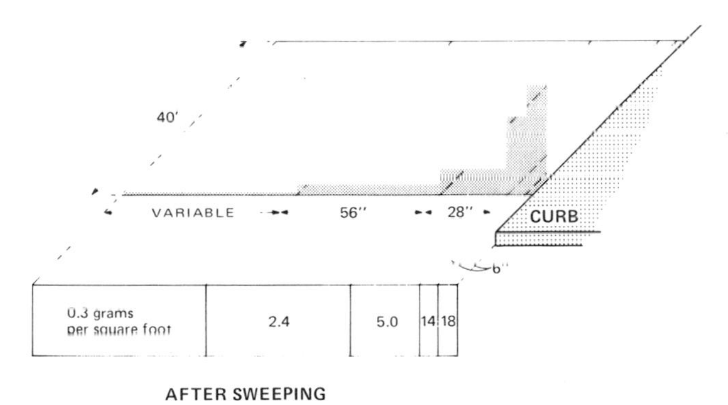

2. Distribution of contaminants.

Control Possibilities

Current street cleaning practices are essentially for aesthetic purposes, and even under well operated and highly efficient street sweeping programs, their efficiency in the removal of the dust and dirt fraction of street surface contaminants is low. The removal efficiency of conventional street sweepers was found to be dependent upon the particle

Table 6
Sweeper Efficiency vs. Particle Size

Particle Size (microns)	*Sweeper Efficiency*
2000	79
840 → 2000	66
246 → 840	60
104 → 246	48
43 → 104	20
< 43	15
Overall	50

size range of street surface contaminants, as shown in Table 6. The overall removal effectiveness for the dust and dirt fraction is 50 percent, that is, one half of the dirt and dust fraction remains on the street. The removal effectiveness for litter and debris, however, ranges from 95 to 100 percent.

To achieve greater removal of the dust and dirt fraction of street surface contaminants, several times the normal street cleaning effort, measured by equipment minutes per 1,000 sq. ft. of area swept, would have to be expended, as indicated in Table 7.

Increased effectiveness can be achieved by operating at a speed slower than the normal 6 mph or by conducting multiple passes. To achieve an overall effectiveness of 70 percent, two cleaning cycles would be required. Effectiveness values greater than 90 percent are probably not achievable with present state-of-the-art street sweepers.

The following mathematical relationship was used to calculate the removal effectiveness of the dust and dirt fractions by particle size range:

$$M = M^* + (M_o - M^*)\, e^{-kE}$$

where M is the amount of street surface contaminants remaining after sweeping, M_o is the initial amount, and E is the amount of sweeping effort involved in equipment minutes per 1,000 sq ft. (M^* and k are dimensionless empirical constants dependent upon sweeper characteristics, particle size of contaminants, and street surface.)

One of the most serious problems encountered in street sweeping is vehicle parking. In congested areas it is not unusual to find the entire curb sides of streets occupied by parked vehicles. In some large cities, "no parking" regulations have been instituted during scheduled street sweeping hours.

Field studies indicate that catch basins as normally used are rea-

Table 7
Sweeper Effectiveness vs. Effort

Effectiveness (%)	*Effort (equip min/1,000 sq ft)*	*Increase over Normal (0.237)*
95	1.5	6.3
90	.85	3.6
70	.50	2.1

sonably effective in removing coarse, inorganic solids, such as coarse sand and gravel, from storm runoff but are ineffective in removing fine solids and most organic matter. This fact is important because it is these latter materials which contribute most heavily to water pollution. Sample analyses have indicated that much of the material found in urban and suburban catch basins consists of litter, leaves, used oil, etc. from sources other than surface runoff. Upon decomposition, the contents of catch basins become even more threatening to receiving-water quality. This subject is presently under study by EPA and the American Public Works Association.

Summary

This study not only represents the most comprehensive summary of pollutant characteristics developed to date on street surface characteristics, but is also more extensive and complete than much of the information presently available on the pollutional characteristics of municipal and industrial waste discharges. A more detailed discussion is presented in the report *Water Pollution Aspects of Street Surface Contaminants*, Sartor and Boyd, available through U.S. Environmental Protection Agency (EPA-R2-72-081).

This research has led to the following conclusions:

1. The water which drains from street surfaces during and following storms is generally highly contaminated.

2. Most of the contaminant material was found to be inorganic, mineral-like matter, similar to common sand and silt.

3. The origin of this material was extremely diverse and could not be established with certainty. On both a weight and volume basis, most of it appears to be soil that has been transported and deposited on the street. Fragments of glass, plastic, paper, and plant foliage are also abundant, as are rubber, metals, and asbestos.

4. A great portion of the overall pollutional potential is associated with the fine solids fraction of the street surface contaminants.
5. Chemical oxygen demand (COD) tests provide a good basis for estimating the oxygen demand potential.
6. The quantity of contaminant material at a given test site was found to depend upon the length of time which had elapsed since the site was last cleaned.
7. Street surface contaminants are not distributed uniformly across the streets.
8. Current street-cleaning practices are essentially for aesthetic purposes; their efficiency in the removal of the dust and dirt fraction of street surface contaminants is low even under well operated street-sweeping programs.
9. To achieve better than 50 percent removal of the dust and dirt fraction of street surface contaminants, several times the normal street-cleaning effort would have to be expended.
10. Specially conducted field studies indicate that catch basins as normally used are reasonably effective in removing coarse inorganic solids from storm runoff, but are ineffective in removing the fine solids which contribute most heavily to water pollution.

Recommendations

This study has shown that street surface contaminants contain high concentrations of materials known to cause problems of water pollution, including heavy metals, organic toxicants, suspended solids, oxygen demanding compounds, etc. On the basis of these findings, the following recommendations are offered:
1. Street-cleaning equipment operators should be trained not only in how their equipment can best be operated (i.e., vehicle speed, broom speed, broom position, etc.), but also in what material needs to be removed and where it is commonly located. Much of the fine material that normally lies in the gutters could be picked up if the operators appreciated its importance relative to water pollution.
2. Increased effort should be expended on street-cleaning operations. Operating speeds should not exceed 5 miles per hour unless on high-speed arterials. Additional cleaning cycles should be scheduled on streets that are the principal vehicular arterials.
3. Public works departments should maintain accurate and detailed records of street-cleaning operations, including manpower utilization, equipment utilization, and equipment maintenance. The American

Public Works Association (1969) has created guidelines for standard procedures to be used in collecting and reporting statistics and in measuring and evaluating equipment performance. The procedures outlined in these reports should be utilized in providing the necessary input data to a cost-effective street-cleaning program.

4. Public works departments should pay increased attention to maintaining pavements in good condition. When the material for paving is being selected, it is recommended that the increased contaminant loadings with asphalt be taken into consideration.

5. Cities should give special consideration to ways of restricting on-street parking on the days that sweepers or flushers make their regular rounds. An effective approach employed in Baltimore was to pass an ordinance restricting on-street parking, send public works crews out to educate local residents about their street cleaning programs, post signs along the streets, and enforce the ordinance through citations and/or tow-away of vehicles. Through this program, the city has achieved substantially better control of all forms of street contaminants and debris as well as broad public support for their work.

6. Routine maintenance schedules should include proper adjustments to sweeper-operating parameters as specified in manufacturer's, owner's, and operating manuals.

7. The role of gutter brooms in street cleaning should be further evaluated and research directed toward the development of new techniques for the efficient cleaning of gutters.

8. Public works departments should give serious consideration to the value of catch basins in their particular system. Where a simple storm-water inlet structure would suffice, it is probably desirable to get rid of the catch basin either by replacing it or filling it in. An interim response of considerable value in most communities would be to clean out dirty catch basins regularly. This would be particularly effective if they were cleaned just before periods of major rainfall.

9. The design of future systems for controlling pollutional effects of street runoff should take into account the fact that particulate contaminants arrive at the point of entry to the sewer system in a quite regular and predictable manner.

10. Further consideration should be given to the desirability of separating storm and sanitary sewers.

11. Special studies should be conducted to determine the amount and nature of materials that can wash off freeways during storms and to identify means for controlling this source of water pollutants. It would be important to conduct this study on well-defined test areas with existing drainage systems. Since traffic characteristics and aerial transport of fine solids both have a pronounced effect on contam-

inant loadings, it is imperative that these be studied concurrently with the freeway surface itself.

12. Special tests should be conducted to evaluate the technical, operational, and economic feasibility of vacuum wand units in collecting street surface contaminants.

13. Research and development should be conducted to explore special curb/gutter configurations that would allow free flow of water and flushed debris to a pickup point, even in the presence of parked cars. Other aspects of this study would be to develop special mobile flushers (probably evolutionary extensions of the low volume/high velocity units developed by the U.S. Naval Radiological Defense Laboratory for removing radioactive fallout materials from street surfaces), as well as special equipment and techniques for picking up the water and contaminants after flushing.

14. A full-scale test program should be conducted in cooperation with the municipal public works departments to examine the overall effectiveness of street cleaning operations and the feasibility of a cost-effectiveness model that could be utilized by municipalities to upgrade current street cleaning practices. The program should include the evaluation of newly developed street cleaning equipment such as vacuumized sweepers and broom sweepers and the general feasibility of adopting special public works practices involving the use of special flushing units, modified gutter and inlet designs, catch basins, and extra cleaning cycles for both catch basins and streets.

15. A study should be conducted in several snow-belt cities located near bodies of water to determine the extent and severity of the problem of snow disposal. The results of such a study should serve to define possible requirements for modifying current snow dumping practices and developing safer means of ultimate snow disposal.

References

American Public Works Association. 1969. *Water pollution aspects of urban runoff.* Report no. 11030 DNS 01/69, contract no. WA 66-23. Environmental Protection Agency.

Avco Economic Systems Corporation. 1970. *Storm water pollution from urban land activity.* Federal Water Pollution Control Administration, United States Department of the Interior final report, contract no. 14-12-187.

Sartor, J. D., and Boyd, G. B. 1972. *Water pollution aspects of street surface contaminants.* Report no. EPA-R2-72-081. Environmental Protection Agency.

This study was conducted under the sponsorship of the Environmental Protection Agency in fulfillment of Project No. 11034 FUJ.

22. Water Pollution and Associated Effects from Street Salting

Richard Field,
Edmund J. Struzeski,
Hugh E. Masters
and Anthony N. Tafuri

Introduction

Salt contamination in runoff is generated by storm events. Accordingly, countermeasures for this form of pollution are being investigated by the U.S. Environmental Protection Agency's Storm and Combined Sewer Pollution Control Research, Development, and Demonstration Program.

This report comprises a state-of-the-art review of highway de-icing practices and associated environmental effects and offers a critical summary of the available information on the following:

1. Methods, equipment, and materials used for snow and ice removal;
2. Chlorides found in rainfall and municipal sewage during the winter;
3. Salt runoff from streets and highways;
4. De-icing compounds found in surface streams, public water supplies, groundwater, farm ponds, and lakes;
5. Special nutritious or toxic additives incorporated into de-icing agents;
6. Vehicular corrosion and deterioration of highway structures and pavements attributable to salting; and
7. Effects of de-icing compounds on roadside soils, vegetation, and trees.

Use of De-Icing Compounds

It has been found that the current annual use of highway de-icers is approximately 9 to 10 million tons of sodium chloride, 0.3 million tons of calcium chloride, and about 11 million tons of abrasives. Reported amounts of these materials deployed for highway de-icing by individual states and regions during the winter of 1966–1967 are presented in Table 1. Twenty-one states in the eastern and north-

Table 1
Reported Use (Tons) of Sodium Chloride, Calcium Chloride, and Abrasives by States and Regions in the United States, Winter of 1966–1967[a] [b]

State	*Sodium Chloride*	*Calcium Chloride*	*Abrasives*
EASTERN STATES			
Maine	99,000	1,000	324,000
New Hampshire	118,000	—	26,000
Vermont	89,000	1,000	89,000
Massachusetts	190,000	6,000	423,000
Connecticut	101,000	3,000	335,000
Rhode Island	47,000	1,000	86,000
New York	472,000	5,000	1,694,000
Pennsylvania	592,000	45,000	1,162,000
New Jersey	51,000	6,000	70,000
Delaware	7,000	1,000	2,000
Maryland	132,000	1,000	40,000
Virginia	77,000	22,000	204,000
	1,975,000	92,000	4,455,000
NORTH CENTRAL STATES			
Ohio	511,000	12,000	43,000
West Virginia	55,000	9,000	230,000
Kentucky	60,000	1,000	—
Indiana	237,000	6,000	77,000
Illinois	249,000	10,000	60,000
Michigan	409,000	7,000	6,000
Wisconsin	225,000	3,000	102,000
Minnesota	398,000	14,000	84,000
North Dakota	2,000	1,000	13,000
	2,146,000	63,000	615,000
SOUTHERN STATES			
Arkansas	1,000	—	—
Tennessee	—	—	—
North Carolina	17,000	2,000	75,000
Mississippi	—	—	—
Alabama	—	—	—
Georgia	—	—	—
South Carolina	—	—	—
Louisiana	—	—	—
Florida	—	—	—
	18,000	2,000	75,000

Table 1 continued

State	*Sodium Chloride*	*Calcium Chloride*	*Abrasives*
WEST CENTRAL STATES			
Iowa	54,000	2,000	291,000
Missouri	34,000	3,000	—
Kansas	25,000	2,000	31,000
South Dakota	2,000	1,000	36,000
Nebraska	10,000	—	6,000
Colorado	7,000	—	150,000
	132,000	8,000	291,000
SOUTHWEST STATES			
Oklahoma	7,000	—	2,000
New Mexico	7,000	—	—
Texas	3,000	—	1,000
	17,000	—	3,000
WESTERN STATES			
Washington	2,000		155,000
Idaho	1,000	—	47,000
Montana	4,000	—	80,000
Oregon	1,000	—	200,000
Wyoming	1,000	—	43,000
California	11,000	—	94,000
Nevada	4,000	—	50,000
Utah	28,000	—	56,000
Arizona		—	—
	52,000	—	725,000
District of Columbia	36,000		—
1966–1967 Reported *Totals*[c]	4,376,000	165,000	6,164,000

[a]Data taken from Salt Institute 1966–1967 Survey for U.S. and Canada.

[b]Represents data by all governmental authorities reporting within each state.

[c]Overall values given in Table 1 represent about 75 percent of true values (reported and unreported) of salts and abrasives used in 1966–1967. With confidential data and appropriate adjustments, the Salt Institute estimates that U.S. total consumption for the winter 1966–1967 was 6,320,000 tons sodium chloride, 247,000 tons calcium chloride, and 8,400 tons abrasives.

central sectors of the country use more than 90 percent of all chloride compound de-icers. Leading states in de-icer use are Pennsylvania, Ohio, New York, Michigan, and Minnesota. The state of New Hampshire has used highway salts since the mid-1940's, and over this period the cumulative use of highway salts in the small state alone has probably exceeded 2.3 million tons.

The demand that roads be safe and usable at all times and that June driving conditions be provided in January has in recent years led to adoption of a "bare-pavement" policy by practically all highway departments in the snow-belt region. As a result, the use of de-icing salts has greatly increased; in many cases they have replaced the abrasives previously used. Unfortunately, the more damaging chlorides are more efficient in melting snow and ice, are not blown off the road as easily by wind and traffic, require less application time, and are less costly both in application and in cleanup. At the end of the winter large amounts of abrasives must be retrieved from shoulder areas, catch basins, and conduits in order to reestablish proper road drainage (Mass. Legislative Res. Council, 1965), whereas chemical de-icers attack the ice and packed snow surfaces directly, melt them, and are themselves dissolved. The salt dissolves the ice and, most importantly, causes a break in the tight bonding of ice to pavement. Chemicals also prevent the formation of new ice. The resulting salt residue is then readily washed off the pavement.

Marine salts have been shown to be comparable in cost-effectiveness to rock salt and are being used for highway de-icing (Mass. Legislative Res. Council, 1965). These salts are probably being sold separately or mixed together with commercial rock salt. It is known that the major constituents in sea water approximate 30.5 percent sodium, 55.1 percent chlorides, 3.7 percent magnesium, 7.7 percent sulfates, 1.2 percent calcium, 1.1 percent potassium, 0.2 percent bromides, and 0.4 percent bicarbonates and carbonates (Fairbridge, 1966; Riley and Skirrow, ed., 1965; Harvey, 1963). Although commercial marine salts exclude impurities to a large extent, information on product composition is not readily available. Further marine salt data are needed on composition, comparative de-icing efficiency, and the potential contribution of sulfates, magnesium, potassium, and other available constituents to environmental pollution.

Highway salting rates are generally in the range of 400 to 1,200 pounds of salt per mile of highway per application (Hanes, et. al., 1970; Schraufnagel, 1965). Over the winter season, many roads and highways in the U.S. may receive more than 20 tons of salt per lane mile. Considerable wasting of highway de-icers undoubtedly occurs because of excessive application, misdirected spreading, and general wintertime difficulties. In some cases, salts are applied as soon as or even before snow occurs based upon weather forecasts. It is believed that highway salts have frequently been used when no snow followed.

Environmental problems are minimized by deploying chemicals as sparingly as possible to maintain "bare-pavement" conditions. The proper application and spreading of highway salts have generated some studies, but this area is still neglected. The Highway Research Board in a 1967 report indicates several challenges presented by previous research findings, including the improvement of present maintenance practices to eliminate the overapplication of highway salts where conditions do not warrant, improved regulation of spreading equipment to prevent distribution of salt material beyond the pavement break, and improved site selection and protection for stockpiles of chloride salts. Greene (1968) notes that improper calibration of salt spreaders is extremely common. This along with improper operation of equipment leads to excessive salt application rates, which not only increases over all costs but also contributes to the damage of vegetation and water supplies and the deterioration of concrete pavement and structures. In Ontario, it is claimed, $1 million per year is saved by improved application of highway salts (Greene, 1968). Over recent years, road-salt use in the state of Maine is reported to have been reduced some 30,000 tons annually due to improved practices.* Greene estimates operational savings of several million dollars per year are possible nationwide without reducing the quality of wintertime road maintenance (1968). Significant improvements in wintertime road maintenance practices would be derived from better field testing and control, good equipment with good maintenance schedules, greater use of mechanized equipment, frequent calibration, increased reliance and improvement of salt-metering instrumentation, education and awareness through the ranks particularly at the working level, concerted effort and increased training carried forth at the state highway department level, and due consideration to environmental protection.

Chloride Salt Reduction Possibilities

Various alternatives to chemical de-icing may become more prominent in the future, especially when communities realize that a price must

*F. E. Hutchinson, 1970. Personal communication.

be paid to alleviate the pollutional effects of wintertime salting. Some of these alternatives are as follows:

1. External and in-slab thermal melting systems;
2. Stationary (or pit) and mobile thermal "snow melters";
3. Substitute de-icing compounds;
4. Compressed-air or high-speed fluid streams in conjunction with snowplow blade or sweepers to loosen pavement bond and lift snow;
5. Snow-adhesion reducing substances in pavement;
6. Pavement substances that store and release solar energy for melting;
7. Electromagnetic energy to shatter ice;
8. Road and drainage design modifications to enhance runoff, reduce wintertime accidents, and capture snowmelt for treatment or control;
9. Salt retrieval or treatment possibly enhanced by the addition of chelating agents;
10. Improved tire or vehicular design to reduce de-icer requirements.

It is recognized that power, maintenance, and chemical costs for the above systems are high when compared to rock salt. However, municipalities such as Burlington, Massachusetts, have expressed a willingness to explore new methods regardless of cost.* Burlington has recently suspended roadway salting practices when a study indicated that their well-water chloride concentrations could exceed the recommended limit of 250 mg/l if salting was continued (Whitman and Howard, Inc., 1971; Huling and Hollocher, 1972; Public Health Service, 1962; U.S.D.I., 1968).

Salt Storage

Salt storage is needed for sustaining highway de-icing operations, but too frequently materials have been stockpiled unprotected in open areas or located on marginal lands adjacent to streams and rivers and have become chronic sources of ground and surface water pollution. To prevent water supply contamination, many communities have turned to covered salt piles, enclosed structures, and the diversion of salt-laden drainage (The Salt Institute, 1967 and 1968).

Fitzpatrick (1970) of the Ontario Department of Highways has

*William H. Burling, Jr., 1972. Personal communications.

1. The "Beehive"—salt storage structure with air vents at top. Courtesy of The Salt Institute, Alexandria, Virginia.

described a dome-like structure or "beehive," as shown in Figure 1, which is now being used to store large quantities of sand-salt mixtures in the Province. The beehive is unique in that it can store up to 5,000 cubic yards of sand-salt under one roof with a clear span free of posts, poles, or pillars. The structure has a 100-foot base diameter and is 50 feet high. Trucks, front-end loaders, and other equipment can easily move about the structure for loading and unloading, thus alleviating the pollution effects from spillage in open areas (Figure 2). Costs are reported around $5.00–$6.00 per ton of sand-salt mixture stored (approximately equal to $3.50 per square foot of floor area) (Fitzpatrick, 1970).

Environmental Effects of De-Icing Compounds

Runoff, Sewage, and Surface Streams

Street runoff from the melting of ice and snow, mixed with chloride salts, finds its way via combined and sanitary sewers to the local sewage treatment plant and then to the streams and also via storm sewers

2. The "Beehive"—method of loading; second stage by conveyor. Courtesy of The Salt Institute, Alexandria, Virginia.

to nearby receiving waters. Daily chloride loads were shown to be 40 to 50 percent higher for winter months as compared to summer months in municipal sewage at Milwaukee, Wisconsin* (Schraufnagel, 1965). During days of heavy snow melt, daily chloride loads were three times the normal summertime loads. Calculations (Hanes, et al., 1970) show that 600 pounds of salt, when applied to a one-mile section of roadway 20 feet wide containing 0.2 inches of ice, will produce an initial salt solution of 69,000 mg/l (at 10° F) to 200,000 mg/l (at 25° F). Street runoff samples collected from a downtown Chicago expressway in the winter of 1967 showed chloride contents from 11,000 to 25,000 mg/l (Sullivan, 1967; A.P.W.A., 1969). Table 2 illustrates some high chloride concentration values found in runoff.

At Milwaukee, on Janurary 16, 1969, extremely high chloride levels of 1,510 to 2,730 mg/l found in the Milwaukee, Menomonee, and Kinnickinnic rivers were believed directly attributable to de-icing salts entering these streams from highway snow melt (The Salt Institute, 1968). Table 3 contains the Milwaukee results. Meadow Brook in Syracuse, New York, contained chloride concentrations usually in the range of 200 to 1,000 mg/l, but frequently above a few thousand mg/l

*Members of the Milwaukee Sewerage Commission and Jones Island sewage treatment plant personnel, 1970. Personal communication.

Table 2
High Chloride Values in Runoff

Location	*Source*	*Date*	*Chlorides (mg/1)*	*Reference*
Chippewa Falls, Wisc.	Highway	1956-1957	10,250	Schraufnagel, 1965
Madison, Wisc.	Street	1956-1957	3,275	Schraufnagel, 1965
Lake Monona, Wisc.	Snow Pile	1956-1957	1,130	Schraufnagel, 1965
Chicago, Ill.	JFK Expressway	1966-1967	25,100	Schraufnagel, 1965
Des Moines, Iowa	Cummins Pkwy. Storm Drain	1958-1969	2,720	Henningson, et al., 1970

(Hawkins, 1971). For example, a sample in December showed about 11,000 mg/l chlorides in the Meadow Brook watershed (A.P.W.A., 1969). From the limited data available on streams, increasing chloride trends are evident for some large rivers in the U.S. (Hanes et al., 1970). Wintertime highway runoff eventually running into freshwater streams and natural or man-made lakes or ponding areas may have adverse effects upon water life in the future (Hanes, et. al., 1970; Fitzpatrick, 1970).

Table 3
Special River Sampling, Milwaukee Sewerage Commission, January 16, 1969[a]

Location	*Water Temperature (°C)*	*Chlorides (mg/1)*
Kinnickinnic River at Chase Avenue	10.0	2,005
Menomonee River at 13th St. and Muskego	10.5	200
Menomonee River at 70th and Honey Creek Pkwy.	5.0	2,730
Milwaukee River at Silver Spring Road	4.0	2,680
Milwaukee River at Port Washington Road	6.5	1,510

[a]Records received from the Milwaukee Sewerage Commission, May 1970.

The dumping of extremely large amounts of accumulated snow and ice from streets and highways either directly or indirectly into nearby water bodies could constitute a serious pollution problem. These deposits have been shown (Soderlund et al., 1970) to contain up to 10,000 mg/l sodium chloride, 100 mg/l oils, and 100 mg/l lead. The latter two constituents are attributable to automotive exhaust.

Farm Ponds and Lakes

Effects of highway salts upon farm ponds and lakes have been described by various investigators. A 1966-1968 survey of twenty-seven farm ponds along various highways in the state of Maine showed that road salts have strong seasonal influence on the chloride level of these waters and that salt concentrations are increasing yearly (Hanes et al., 1970; Sullivan, 1967; Hutchinson, 1967; Hutchinson, 1969). Density stratification of chlorides was observed in Beaver Dam Lake at Cumberland, Wisconsin (Schraufnagel, 1965) in First Sister Lake near Ann Arbor, Michigan (Judd, 1970), and in Irondequoit Bay at Rochester, New York (Dimend and Bubeck, 1971)—all attributed to salt runoff from nearby streets. It has also been estimated that highway salts contribute 11 percent of the total input of waste chlorides entering Lake Erie annually (Hanes, et al., 1970; Owmbey and Kee, 1967). Sodium from road salts entering streams and lakes may additionally serve to increase existing levels of one of the monovalent ions essential for optimum growth of blue-green algae, thereby stimulating nuisance algal blooms* (Sharp, 1970).

Recent investigations (Feick, et al., 1972) have brought attention to the hazardous potential of sodium and calcium ion exchange with mercury tied up in bottom muds. This could release highly toxic mercury to the overlying fresh waters. Other poisonous heavy metals may also be released in this manner.

De-icing Additives

Special additives present within most of the highway salts sold today may create pollutional problems even more severe than those caused by the chloride salts (Cargill, Inc., 1967; Smith, 1968). Ferric ferrocyanide and sodium ferrocyanide are commonly used to minimize the caking of salt stocks (Schraufnagel, 1965). The sodium form in particular is soluble in water and will generate cyanide in the presence of

*L. E. Keup, 1971. Personal communication.

sunlight (Hanes et al., 1970). Tests by the state of Wisconsin showed that 15.5 mg/l of the sodium salt can produce 3.8 mg/l cyanide after 30 minutes (Hanes et al., 1970; Schraufnagel, 1965). Maximum levels of cyanide allowed in public water supplies range from 0.2 to 0.1 mg/l (Public Health Service, 1962). Chromate and nutritious phosphate additives are used in de-icers as corrosion inhibitors (Hanes, 1970; Schraufnagel, 1965; Cargill, Inc., 1967). As with cyanide, chromium is a highly toxic ion, and limits permitted in drinking and other waters are in the same low range (Public Health Service, 1962). During the winter of 1965–66 in the Minneapolis–St. Paul area, snow melt samples showed maximum levels of 24 mg/l hexavalent chromium, and 3.9 mg/l total chromium (Cargill, Inc., 1967).

Ground and Surface Water Supply Contamination

Serious groundwater pollution has occurred in many locations because of the heavy application of salts onto highways and inadequate protection given to salt storage areas (Mass. Legislative Res. Council, 1965; Hanes et al., 1970; Whitman and Howard, Inc., 1971; Huling and Hollocher, 1972; Hutchinson, 1969; Walker, 1970; Rahn, 1968; Schraufnagel, 1967; Scheidt, 1967). By 1965 New Hampshire had replaced more than 200 roadside wells contaminated by road salts. Some of these wells contained more than 3,500 mg/l chlorides (Hanes, 1970; Schraufnagel, 1967). In Manistee County, Michigan, a roadside well located 300 feet from a highway department salt storage pile was found to contain 4,400 mg/l chlorides (Hanes, 1970; Deutsch, 1963). Tastes and odors in domestic water supplies in Connecticut have been traced to chlorides and sodium ferrocyanide originating from salt storage areas (Scheidt, 1967). Within Massachusetts, salt increases have been noted in the water supplies of some sixty-three communities, and various supplies have been abandoned at least in part because of road salting and salt storage piles (Whitman and Howard, Inc., 1971; Huling and Hollocher, 1972; *Boston Globe*, May 1970; *American City*, August 1965; *Reclamation News*, April 1965; *Boston Globe*, August 1971).

As previously cited, Burlington, Massachusetts, conducted a study indicating that the chloride content of area wells was becoming increasingly high (Figure 3). If this chloride concentration were to increase at the rate shown, water supplied by these Burlington wells could soon exceed the upper limit of 250 mg/l chloride established by the U.S. Public Health Service (1962), a condition that could force closing of wells. It should be emphasized that a desirable concentra-

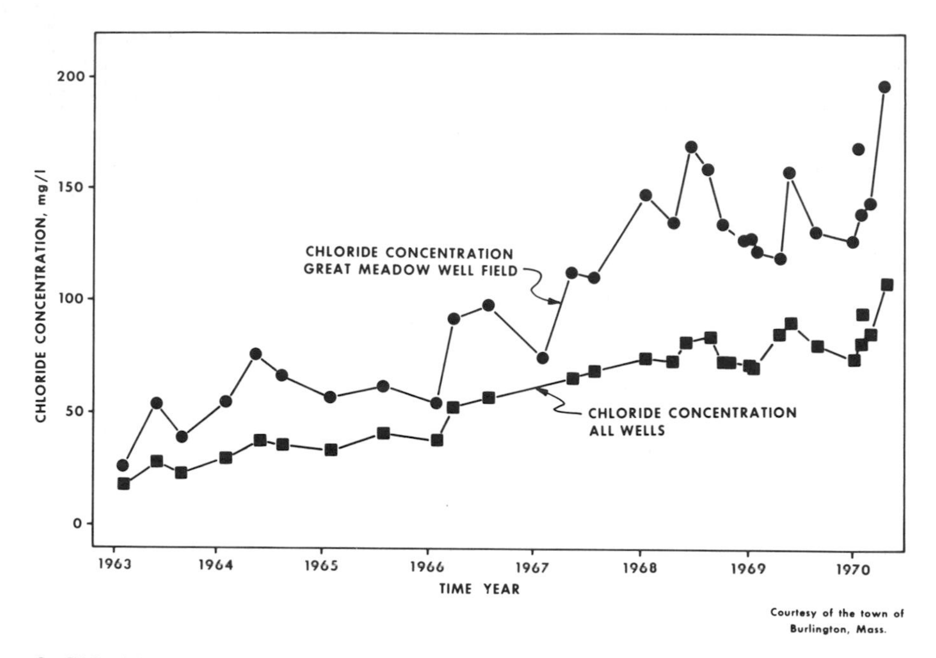

3. Chloride concentration of wells in the Burlington, Massachusetts, area, 1963-1970.

tion of chlorides is considered to be 25 mg/l or less (U.S.D.I., 1968) and that the American Heart Association recommends that water containing more than 20–22 mg/l of sodium (59 mg/l sodium chloride) not be used when patients are on diets with an intake of sodium restricted to less than 1,000 mg/day (Amer. Heart Assoc., 1957; Hanes, 1970; Huling and Hollocher, 1972). The normal adult intake of sodium is about 4,000 mg/day.

Other specific cases of water supply contamination in Massachusetts merit attention. In 1951 the town of Becket found that the chloride content of water in a well located downhill from a salt storage pile had increased dramatically to about 1,360 mg/l (*Boston Globe*, April 1970). Private and public water supplies in the Weymouth, Braintree, Randolph, Holbrook, Auburn, and Springfield areas were among those believed to be affected by highway salts. Large salt storage piles located at Routes 128 and 28 in Randolph and alongside the Blue Hill River were suspected of introducing contamination into Great Pond, which serves as water supply for Braintree, Randolph, and Holbrook. Water supplies in Tyngsboro and Charlton were similarly experiencing salt increases, and two wells in Charlton were likely to be abandoned. Also, in the general Boston area, snow removal and disposal practices reportedly contributed to heavy salt content in the Mystic Lakes (*Globe*, May 1970).

On the subject of industrial water use, Schraufnagel (1965, 1967) found that chlorides have been responsible for the corrosion of various metals including stainless steels. Schraufnagel also cited an industry statement that "increasing the salinity average above the then 40 to 50 mg/l, or lengthening the periods of high salinity, would increase corrosion of all metals used in the handling system." McKee and Wolf in 1963, in their extensive review of the literature, summarized chloride tolerances for various industries as follows: food canning and freezing, 760 mg/l; carbonated beverages, food equipment washing, and paper manufacturing, 200 to 250 mg/l; steel manufacturing, 175 mg/l; textiles, brewing and paper manufacturing (soda and sulfate pulp), 60 to 100 mg/l; and dairy processing, photography, and sugar production, 20 to 30 mg/l (Hanes, 1970; McKee and Wolf, 1963). Other reviews on water quality needs for industry including chloride limits are also available (Moore, 1950; American Society for Testing and Materials, 1953).

Vehicular and Roadway Damage

Road salts not only promote vehicular corrosion but may also affect structural steel, house sidings, and other property (Mass. Legislative Res. Council, 1965; Schraufnagel, 1967; Lockwood, 1965; Kallen, 1956; *Milwaukee Journal*, May 1970; APWA Reporter, October 1970; *Automotive Industries*, February 1964; *Steel*, March 1964; U.S. Highway Research, 1962; Dickinson, 1961; Portland Cement Association, 1958; Hamman and Mantes, 1966). It has been estimated that the private car owner pays about $100 per year for corrosion (*Milwaukee Journal*, May 1970). De-icers may cause appreciable damage to highway structures and pavements, particularly those constructed of Portland cement. Though air-entrained concrete is reportedly superior to non-air-entrained concrete in its resistance to salts, neither form should be exposed to road salts for at least one year after being poured (Mass. Legislative Res. Council, 1965; U.S. Highway Research, 1962; Dickinson, 1961). Detrimental effects from de-icing salts have been reported (Schraufnagel, 1967; Hamman and Mantes, 1966) on various underground utilities, such as cables and water mains.

Soil, Vegetation, and Trees

Widespread damage of roadside soils, vegetation, and trees has been observed where there has been liberal application of road salts (e.g., Hanes, 1970; Hutchinson, 1969; Roberts and Zybura, 1967; Prior, 1968; Hutchinson and Olson, 1967; Roberts, 1968; Verghese et al., 1969; Rich, 1968; Westing, 1966; Holmes and Baker, 1966; Zelazny et al., 1970; Sauer, 1967). Most of the studies dealing with plant injury and death have focused on the sugar maple decline that has occurred over a sixteen-state area, mostly in New England. Figures 4 through 7 taken from a study conducted by the Connecticut State Highway Department (Button, 1964 and 1965) show the progressive deterioration of sugar maples and associated leaf damage. Figure 4, taken in 1960, depicts the condition of sugar maples on two sides of a road where longitudinal drainage is from right to left. Figure 5 shows the condition of these maples five years later. Leaf-margin burn, limb die-back, and varying degrees of defoliation are pronounced on the trees on the left, which received the impact from salt-laden drainage. These effects are more conspicuously illustrated in Figures 6 and 7. The tree in Figure 6 on the left side of the roadway shows severe stages of deterioration, and the leaves in Figure 7 demonstrate severe leaf-margin burn. It is important to realize here that highway maintenance departments relied principally on abrasives as opposed to salt until the 1960's (Storm and Combined Sewer Technology Branch, 1971).

Tables 4 through 8 give relative salt tolerances of various fruit crops, vegetable crops, field crops, grasses, forage legumes, trees, and ornamentals. For more complete lists, see the sources noted. It is hoped that this information may be used by highway authorities and others in selecting roadside plants and vegetation.

Conclusions

Some important conclusions drawn from this study are as follows:

1. De-icing salts are found in high concentrations in highway runoff. These salts have caused certain injury and damage across a wide environmental spectrum, and their potential dangers are severe.

2. Practically all highway authorities in the U.S. believe that "bare pavement" conditions are necessary, and this attitude often results in excessive salt application.

4. Relatively healthy sugar maples, photographed 1960, Route 17, Durham-Middletown line, Connecticut.

5. Same trees in 1965; those on left exhibit salt damage.

6. Closeup of sugar maple with pronounced salt damage, Route 17, Durham-Middletown line, Connecticut.

7. Healthy (right) vs. damaged sugar maple leaves, from right and left sides of Route 17 respectively, Durham-Middletown line, Connecticut.

Table 4
Salt Tolerance of Fruit Crops[a]

Tolerant	*Moderately Tolerant*	*Sensitive*
Date palm	Pomegranate	Pear
	Fig	Apple
	Olive	Orange
	Grape	Grapefruit
	Cantaloup	Prune

[a]Bernstein, 1965.

Table 5
Salt Tolerance of Vegetable Crops[a]

Tolerant	*Moderately Tolerant*	*Sensitive*
Garden beet	Tomato	Radish
Kale	Broccoli	Celery
Asparagus	Cabbage	Green bean
Spinach	Cauliflower	
	Lettuce	
	Sweet corn	
	Potato	

[a]Bernstein, 1959.

Table 6
Salt Tolerance of Field Crops[a]

Tolerant	*Moderately Tolerant*	*Sensitive*
Barley	Rye	Field bean
Sugar beet	Wheat	
Rape	Oats	
Cotton	Sorghum	
	Sorgo (sugar)	
	Soybean	

[a]Bernstein, 1960.

Table 7
Salt Tolerance of Grasses and Forage Legumes[a]

Tolerant	*Moderately Tolerant*	*Sensitive*
Alkali sacaton	White sweet clover	White dutch clover
Salt grass	Yellow sweet clover	Meadow foxtail
Nuttall alkali grass	Perennial ryegrass	Alsike clover
Bermuda grass	Mountain brome	Red clover
Tall wheatgrass	Harding grass	Ladino clover
Rhodes grass	Beardless wildrye	Burnet
Rescue grass	Oats	
Canada wildrye	Sudan grass	
Western wheatgrass	Reed canary	
Tall fescue	Sourclover	
Barley	Alfalfa	
Birdsfoot trefoil	Wheat	
	Meadow fescue	
	Smooth brome	

[a]Bernstein, 1958.

3. Salt storage sites are persistent and frequent sources of ground and surface water contamination and of vegetation damage.
4. The special additives in road de-icers have latent toxic properties and are therefore particularly dangerous.
5. De-icing salts have raised the chloride content of water supplies and receiving waters to dangerous levels.
6. De-icing salts are a major factor in vehicular corrosion and roadway damage. Rust-inhibiting additives are not effective.
7. Road de-icers can disturb soils, trees, and other vegetation comprising the roadside environment.

Recommendations

EPA Proposals and Projects

Abt Associates, Inc. (1972) has recently completed research on new technology in snow and ice control. The firm investigated the ice-releasing agents used on the exteriors of aircraft, vessels, and outdoor mechanical equipment and recommended the development of a similar hydrophobic or icephobic (water or ice repellant) substance

Table 8
Salt Tolerance of Trees and Ornamentals[a]

Tolerant	*Moderately Tolerant*	*Poorly Tolerant*
Common matrimony vine	Silver buffalo berry	Black walnut
Oleander	Arbor vitae	Little leaf linden
Bottlebrush	Spreading Juniper	Barberry
White acacia	Lantona	Winged euonymus
English oak	Golden willow	Multiflora rose
Silver poplar	Ponderosa pine	Spiraea
Gray poplar	Green ash	Artic blue willow
Black locust	Eastern red cedar	Viburnum
Honey locust	Japanese honeysuckle	Pineapple guava
Osier willow	Boxelder maple	Rose
White poplar	Siberian crab	European hornbeam
Scotch elm	European black currant	European beech
Russian olive	Pyracantha	Italian popular
Squaw bush	Pittosporum	Black alder
Tamarix	Xylosma	Larch
Hawthorne	Texas privet	Sycamore maple
Red oak	Blue spruce	Speckled alder
White oak	Douglas fir	Lombardy poplar
Apricot	Balsam fir	Red maple
Mulberry	White spruce	Sugar maple
	Beech	Compact boxwood
	Cottonwood	Filbert
	Aspen	
	Birch	

[a]Zolazny, L., 1968.

for ice control on pavements. The work of this firm has stimulated two EPA proposals for further research.

A larger project (see Arthur D. Little, Inc., 1972) will provide two manuals: (1) *A de-icer users' manual* to describe snow and ice removal practices and the best systems of applying de-icing chemicals to streets and highways. The manual will describe the absolute minimum amounts of de-icing chemicals for maintaining safe traffic flows, critical placement points for salt applications, improved calibration, etc. (2) *A manual of design and recommended practices for storage facilities and methods of handling de-icing materials throughout storage* will describe proper siting of facilities, adequate covering of storage sites to protect materials and prevent surface drainage, adequate foundation and footing, physical, mechanical, and chemical techniques for preventing salt caking, etc. Instructional materials currently available from highway agencies and the salt industry have in general not given adequate emphasis to the pollution problems associated with materials storage.

Additional Needs

The following additional steps are recommended:

1. Base-line data should be obtained on long-term environmental changes that may be taking place because of de-icing chemicals. Data are especially needed on de-icing chemicals in surface water, groundwater, selected soils, and vegetation, and on the prevailing deterioration levels of salt-affected vehicular traffic, highway pavements and structures, and underground utilities.
2. Findings of past studies on the vehicular corrosion and deterioration of highway pavements, structures, and utilities caused by road de-icers should be made readily available for further use.
3. Governmental authorities should give consideration in roadway design to reducing de-icing requirements and enhancing the control, collection, and treatment of ensuing salt runoff.
4. Detailed studies should be undertaken both in the laboratory and field on the various toxic and nutrient additives mixed with de-icing materials so as to determine their potential hazards and safe levels of use.
5. Various suppliers and highway authorities should make available full information on marine salts: their current and future expected use in highway de-icing, chemical composition, physical properties including melting efficiencies, and comparison with the common chloride salts.
6. The merits and demerits of various substitutes for the common chloride salts should be evaluated. Those de-icers with high efficiency and minimum side effects should be identified.
7. The actual contribution of de-icing salts to the safety of winter driving for which they were intended should be scientifically determined.
8. Information should be compiled, disseminated, and in some cases developed on the best selection of roadside plantings and the various remedial measures for restoring roadside soils and vegetation damaged by de-icing chemicals.

References

Abt Associates, Inc. 1972. *Highway snow and ice control: a search for innovative technological alternatives.* EPA contract no. 681-01-0706. Washington, D.C.: Environmental Protection Agency.

Allison, L. E. 1964. Salinity in relation to irrigation. *Advances in agronomy* 16: 139-180.

American city. 1965. Side effects of salting for ice control. *American city* 80: 33.

American Heart Association. 1957. *Your 1000-milligram sodium diet.* New York: American Heart Association.

American Public Works Association. 1969. *Water pollution aspects of urban runoff.* Report for EPA no. 11030 DNS. Washington, D.C.: Environmental Protection Agency.

American Public Works Association. 1970. Vehicle corrosion caused by deicing salts. *APWA Reporter,* Special Report no. 34, September, 1970.

American Society for Testing and Materials. 1953. *Manual on industrial water.* Special technical publication no. 148. Philadelphia: ASTM Committee D-19 on Industrial Water.

Arthur D. Little, Inc. 1972. *Study of the environmental impact of highway deicing.* EPA contract no. 68-03-0154. Washington, D.C.: Environmental Protection Agency.

Bernstein, L. 1958. *Salt tolerance of grasses and forage legumes.* Washington, D.C.: Department of Agriculture bulletin no. 194.

Bernstein, L. 1959. *Salt tolerance of vegetable crops in the West.* Washington, D.C.: Department of Agriculture bulletin no. 205.

Bernstein, L. 1960. *Salt tolerance of field crops.* Washington, D.C.: Department of Agriculture bulletin no. 217.

Bernstein, L. 1965. *Salt tolerance of fruit crops.* Washington, D.C.: Department of Agriculture bulletin no. 292.

Boston Globe. 1969. Road salt blamed for souring water. *Boston Globe,* April 24, 1969, p. 60.

Boston Globe. 1970. Alewife brook polluted. *Boston Globe,* May 3, 1970, p. 15.

Boston Globe. 1970. Salt buildup in drinking water a danger to some Bay Staters. *Boston Globe,* May 8, 1970, p. 3.

Button, E. F. 1964. *Influence of rock salt used for highway ice control on natural sugar maples at one location in central Connecticut.* Hartford: Connecticut State Highway Department report no. 3-A.

Button, E. F. 1965. Ice control chlorides and tree damage. *Public Works* 93: 136-137.

Cargill, Inc. 1966. *Toxicity and pollution study of carguard chemicals, 1965-1966.* Minneapolis: Cargill, Inc.

Commonwealth of Massachusetts. 1965. *Legislative Research Council report relative to the use and effects of highway deicing salts.* Boston: State Legislature.

Deutsch, M. 1963. *Ground water contamination and legal controls in Michigan.* Water supply paper no. 1691. Washington, D.C.: U.S. Geological Survey.

Dickinson, W. E. 1961. *Problems of ice removal from pavements.* Publication no. 98. Washington, D.C.: National Ready Mixed Concrete Association.

Diment, W. H., and Bubeck, R. C. 1971. Runoff of deicing salt: effect on Irondequoit Bay, Rochester, New York. Paper presented at the Street Salting Urban Water Quality Workshop, SUNY Water Resources

Center, Syracuse University, Syracuse, New York, May 6, 1971.
Fairbridge, R. W., et al. 1966. *The Encyclopedia of oceanography*. Vol. 1 in *Encyclopedia of earth sciences*. New York: Reinhold Publishing Company.
Feick, G., et al. 1972. Release of mercury from contaminated freshwater sediments by the runoff of road deicing salt. *Science* 175: 1142–43.
Fitzpatrick, J. R. 1970. Beehives protect snow-removal salt and prevent water pollution. *American city* 85: 81–84.
Geschelin, Joseph. 1964. Calcium chloride versus sodium chloride. *Automotive industries* 130: 47.
Greene, W. C. 1968. "What are the problems?" Paper presented in Bridgeport, Connecticut, at the University of Connecticut Symposium on Pollutants in the Roadside Environment, February 29, 1968.
Hamman, W., and Mantes, A. J. 1966. Corrosive effects of deicing salts. *Journal of American Water Works Association* 58: 1457–61.
Hanes, R. E., et al. 1970. *Effects of deicing salts on water quality and biota-literature review and recommended research*. National Cooperative Highway Research Program report no. 91. Blacksburg: Virginia Polytechnic Institute and Highway Research Board.
Harvey, H. W. 1963. *The chemistry and fertility of sea waters*. London: Cambridge University Press.
Hawkins, R. H. 1971. Street salting and water quality in Meadow Brook, Syracuse, New York. Paper presented at the Street Salting Urban Water Quality Workshop, SUNY Water Resources Center, Syracuse University, Syracuse, New York, May 6, 1971.
Henningson, Durham and Richardson, Inc. 1970. *Rainfall–runoff and combined sewer overflow*. Final draft report, contract no. 14–12–402. Washington, D.C.: EPA.
Holmes, F. W., and Baker, J. H. 1966. Salt injury to trees, II. Sodium and chloride in roadside sugar maples in Massachusetts. *Phytopathology* 56: 633–36.
Huling, E. E., and Hollocher, T. C. 1972. Groundwater contamination by road salt: steady-state concentrations in east central Massachusetts. *Science* 176: 288–290.
Hutchinson, F. E. 1967. *The influence of salts applied to highways on the levels of sodium and chloride ions present in water and soil samples*. Project no. R1086-8, Progress reports nos. I and II. Washington, D.C.: Department of the Interior, Office of Water Resources Research.
Hutchinson, F. E. 1969. Effects of highway salting on the concentration of sodium and chloride in private water supplies. In *Research in the life sciences* 17: 15–19.
Hutchinson, F. E. 1969. *The influence of salts applied to highways on the levels of sodium and chloride ions present in water and soil samples*. Project no. A-007-ME. Washington, D.C.: Department of the Interior, Office of Water Resources Research.
Hutchinson, F. E., and Olson, B. E. 1967. *The relationship of road salt applications to sodium and chloride ion levels in the soil bordering major highways*. HRB report no. 193. Washington, D.C.: Highway Research Board.
Judd, J. H. 1970. Effect of salt runoff from street deicing on a small lake. Ph.D. dissertation. University of Wisconsin, Madison.
Kallen, H. P. 1956. Corrosion. *Power* 100: 73–108.
Lockwood, R. K. 1965. *Snow removal and ice control in urban areas*. Research project no. 114. Chicago: American Public Works Association.

McKee, J. E., and Wolf, H. W. 1963. *Water quality criteria.* Sacramento, California: State Water Quality Control Board.

Milwaukee Journal, "What is highway salt doing to us?" *Milwaukee Journal*, May 4, 1970, p. 1.

Monk, R. W., and Peterson, H. B. 1962. Tolerance of some trees and shrubs to saline conditions. *Proceedings of American Society of Horticultural Science* 81: 556–61.

National Technical Advisory Committee to the Secretary of the Interior. 1968. *Water quality criteria.* Washington, D.C.: Department of the Interior.

Ownbey, C. R., and Kee, D. A. 1967. Chlorides in Lake Erie. Paper read at Tenth Conference of Great Lakes Research, Toronto, Canada, April, 1967.

Portland Cement Association. 1945. *The elimination of pavement sealing by use of air entraining Portland cement.* Highway bulletin 10.0. Chicago: Portland Cement Association.

Portland Cement Association. 1958. *Protection of existing concrete pavements from salt and calcium chloride.* Highway bulletin 1-2. Chicago: Portland Cement Association.

Prior, G. A. 1968. Salt migration in soil. Paper presented at the University of Connecticut Symposium on Pollutants in the Roadside Environment, Bridgeport, February 29, 1968.

Rahn, P. H. 1968. Movement of dissolved salts in groundwater systems. Paper presented at the University of Connecticut Symposium on Pollutants in the Roadside Environment, Bridgeport, February 29, 1968.

Rich, A. E. 1968. Effect of deicing chemicals on woody plants. Paper presented at the University of Connecticut Symposium on Pollutants in the Roadside Environment, Bridgeport, February 29, 1968.

Riley, J. P., and Skirrow, G., eds. 1965. *Chemical oceanography*, vol. 1. New York: Academic Press.

Roberts, E. C., and Zybura, E. L. 1967. *Effect of sodium chloride on grasses for roadside use.* HRB report no. 193. Washington, D.C.: Highway Research Board.

Rudolfs, W. 1919. Influence of sodium chloride upon the physiological changes of living trees. *Soil science* 8: 397–425.

Salt Institute. undated. *Survey of salt, calcium chloride and abrasive use for street and highway deicing in the United States and in Canada for 1966–1967.* Alexandria, Virginia: The Salt Institute.

Salt Institute. 1967. *Storing road deicing salt.* Alexandria: The Salt Institute.

Salt Institute. 1968. *The snowfighter's salt storage handbook.* Alexandria: The Salt Institute.

Salt Institute. undated. *Survey of salt, calcium chloride and abrasive use for street and highway deicing in the United States and Canada for 1969–1970.* Alexandria: The Salt Institute.

Scheidt, M. E. 1967. Environmental effects of highways. *Journal of the Sanitary Engineering Division, proceedings of the American Society of Civil Engineers* 93, no. SA5, paper no. 5509: 17–25.

Schraufnagel, F. M. 1965. *Chlorides.* Madison, Wisconsin: Commission on Water Pollution.

Schraufnagel, F. M. 1967. *Pollution aspects associated with chemical deicing.* HRB report no. 193. Washington, D.C.: Highway Research Board.

Sharp, R. W. 1970. *Road salt as a polluting element.* Special environmental release no. 3. Washington, D.C.: Department of the Interior, Bureau of Sport Fisheries and Wildlife.

Smith, H. A. 1968. Progress report on NCHRP project 16-1; effects of deicing compounds on vegetation and water supplies. Paper presented at the 54th meeting of the American Association of State Highway Officials, Minneapolis, Minnesota, December 5, 1968.

Soderlund, G., et al. 1970. Physicochemical and microbiological properties of urban stormwater runoff. Paper presented at the Fifth International Water Pollution Research Conference, San Francisco, California, July 29-August 3, 1970.

Storm and Combined Sewer Technology Branch, Edison Water Quality Research Lab. 1971. *Environmental impact of highway deicing*. EPA report no. 11040 GKK 06/71.

Sullivan, R. H. 1967. Effects on winter storm runoff of vegetation and as a factor in stream pollution. Paper presented at the Seventh Annual Snow Conference, Milwaukee, Wisconsin, April 12, 1967.

U.S. Highway Research Board. 1962. *Effects of deicing chemicals on structures: a symposium*. Publication no. 100. Washington, D.C.: Highway Research Board.

U.S. Highway Research Board. 1967. *Highway research record report no. 193 on environmental considerations in the use of deicing chemicals*. Washington, D.C.: Highway Research Board.

U.S. Public Health Service. 1962. *Public Health Service drinking water standards—1962*. PHS publication no. 956.

Verghese, K. G., et al. 1969. *Sodium chloride uptake and distribution in grasses as influenced by fertility interaction and complementary anion competition*. Unpublished report. Blacksburg: Virginia Polytechnic Institute Research Division.

Walker, W. H. 1970. Salt piling—a source of water supply pollution. *Pollution engineering* 2: 30-33.

Westing, A. H. 1966. Sugar maple decline: an evaluation. *Economic botany* 20: 196-212.

Whitman and Howard, Inc. 1971. *Salt contamination of existing well supplies*. Report for town of Burlington, Massachusetts. Boston: Whitman and Howard, Inc.

Zelazny, L. W. 1968. Salt tolerance of roadside vegetation. Paper presented at the University of Connecticut Symposium on Pollutants in the Roadside Environment, Bridgeport, February 29, 1968.

Zelazny, L. W., et al. 1970. Effects of deicing salts on roadside soils and vegetation, II. Effects on silver maples (Acre saccharinum L). Unpublished report. Blacksburg: Virginia Polytechnic Institute, Research Division.

23. Livestock Grazing—A Non-Point Source of Water Pollution in Rural Areas?

George B. Coltharp and Leslie A. Darling

Introduction

Millions of acres of "wildland watersheds"* throughout the U.S. are strategically located in terms of their influence on water production. Society should be concerned not only with the *quantity* of water produced from wildlands but also with the *quality*. Many of these wildland watersheds were formerly closed to intensive uses, but more recently have come under the "multiple use" concept of land management and are being increasingly used for recreation, logging, grazing, etc. Certain types of land management activities have been shown to affect dramatically the quantity of water produced from a given watershed (Boughton, 1970; Ward, 1971). These same management or land use activities have also been shown to affect the quality of water produced. In most cases the quality parameters studied have been confined to physical values, such as sediment production and temperature regimes (Brown, 1972). Those aspects of wildland water quality more closely allied to human health, e.g. the bacteriological and chemical, have largely been ignored.

It has been estimated that grazed areas (pasture, grassland, and range) constitute over one third of the land area of the conterminous 48 states, and forested areas, which are grazed in many instances, constitute an additional one third of this land area (Wooten, 1959). Because these grazed wildlands make up such a large part of our natural headwater areas, it is important that information relative to grazing and water quality be made available.

The recent concern for a quality environment has prompted many individuals and organizations to make rash judgments on natural resource uses. There have been suggestions made that livestock be prohibited from grazing on important water-yielding areas. Others suggest that, regardless of the magnitude of water-quality deterioration due to livestock grazing, most municipalities routinely treat their water supplies; hence there should be no real concern. Such suggestions,

*Wildlands refer, in the context of this paper, to natural forest and rangelands not subjected to farming, urbanization, or industrialization.

although they may have merit, are based largely on lack of information or on emotion. There are, however, some well justified concerns pertaining to livestock grazing and rural water quality. There are many rural uses of untreated stream water, such as culinary, bathing, and dairy barn uses. If livestock grazing does, in fact, adversely affect wildland or rural water quality, people using these untreated water supplies may be subject to various water-borne infections. Likewise, the millions of recreationists who unsuspectingly drink, prepare food, and bathe in wildland streams could also contract infectious diseases.

Background

What is known at this point about the effects of domestic livestock grazing on wildland water quality? The problem of animal waste disposal has been of increasing interest during the last decade, particularly in reference to confined concentrations of domestic animals. Numerous conferences, symposia, and publications attest to the concern for this area of environmental awareness (Loehr, 1969; Webber, 1971; Lin, 1972). It has been well documented that animal wastes contribute significantly to water pollution (Loehr, 1968). Many studies of runoff from livestock feedlots and intensive agricultural areas have shown that domestic animal wastes contain pathogens that can affect human health, and that these pathogens can be transmitted to man via water (Diesch, 1970). Most of the literature dealing with livestock waste disposal and livestock contribution to water quality are related to feedlot, barnyard, and closely confined pasture situations. Very little information is available on the contribution of livestock grazing to wildland water quality. In fact, it has been noted that little interest has been shown by the water pollution control agencies in the pollutional effects of animals grazing vegetated land areas (range or pasture) (Miner and Willrich, 1970).

It is important to know the background quality of undisturbed or pristine streams in order to evaluate any imposed use effects on these streams. Several studies of undisturbed streams in Montana, Colorado, and Utah have revealed appreciable background levels of bacteria in them, attributable to indigenous wildlife populations (Bates, 1963; Walter and Bottman, 1967; Bissonette et al., 1970; Goodrich et al., 1970; Stuart et al., 1971). In fact, it appears that watersheds fully protected from all human activities frequently act as wildlife sanctuaries and heavy wildlife populations pollute the streams draining from these areas. As a result of this observation, Goodrich et al. (1970)

concluded that watersheds untouched by man may produce water with high coliform populations, and by trying to meet rigid water quality standards, man may be attempting to keep water cleaner than it naturally occurs.

The effects of wildland grazing or unconfined grazing on wildland water quality is less well known. A study conducted by Meiman and Kunkle (1967) in Colorado is the only direct reference to the effects of wildland grazing and water quality that could be found in the literature. This study indicated livestock do affect bacteriological counts of streams draining grazed areas. The authors suggest that bacterial counts are better indicators of land-use impacts on water quality than suspended sediment or turbidity. Another study in Colorado which investigated pastureland adjacent to a stream also indicated a contaminating influence of grazing animals (Morrison and Fair, 1966). Other investigators have alluded to the effects of livestock grazing on wildland water quality (Peterson and Boring, 1960; Teller, 1963; Lee, Symons, and Robeck, 1970).

Kunkle (1970) investigated the effects of various land-use practices on water quality in northern Vermont. He found that grazing, in a pasture situation, did significantly increase bacteriological-indicator counts if the livestock were allowed direct access to the stream and/or the immediate bank area.

A common problem of evaluating land-use effects on water quality is that of distinguishing which uses contribute to the deterioration of the water quality. Because many of the grazed wildland watersheds are subject to outdoor recreation pressure also, differentiating bacterial contamination in stream-flow in terms of either man or animal origin is a problem. Geldreich and Kenner (1969) suggest that the ratio of fecal coliform count to fecal streptococci count indicates the origin of fecal contamination because of the relatively larger numbers of fecal strep produced by animals. They suggest that a ratio of 4:4 or higher indicates human origin, and a ratio of 0:7 or less would be animal origin. Values falling between these limits would be less indicative of the source. A qualification is that the water samples should be obtained within twenty-four hours of fecal input to the stream or the more rapid dieoff rate of the fecal strep will invalidate the ratio. Another possible indicator of the origin of fecal contamination is the presence of *Streptococcus bovis* and *S. equinus*, which are significant in fecal material from cattle, horses, and other forms of livestock but notably absent in human feces (Goldreich and Kenner, 1969).

It seems apparent from the preceding discussion that there is a critical need for studies to define the effects of various land uses on wildland water quality. Because of the areal extent and monetary importance of livestock grazing in the U.S., investigation of the effects of this land use on water quality should receive high priority.

A Study

In order to ascertain the relative contribution of wildland grazing to rural water quality, selected water quality parameters were investigated on three small watersheds within the Logan River drainage of northern Utah during 1971 and 1972. The study was designed to

1. Determine differences in the quality of water draining from grazed and ungrazed watersheds;
2. Determine changes in water quality as streams flow through grazed areas;
3. Determine differences in water quality due to the type of livestock grazing an area.

Methods

Preliminary water quality sampling was initiated in June of 1971 on streams draining both grazed and ungrazed watersheds in northern Utah. This survey of the general area indicated that grazing activity did affect the counts of traditional bacteriological indicators, but the magnitude and significance of these effects were confounded by the type of livestock grazed, indigenous wildlife populations, and a host of other factors difficult to control (Coltharp and Darling, 1973). Beginning with the winter of 1972, this study was delimited to three small watersheds of fairly close proximity in an effort to control as many of these confounding factors as possible.

General Area

The criteria employed in selecting the three watersheds studied were existence of one type of livestock grazing or activity over the entire watershed area, drainage by a perennial stream, and accessibility of the stream for sampling. The three watersheds selected were Twin Creek, Little Bear Creek, and Woodcamp Hollow Creek; they range in area from 1.7 to 5.8 square miles and are all located within the Logan River drainage (Figure 1).

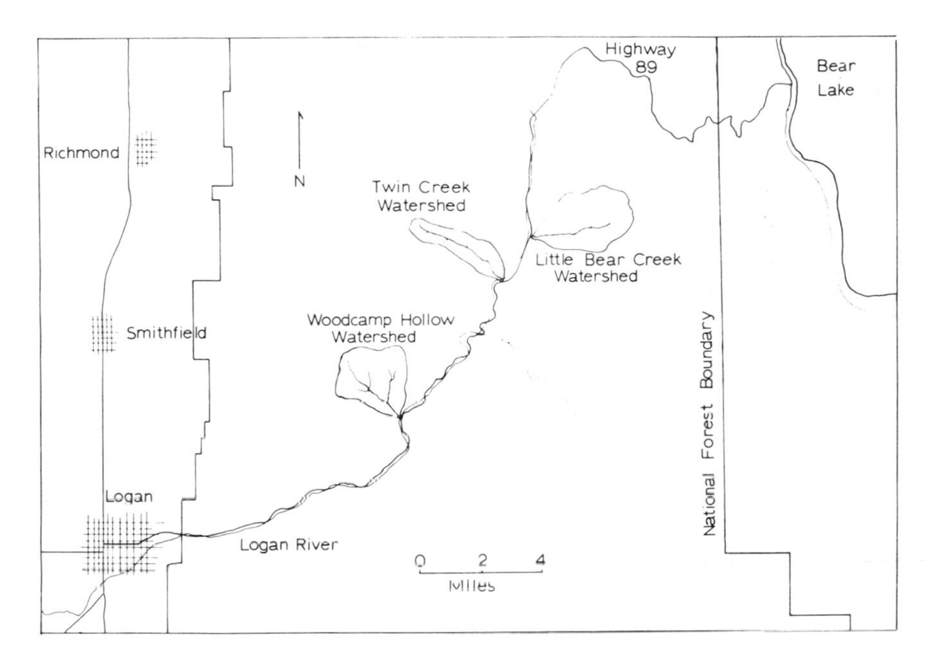

1. Map of Logan Canyon study area.

Logan River drains a large area of the Bear River Range, which is composed of dolomitic limestone residuals of the Laramide Orogeny. This mountain range extends north and south, crossing the border between northern Utah and southern Idaho. The general vegetation cover on all three watersheds includes extensive areas of quaking aspen (*Populus tremuloides*), Englemen spruce (*Picea engelmannii*), Douglas fir (*Pseudotsuga menziesii*), subalpine fir (*Abies lasiocarpa*) on the higher elevations, while the lower elevations are covered with sagebrush (*Artemisia tridentata*), Rocky Mountain juniper (*Juniperus scopulorum*), miscellaneous grasses (*Bromus spp.*, *Agropyron spp.*, *Festuca spp.*, etc.) and shrubs (Figure 2).

Watersheds

The Twin Creek watershed lies entirely within the Logan Canyon Cattle Allotment and was subjected to cattle grazing from August 11 to October 6, 1972, by an undetermined number of the 1,506 head of cattle that are permitted on the allotment. (The Logan Canyon Cattle Allotment comprises an area of 39 square miles; the Little Bear

2. General view of study area.

Sheep Allotment comprises approximately 12 square miles.) The live stream length is approximately 2.5 miles and drains land from elevations of 8,500 feet at the watershed headwater divide to 6,000 feet where it joins the Logan River (Figure 3a). Twin Creek was sampled at five locations (Stations 1–5), with three sites at springs, the fourth at a point below heavy grazing activity, and the fifth at the watershed outlet.

The Little Bear Creek watershed is located on the east side of Logan River approximately two miles upstream from Twin Creek. Elevations range from 8,700 feet at the upper divide to 6,200 feet at the mouth. The watershed is within the Little Bear Sheep Allotment, which permits grazing of approximately 1,057 sheep for a period of 75 consecutive days. From July 27 to August 4, 1972, the sheep grazed the portion of the watershed through which the live stream flows. Along the perennial portion of Little Bear Creek, three fourths of a mile long, three sampling stations were established: Station 6 was located at the headwater springs; Station 7 was approximately midway between the springs and the watershed outlet; and Station 8 was at the outlet where the Little Bear joins Logan River (Figure 3b).

Woodcamp Hollow watershed was included in the study because it was representative of those watersheds in Logan Canyon that were not grazed during the 1972 sampling season. In fact, this area had not been grazed for several years prior to the study. The watershed is located approximately six miles south of Twin Creek on the west side of Logan Canyon. Elevations range from a little over 9,000 feet down to approximately 5,500 feet at the outlet. Three sampling stations were established along this stream (Stations 9, 10, and 11), but because of a lack of perennial streamflow at Stations 9 and 10, only Station 11 provided a consistent record for the entire season (Figure 3c).

Techniques

Water samples were collected at least twice a month from December 1971 through April 1972 on Twin Creek and approximately twice a week during the intensive sampling period of May through November 1972 for all three watersheds. The water quality parameters measured were temperature, turbidity, pH, nitrates, phosphates, total coliform, fecal coliform, and fecal streptococci. Bacteriological analyses were conducted by means of the membrane filter technique as described in *Standard Methods* (APHA, 1972), while chemical and physical values were obtained in the laboratory using a Hach DR-EL Engineers Laboratory (Hach Kit).* Streamflow was measured on Twin Creek by means of a Parshall flume equipped with a waterstage recorder.

Results and Discussion

Bacteriological

Bacterial counts in Twin Creek reflected both a seasonal variation and a response to cattle grazing. Figure 4 demonstrates this response for total coliform counts for Stations 1 through 4. Stations 1, 2, and 3 are springs and therefore exhibit low bacterial counts. Station 4, which is downstream of a heavily grazed area, shows a strong response to snowmelt runoff in May, then pronounced high counts during the grazing period, with a drastic reduction in counts during late fall.

The bacterial counts at Station 5 (outlet of Twin Creek) are presented in Figure 5. Total coliform counts at this station follow the

*The use of trade names does not imply endorsement of specific products by Utah State University.

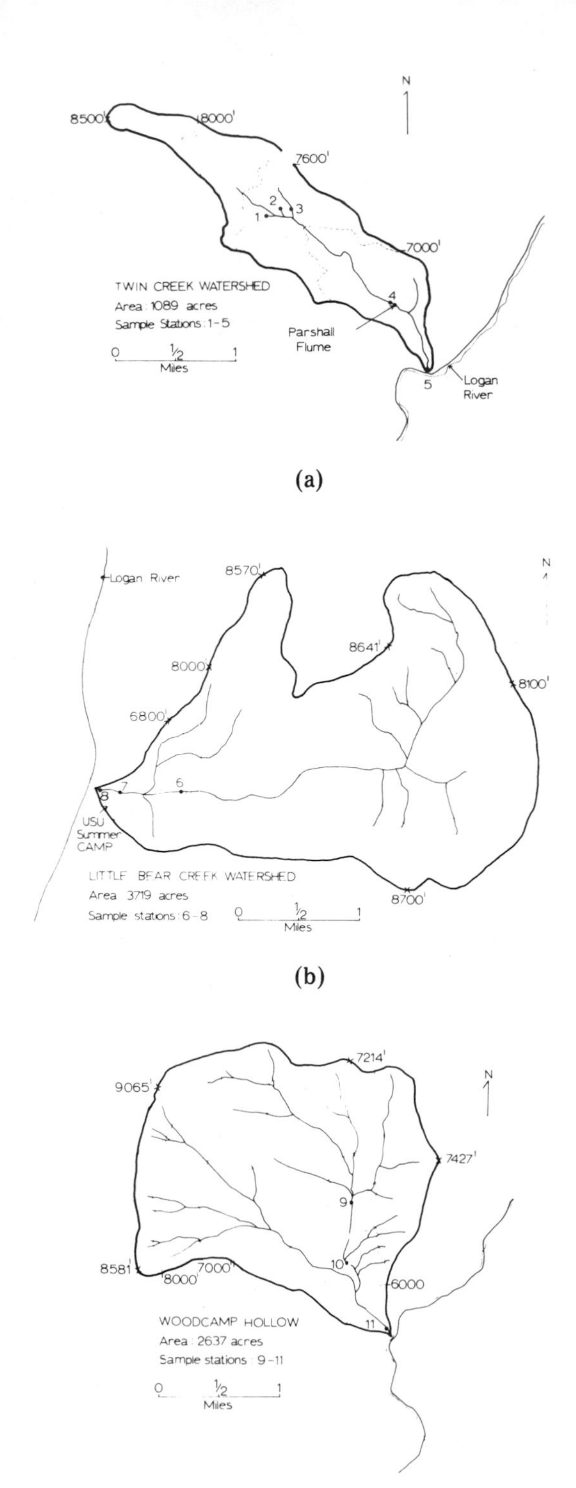

3. Maps of study watersheds: (a) Twin Creek; (b) Little Bear Creek; (c) Woodcamp Hollow Creek.

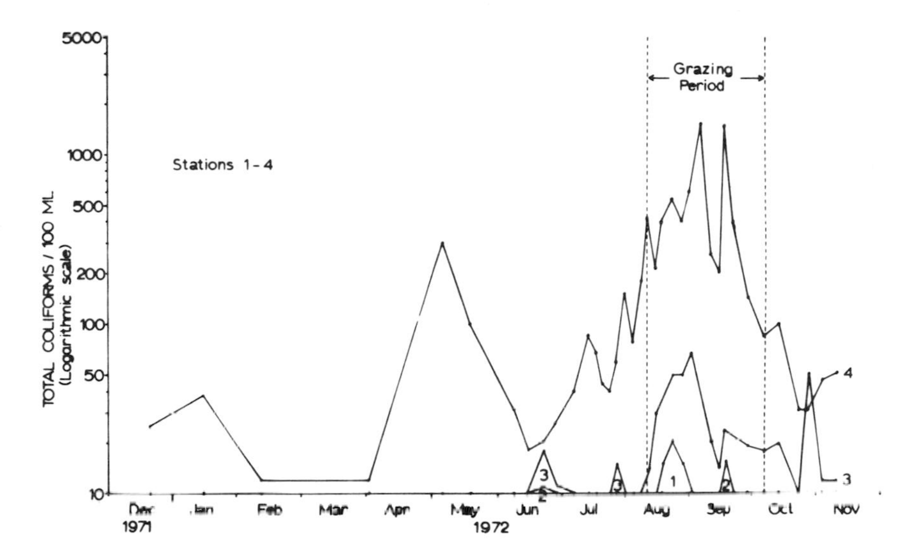

4. Total coliform counts for Stations 1, 2, 3, and 4 on Twin Creek Watershed.

seasonal variation demonstrated at Station 4. Fecal coliform counts demonstrate variation similar to total coliform, but with lesser values. Fecal streptococci counts at Station 5 generally follow similar seasonal trends, but demonstrate relatively high counts in early summer that are not readily explainable. These high pregrazing-period counts may be attributable to an elk herd which roams over this general area during early summer.

This visual indication of grazing influence on bacterial counts was statistically evaluated by dividing the intensive sampling interval (April–November 1972) into three segments (before, during, and after grazing) and testing for differences between the means of these segments. The total coliform, fecal coliform, and fecal streptococci means for these three segments are presented in Figure 6 for three stations in Twin Creek. Total coliform means at Station 2 (a spring) were all less than 10 and not significantly different. The means at Station 4 reflected a significant ($\geqslant$ .05 level) increase in total coliform counts during the grazing period, but were similar before and after grazing. Station 5 also exhibited a significant grazing impact with a lower during-grazing mean than at Station 4. This lower value is probably due to bacterial dieoff in the stream, as most of the grazing impact and thus the bacterial input was above Station 4. Fecal coliform and fecal streptococci means reflected similar comparisons to the total coliform means with the means during grazing being significantly greater than the means before or after grazing.

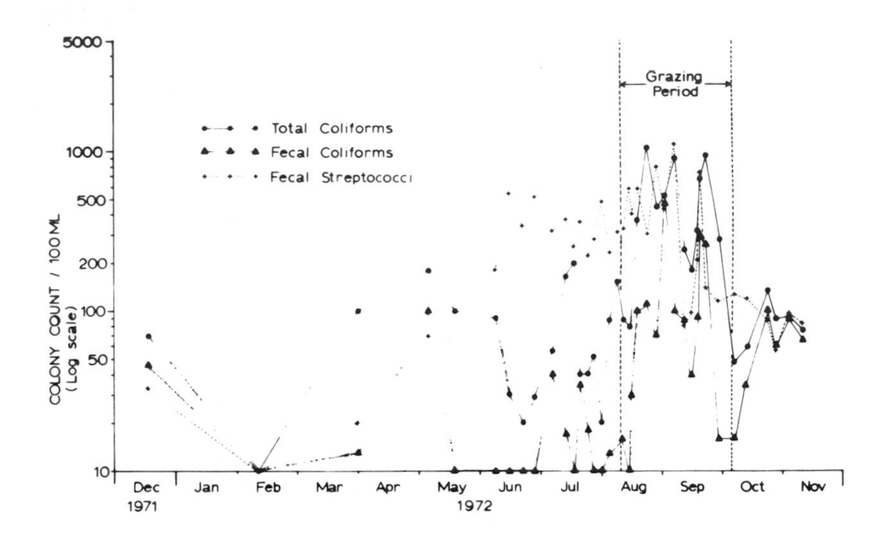

5. Total coliform, fecal coliform, and fecal streptococci counts at Station 5 on Twin Creek Watershed.

Total coliform values for the sheep-grazed Little Bear Creek are presented in Figure 7. These counts showed a tenfold increase during the sheep-grazing period, and the mean count for this grazing period was significantly greater than the mean counts for either the before or after grazing periods. Fecal coliform and fecal streptococci counts followed total coliform count-trends in a manner similar to the results shown for Twin Creek (Figure 5). That is, fecal coliform was correlated with total coliform, and fecal streptococci had relatively high before-grazing counts.

Total coliform, fecal coliform, and fecal streptococci counts for the ungrazed Woodcamp Hollow Creek are presented in Figure 8. These counts display a somewhat erratic trend over the sampling season and are conspicuously absent of any midsummer peaks that would coincide with the peaks associated with grazing on the other watersheds. The values obtained are mostly less than 100 and the variation appears to be random.

Chemical and Physical

Analyses of the chemical and physical values did not conclusively determine the grazing impacts on the three streams studied. The mean values for these parameters are presented, by grazing period segment,

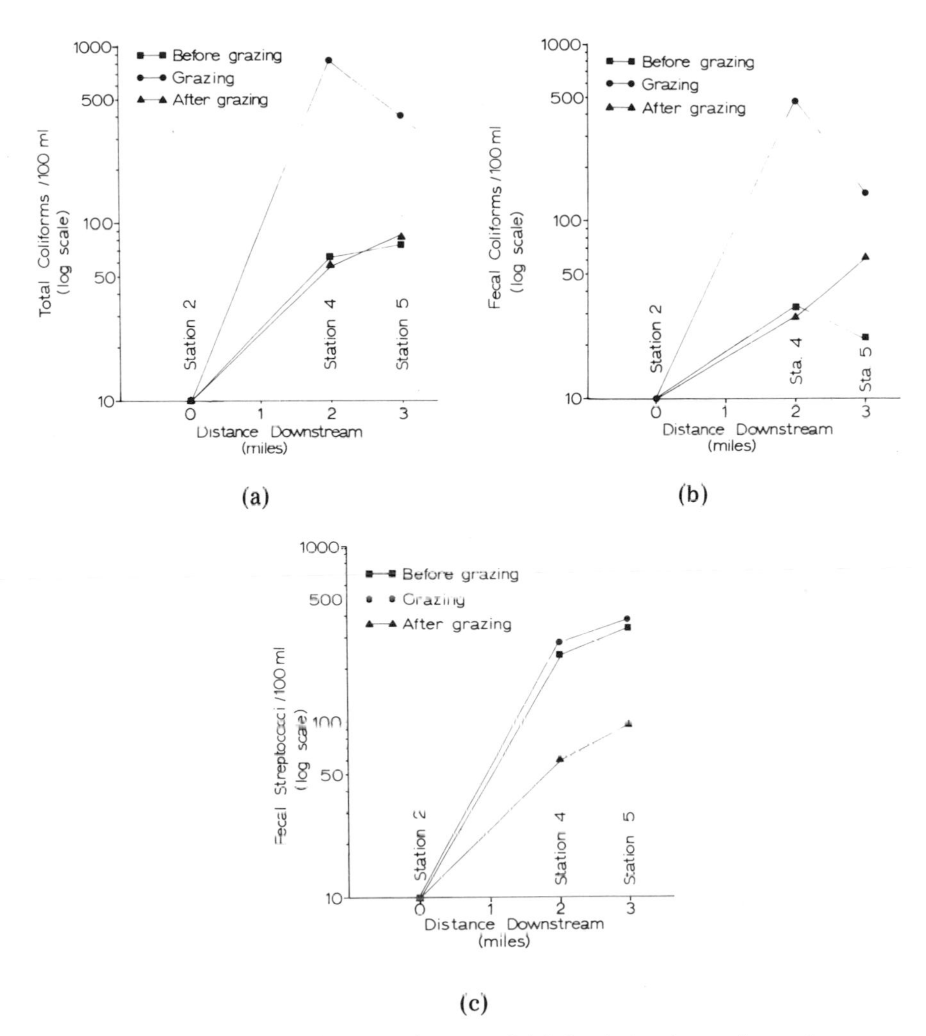

6. (a) Total coliform, (b) fecal coliform, and (c) fecal streptococci count means before, during, and after grazing periods at three locations on Twin Creek Watershed.

in Table 1, along with indicated statistical significance. Temperature means over the three treatments did show significant differences at five sample stations, but these differences appear to be more related to seasonal fluctuations than to grazing activity. The pH values exhibited three means that were significantly different, but, again, they were apparently not related to watershed activity. Turbidity values

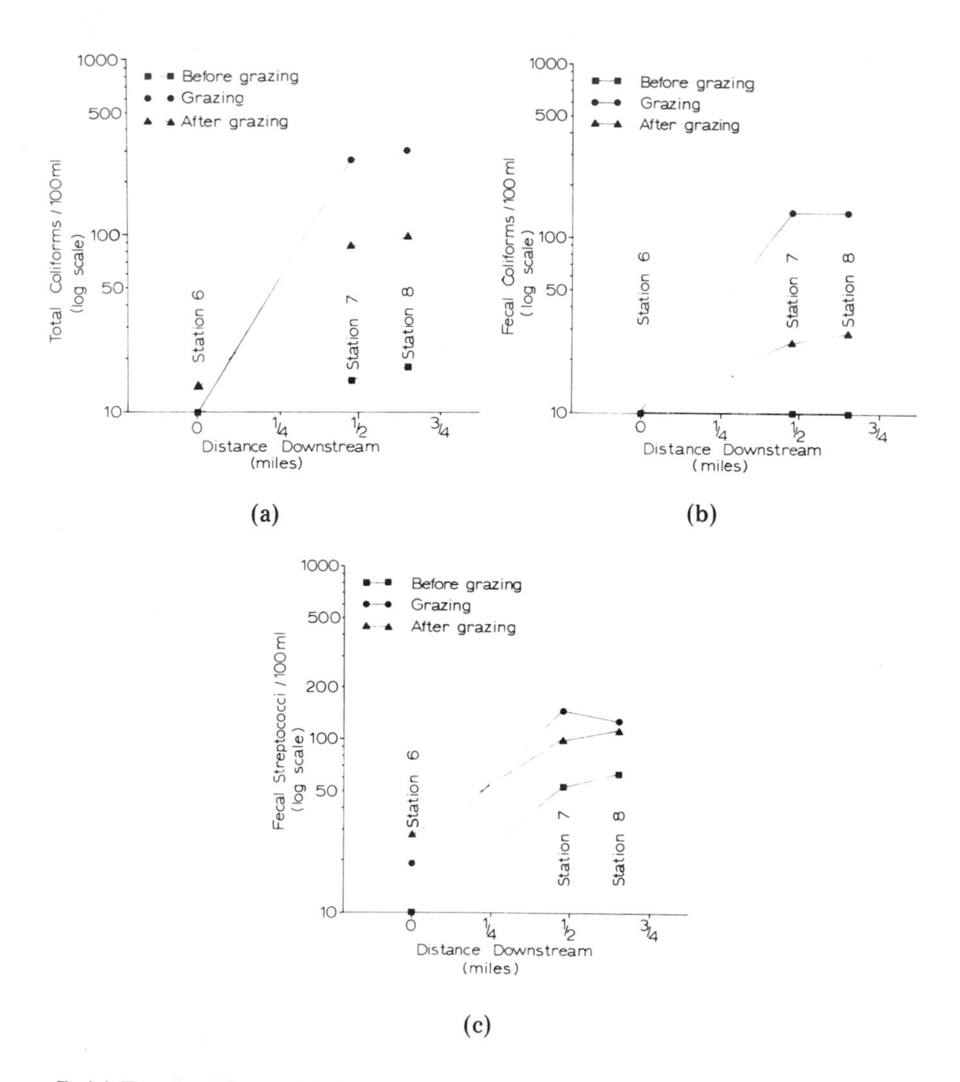

7. (a) Total coliform, (b) fecal coliform, and (c) fecal streptococci count means before, during, and after grazing periods on Little Bear Creek Watershed.

were generally quite low and reflected only sheep-grazing activity at Stations 7 and 8. Nitrate values showed significant differences at four stations. Station 4, however, showed the only significant increase that might be related to grazing, while Stations 10 and 11, which were not grazed, also showed increases for the same time period. Phosphate values showed no significant changes over the sampling season.

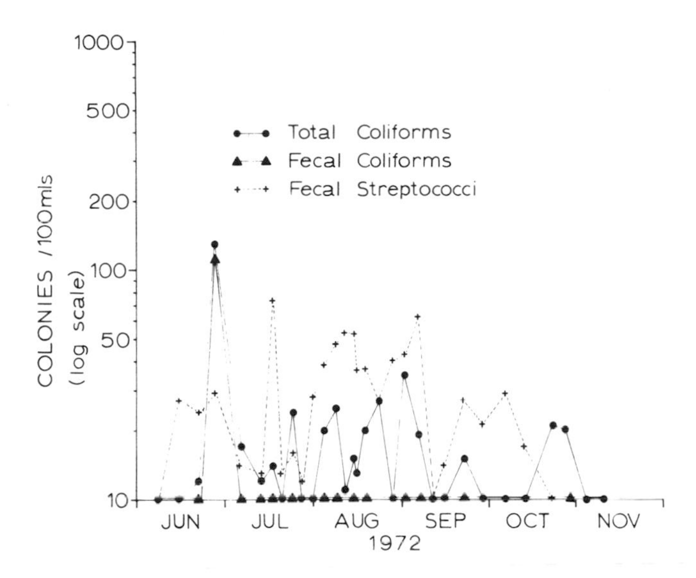

8. Total coliform, fecal coliform, and fecal streptococci counts at Station 11 on Woodcamp Hollow Watershed.

General

Mean values for the measured parameters at the most downstream station of each watershed (Stations 5, 8, and 11) were statistically compared (ANOV) and the results are presented in Table 2. Several values that were significantly different (temperature, pH, and nitrates) are not clearly related to grazing effects. The bacterial indicators, however, clearly reflect the influence of grazing.

Multiple regression analyses, using stepwise deletion procedures, were applied to the chemical, physical, and bacteriological data in order to investigate the relationship of the variables to the predictability of total coliform counts (the most dramatically influenced parameter) and to rate these variables in the order they were deleted from the regression models. All stations which were directly affected by grazing activity (Stations 4, 5, 7, and 8) were combined and a stepwise deletion was performed using total coliform as the dependent variable (y). Fecal coliform was shown to be the most important predictive independent variable (Xn), since it is a component of total coliform, while grazing, a qualitative variable, was the second most important variable out of ten. Lumping those stations together that were not affected by grazing (Stations 1, 2, 3, 6, 10, and 11) we

Table 1
Sampling Period Means for Physical and Chemical Parameters by Sample Stations

Station		Temperature (°C)			pH			Turbidity (JTU's)			Nitrates (mg/1 N-NO_3)			Phosphates (mg/1 PO_4)		
		B[a]	D	A	B	D	A	B	D	A	B	D	A	B	D	A
Twin Creek	2	5.5	5.5	5.2[b]	7.6	7.5	7.6	.9	0	0	1.32	1.28	.96[b]	.35	.28	.17
	4	11.3	10.7	4.1[b]	8.3	8.4	8.4	2.9	4.4	.8	.19	.34	.21[b]	.23	.19	.13
	5	14.2	11.7	3.9	8.3	8.4	8.4[b]	3.6	2.4	1.0	.10	.16	.14	.26	.30	.09
Little Bear	6	7.3	7.9	7.3	8.1	8.2	8.2	1.1	0	0[b]	.27	.33	.30	.26	.25	.21
	7	9.2	10.7	7.8[b]	8.1	8.4	8.3	1.7	4.8	.2[b]	.23	.27	.21	.25	.12	.17
	8	9.9	10.8	8.4	8.3	8.4	8.3	2.1	3.0	.3[b]	.21	.23	.22	.21	.11	.15
Wood-camp	10	8.2	8.1	6.2[b]	8.0	8.1	8.3[b]	.4	0	0	.16	.25	.15[b]	.15	.16	.10
	11	9.0	8.4	5.3[b]	8.2	8.4	8.4[b]	.5	0	0	.12	.22	.07[b]	.18	.16	.13

[a]B = before grazing, D = during grazing, A = after grazing.

[b]Significant differences between B, D, and A mean at the 95% level of confidence.

Table 2
Comparisons of Parameter Mean Values for the Most Downstream Stations on Twin Creek, Little Bear Creek, and Woodcamp Hollow

Parameter	*Twin Creek (Station 5)*	*Little Bear Creek (Station 8)*	*Woodcamp Hollow (Station 11)*
(n) =	(38)	(35)	(31)
Temperature (°C)	11.3	9.1	8.0[a]
pH	8.39	8.32	8.31[a]
Nitrates (mg/1 N - NO_3)	.14	.22	.15[a]
Phosphates (mg/1 PO_4)	.25	.16	.16
Turbidity (JTU's)	2.6	1.2	.23
Total Coliform (counts/100 ml)	240.00	103.00	17.70[a]
Fecal Coliform (counts/100 ml)	88.20	37.70	4.45[a]
Fecal Streptococci (counts/100 ml)	325.00	101.00	27.00[a]
Total Heterotrophic Bacteria (counts/ml)	3780	2780	3990

[a]Significant at 95 percent level.

found by a similar analysis that fecal coliform was the most important variable, but grazing was seventh out of a total of ten variables.

Summary and Recommendations

What does this study tell us about the effect of livestock grazing on rural water quality? It indicates that cattle and sheep grazing on small mountain watersheds with live streams significantly increases the total coliform, fecal coliform, and fecal streptococcus counts in the streams immediately below grazed areas. These indicator organisms were present at all downstream sampling stations (nonspring areas) regardless of the presence or absence of grazing livestock, but they were much more numerous in the streams draining grazed areas.

If the number of bacteriological indicator organisms present in water is an accurate indication of actual pathogen presence, then cattle and sheep grazing a watershed through which streams flow can increase the potential health hazards of that stream to downstream users. Van Donsel and Geldreich (1971) tend to support this thesis. They investigated mud samples from many stream and lake bottoms and recovered salmonella in 19 percent of the samples when the fecal coliform count was 1-200/100 ml, 50 percent when counts were 201-2000, and 80 percent when counts were over 2,000. This research and other similar studies have led the National Technical Advisory Committee on Water Quality Criteria to recommend a fecal coliform limit of 200/100 mls for primary contact recreational waters (FWPCA, 1968). Mean values for fecal coliform reported in this study were well above this figure during grazing periods on Twin and Little Bear Creeks.

It should be acknowledged, however, that the elevated counts measured in this study, which were attributable to grazing, did not persist after mixing with Logan River. Total coliform counts were generally less than 100/100 ml in Logan River below the study watersheds, and fecal coliforms ranged from 5 to 50/100 ml.

Also, livestock grazing had no apparent effects on the chemical and physical values measured in this study. Several of the parameters were slightly elevated during the grazing period, but in most instances they were not significantly different from the values before and after grazing.

Judging from the findings of this study and the scant literature available on the subject, it appears that livestock grazing can impose an additional pollution load on wildland streams (additional to the background load provided by indigenous wildlife). What, then, can be done to alleviate this additional source of contamination? *The most important recommendation at this time is to prevent direct access of livestock to important water supply streams.* This is not a recommendation to eliminate grazing. Water can be provided for livestock by piping it away from live streams and/or constructing small stock-water ponds. Fencing could be provided to create a buffer strip between the livestock and the stream.

Rather than curtailment of livestock grazing, perhaps a concerted effort should be made to advise wildland recreationists of the possible health hazards involved in drinking untreated stream water.

References

American Public Health Assoc. 1971. *Standard methods for the examination of water and wastewater*. 13th ed. New York.

Bates, James W. 1963. Effects of beaver on stream flow and water quality. M.S. thesis, Utah State University.

Bissonette, G. K., Stuart, D. G., Goodrich, T. D., Walter W. G. 1970. Preliminary studies of serological types of enterbacteria occurring in a closed mountain watershed. *Proc. Mont. Acad. Sci.* 30: 66–76.

Boughton, W. C. 1970. *Effects of land management on quantity and quality of available water. A review*. Australian Water Resources Council Research Project 68/2, Report 120.

Brown, George W. 1972. *Forestry and water quality*. Corvallis: Oregon State University Press.

Coltharp, G. B., and Darling, L. L. 1973. Livestock grazing and high quality water yield: Are they compatible? *Soc. for Range Mgt. abst. of papers*: 28–29.

Diesch, S. L. 1970. Disease transmission of waterborne organisms of animal origins. In *Agricultural practices and water quality*, ed. T. L. Willrich and G. E. Smith, pp. 265–285. Ames, Iowa: Iowa State University Press.

Federal Water Pollution Control Assoc. 1968. *Report of the Committee on Water Quality Criteria*. U.S. Dept. Int.

Geldreich, E. E., and Kenner, B. A. 1969. Concepts of fecal streptococci in stream pollution. *Jour. of Water Pollution Control Federation*. 41: R336–352.

Goodrich, T. D., Stuart, D. G., Bissonette, G. K., Walker, W. G. 1970. A bacteriological study of the waters of Bozeman Creek's south fork drainage. *Proc. Mont. Acad. Sci.* 30: 59–65.

Kunkle, S. H. 1970. Sources and transport of bacterial indicators in rural streams. In *Proc. symp. on interdisciplinary aspects of watershed management*, pp. 105–132. New York: A.S.C.E.

Lee, R. D., Symons, J. M., Robeck, G. C. 1970. Watershed human-use level and water quality. *Jour. Amer. Water Works Assoc.* 62: 412–422.

Lin, Sundar. 1972. Nonpoint rural sources of water pollution. Springfield: Illinois Dept. of Registr. and Educ.

Loehr, Raymond C. 1968. *Pollution implications of animal wastes—a forward-oriented review*. F.W.P.C.A., Robt. S. Kerr Water Res. Center.

Loehr, Raymond C., ed. 1969. *Animal waste management: Proc. Cornell Univ. conf. on agric. waste mgt.* Ithaca: Cornell Univ.

Meiman, J. R., and Kunkle, S. H. 1967. Land treatment and water quality control. *Jour. Soil and Water Conserv.* 22: 67–70.

Minor, J. R., and Willrich, T. L. 1970. Livestock operations and fieldspread manure as sources of pollutants. In *Agriculture practices and water quality*, ed. T. L. Willrich and G. E. Smith, pp. 231–240. Ames, Iowa: Iowa State Univ. Press.

Morrison, S. M., and Fair, J. F. 1966. *Influence of environment on stream microbial dynamics*. Hydrology Paper No. 13. Fort Collins: Colo. State Univ. Press.

Peterson, N. J., and Boring, J. R. 1960. A study of coliform densities and escherichia serotypes in two mountain streams. *Amer. jour. of hygiene* 71: 134–140.

Stuart, D. B., Bissonette, G. K., Goodrich, T. D., Walker, W. G. 1971. Effects of multiple-use on water quality of high-mountain watersheds: bacteriological investigations of mountain streams. *Applied microbiol.* 22: 1048–54.

Teller, H. L. 1963. An evaluation of multiple use practices on forested municipal catchments in the Douglas fir region. Ph.D. dissertation, University of Washington.

Van Donsel, D. J., and Geldreich, E. E. 1971. Relationship of salmonella to fecal coliform in bottom sediments. *Water research* 5: 1079–87.

Walter, W. G., and Bottman, R. P. 1967. Microbiological and chemical studies of an open and closed watershed. *Jour. environ. health* 30: 157–163.

Ward, R. C. 1971. *Small watershed experiments: an appraisal of concepts and research developments.* Hull, Quebec: Univ. Hull, Occasional Papers in Geog. No. 18.

Webber, L. R. 1971. Animal wastes. *Jour. soil and water conserv.* 26: 47–50.

Wooten, H. H. 1959. *Major uses of land in the United States: summary for 1959.* Washington, D.C.: Govt. Printing Office.

24. Evaluation of Proposed Animal Manure Handling Practices

George A. Whetstone

It has been calculated that the gross national production of manure exceeds two billion tons per year. Much of this is deposited directly on range or pasture to be recycled in ways which gladden an ecologist's heart. Moreover, the manure which accumulates in barnyards, chicken houses, and milking parlors has traditionally been returned to the land for its fertilizer and soil conditioner values. When virtually all rural residents were engaged in agricultural pursuits, such practices—and public approval of them—developed automatically. Now three significant trends are rendering this bucolic procedure inadequate.

First, the operators of feedlots, finishing pens, or poultry houses frequently have insufficient land under their control for beneficial spreading to be feasible. Depending on the choice of crop, from one to five cattle can supply the assimilable nutrients for an acre of land. For a 64,000-head feedlot, this translates into a land requirement of 20 to 100 square miles. Even with free manure, the costs of hauling and spreading during the seasons when those operations are feasible are such that commercial fertilizers have been preferred.

Second, confined feeding operations have tended to locate near the edges of towns where transportation and other facilities were available. They have become surrounded by urban sprawl and by refugees who wish to be in, but not of, the country. This has reduced both the area available for land application of manure and the acceptance by neighbors of the practice.

Third, pollution has become anathema. The horses working in the hills of Oregon now wear disposable diapers. And the deer, it is reported, have become more wary of environmentalists than of hunters. As the industries and the cities attain zero discharge, nonpoint-source agriculture, now thoroughly outvoted in Congress and most state legislatures, is destined to monopolize the villain's role. Will we, then, be permitted to spread manure on pastures and croplands? Perhaps. But the alternatives merit serious consideration.

Thermochemical Processes

If costs be disregarded, the most acceptable alternative to land spreading is the thermochemical alteration of manure to obtain such products as electricity, oil, gas, or building materials. The market for them is essentially unlimited, and the unit value of energy is certain to increase in the future. The oil most easily obtained from manure, while not particularly amenable to gasoline production, makes an excellent, nearly sulphur-free fuel. These processes inherently destroy all biological life forms in the manure, thus reducing public health problems and aesthetic objections.

Costs are formidable, however. Manure as defecated has a moisture content of from 70 to 90 percent. To secure an easily handled material with a fuel value comparable to that of lignite or oil shale, about 3.5 pounds of water per pound of dry matter must be removed in some manner. This may be by natural evaporation in semi-arid areas or by mechanical drying elsewhere, with the costs of drying alone probably exceeding the value of the final product. The 64,000-head feedlot, which required far too much land for spreading, would produce only 225 tons of dry manure per day. An oil shale plant, to be considered viable—with no drying costs—would need a capacity of at least 50,000 tons per day.

Halligan and Sweazy (1972) evaluated potential costs and benefits associated with the production of methane, oil, or synthesis gas from the manure of 600,000 cattle in feedlots within a fifteen-mile radius on the semi-arid high plains of western Texas. The methane, with a calculated value of $8,275 per day, would require the use of oxygen costing $4,275. A substantial increase in gas prices would be required for feasibility. For oil production, temperatures of 380° C and pressures of 6,000 psi are employed. Much development work would be required to bring costs under the estimated $8,356 values of the 4,178 barrels of oil producible daily. The value as synthesis gas, an intermediate in most ammonia-producing sequences, was calculated to be $9,650 per day. This could well prove to be feasible under current conditions, and more detailed investigations are proceeding.

The literature contains many accounts of methane production by means of anaerobic digesters on individual farms. Some 500 to 600 small plants are reported to have been in operation in France and Algeria in the postwar years. Several South Africans utilize methane from manure to supply their power and heating requirements, and a plantation in East Africa is reported to have doubled its coffee production in a decade by employing as fertilizer the residue from an anaerobic digester which produced sufficient methane to power the plantation.

On the other hand, northern Europe reportedly had stopped producing methane from manure by 1966. An intensive investigation at Waterloo University (Costigane et al.) in 1972 concluded that methane production "is not at present considered feasible for animal waste treatment on a small farm due to the high initial equipment cost." A report by Savery and Cruzan, also in 1972, concluded that a 60,000 chicken unit could be self-sufficient in electricity, but at a cost six times that of its present supply. They considered, however, that improved technology and the then-impending shortage of natural gas would reduce the adverse cost-to-benefit ratio.

A bibliography on small-scale methane production, with abstracts, will be appearing in a report of the Environmental Protection Agency by George Whetstone, Harry Parker, and Dan Wells, entitled *Study of current and proposed practices in animal waste management*.

Refeeding

On the old-time family farm the pigs followed the cattle and the chickens followed the pigs. A well-respected text on bovine husbandry (Snapp and Neumann, 1960) referred to the first portion of this cycle as the "pork profit" from cattle raising. Later editions of this still-used book have dropped the term, presumably in recognition of the specialization now usual in agribusiness. Even in feedlots, however, considerable "browsing" over the deposits occurs. Durham et al. (1966) have reported that autopsies on ruminants fed all-concentrate diets have disclosed digestive tracts compacted with heavy deposits of soil and feces.

It seems evident enough that nutritional values may remain in particles of feed which have passed through a first digestive tract unutilized and essentially unaltered. There are, however, additional (and usually greater) values present. Poultry feces have been reported to be rich in riboflavin (vitamin B_2), in vitamin B_{12}, and in an "unidentified growth factor." Growth, egg production, hatchability, and general welfare of chickens appear to be much improved when they have access to feces, either their own or others, as a portion of their diet (Whetstone et al., 1974, Chapter 5). Rabbits have been reported to consume from 54 to 82 percent of their total fecal production (Eden, 1940), and rats deprived of access to their feces have developed deficiencies in pantothenic acid which have led to cessation of growth, loss of weight, and death (Daft et al., 1963).

Utilization of manure as a component of feed is currently forbidden in the United States by edict of the Food and Drug Administration. The practice is legal in Canada and the United Kingdom. Research has been encouraged in the States, and a considerable body of knowledge has been accumulated, though not all of it has been placed in the public domain. A reasonably good summary of some of the methods used for treating manure for refeeding appears in the draft version of the *Development document for effluent limitations guidelines and standards of performance: feedlot industry* which the Hamilton Standard Division of United Aircraft Corporation prepared for the EPA and released in June 1973. Gaps and errors are alleged to exist in the report and doubtless do exist, since several of the processes are undergoing frequent change and in many cases are enshrouded in at least partial secrecy.

The first method discussed in the Hamilton Standard report (after land spreading and composting) is dehydration. This is a well-understood process for which commercial equipment is available. Values of up to $70 per ton have been assigned to dried poultry waste on the basis of its nutritional value. Much of the British production of poultry manure is dehydrated and sold as a specialty fertilizer to gardeners, florists, and vineyardists, or as a feed component. Zindel, Flegal, and coworkers at Michigan State University have recycled dried poultry waste (DPW) as 12.5 percent and as 25 percent of poultry rations through at least 35 cycles. They have concluded that the practice is safe and that, with proper processing and storage, DPW "has a place in the list of ingredients for all animal rations" (1972).

The General Electric Corporation developed a process for the aerobic production of single-cell protein and built a pilot plant at Casa Grande, Arizona, to test the process, determine cost data, and secure a basis for requesting FDA approval. "However [quoting from the Hamilton-Standard report], since late 1972, the facility has been shut down. Available information indicates that the reasons for the closing are complex and include difficulties with maintenance of the pure bacterial cultures utilized in the process."

Several studies have been or are being made of growing yeast on manure products for its protein value. Singh and Anthony (1968) obtained a final yeast product with 40.9 percent crude protein (dry basis) from the soluble portion of manure on a concrete-floored beef feedlot. Rats fed the dried solubles developed diarrhea, which was attributed to the high mineral content. Meller (1969), basing his findings on studies of municipal wastes and crop residues, concluded that a hydrolysis-fermentation approach to waste conversion by yeast culture had promise but was "at best at the high end of the current high protein supplement price range." Thayer (*Beef*, 1971) of Texas Tech is investigating the possibilities, as are others. According to the Hamilton Standard report, laboratory yeast culture systems have not been completely defined and economic analysis is not yet possible.

At least three more years work is anticipated before practical results appear.

Hamilton Standard has been investigating a process for the anaerobic fermentation of cattle waste into a proteinaceous feed ingredient and methane gas. Enough gas can be produced to power the operation, which is considered feasible for feedlots of 5,000 head and larger. The system is currently in the laboratory stage.

Feed Recycling Company has a proprietary process under active study at Blythe, California, which was described in *CALF News* for January 1973. Since then, as stated in the Hamilton Standard report, it has been revised and simplified to some extent. Frank Senior, consulting engineer for Feed Recycling, insists, however, that the Hamilton Standard account has many misstatements.* A first group of ten test cattle have been slaughtered at the Cal-Poly experimental abattoir at Pomona in the presence of accredited pathological veterinarians. The livers of all five cattle on normal feedyard rations were rejected for human consumption while only two from the test group fed a ration containing about 8 percent treated recycled manure were rejected.

Oxidation ditch residue has been included as a liquid component in a number of refeeding tests; they have generally been reported successful in that the residue has acceptable nutritional value and is an effective protein and mineral supplement in a ration for ruminants. Lack of FDA approval has, however, eliminated commercial development of the concept.

A paper describing a "Total Biochemical Recycle Process" (Carlson, 1971) was presented at the International Symposium on Livestock Wastes in 1971. In summarizing this process, the Hamilton Standard report asserted that "available information was not sufficient to substantiate the developer's claims" and that the proprietary process appeared "to be complex and expensive with no demonstrable payback."

Various chemical treatments have been studied for rendering wastes more digestible, and several investigators have succeeded in producing from them feeds that resulted in good weight gains and carcass characteristics. The work of Smith, Goering, and Gordon of the Agricultural Research Service, Beltsville, Maryland (1969, 1970, 1971a, 1971b), should be consulted for details. Some of Smith's other papers (1971, 1973a, 1973b) give an excellent treatment of refeeding in general, with extensive lists of references.

No treatment of refeeding could be considered complete which did

*Frank Senior, August 1973: personal communication.

not cite the highly successful work of Brady Anthony at Auburn University (1962, 1966, 1967, 1968, 1969, 1971a, 1971b, 1973). Briefly, he collects fresh manure from steers confined on a concrete slab, mixes it with coastal bermuda grass, and ensiles the mixture. The ensilage, called "wastelage," is blended with corn and supplements to constitute a complete ration for the cattle. The excess can be fed to sheep or brood cows on pasture with land disposal of their contribution occurring naturally. A modification of the process on a more extensive scale is being studied for sheep feeding in Colorado (Ward and Beede, 1973).

Flies and other Coprophagists

Land disposal of manure is frequently accompanied by a plowing in for the dual purpose of retaining volatile nutrients in the soil, where they may be beneficial and of keeping them out of the air, where they constitute a nuisance. In nature these functions are usually performed by beetles and other insects. Some interesting work has been done on the introduction of appropriate species of dung beetles to pastures.

An example of somewhat more intensive entomological engineering is the use of fly larvae to catabolize manure. The products of the operation are the protein-rich larvae themselves and a dry, reasonably stable, essentially odorless product from which about 80 percent of the original organic matter has been removed. This converted manure has value, nevertheless, as a fertilizer or soil conditioner.

The utilization of nutrients from manure as fish food has been advocated with some enthusiasm (Durham et al., 1966; Hart et al., 1965). However, experiences with fish kills when slug flows from feedlots have been carried by flooding streams suggest that great caution is necessary in regulating quantities.

Water Hyacinths

To end this rather sordid survey on a note of beauty, we consider finally the conversion of manure nutrients to vegetation like water hyacinths. G. C. Dymond (1949) quoted a production figure of 1,100 tons per acre (66 tons of dry matter) for the plants and suggested they be harvested and composted. C. E. Boyd (1969, 1970) observed that water hyacinths, water lettuce, and hydrilla have mean crude protein levels as high as those of many high quality forages. Problems associated with this solution to the utilization of manure are costs of harvesting and drying, mosquito control, and control of undesirable spreading of the plants as noxious weeds.

Conclusions

The manure disposal problem has a multiplicity of solutions in various states of development and with varying degrees of acceptability in different climatic and cultural settings. Much remains to be done in establishing guidelines for selecting the least costly (using the term in the widest sense) or, hopefully, most beneficial means of utilizing the product in various settings. It is to be anticipated that the choice will vary from site to site and from time to time.

For a more comprehensive discussion the reader is referred to a state-of-the-art study prepared by R. C. Loehr, which emphasized land-spreading, to two recent annotated bibliographies (McQuitty and Barber, 1972; Whetstone et al., 1974), and to the frequently cited Hamilton Standard draft report.

References

Anthony, W. B. 1966. Utilization of animal waste as a feed for ruminants. In *Proceedings of a symposium on management of farm animal wastes*, pp. 109–112. Chicago: American Society of Agricultural Engineers publ. no. SP–0366.

Anthony, W. B. 1967. Manure-containing silage—production and nutritive value. Abstract. *Journal of animal science* 26: 217.

Anthony, W. B. 1968. Wastelage—a new concept in cattle feeding. Abstract. *Journal of animal science* 27: 289.

Anthony, W. B. 1969. Cattle manure: re-use through wastelage feeding. *Proceedings of the conference on agricultural waste management*, pp. 105–113. Ithaca: Cornell University.

Anthony, W. B. 1971a. Animal Waste Value—nutrient recovery and utilization. *Journal of animal science* 32: 799–802.

Anthony, W. B. 1971b. Cattle manure as feed for cattle. In *Proceedings of international symposium on livestock wastes*, pp. 293–296. Chicago: American Society of Agricultural Engineers publ. no. PROC–271.

Anthony, W. B., Cunningham, J. P., Jr., Renfroe, J. C. 1973. Ensiling characteristics of various feedstuffs and animal wastes. Abstract.

Journal of animal science 36: 208.

Anthony, W. B., and Nix, R. 1962. Feeding potential of reclaimed fecal residue. *Journal of dairy science* 45: 1538-39.

Beef. 1971. Feed 'em trash, cut pollution. *Beef* 7: 12.

Boyd, C. E. 1969. The nutritive value of three species of water weeds. *Economic botany* 23: 123-127.

Boyd, C. E. 1970. Vascular aquatic plants for mineral nutrient removal from polluted waters. *Economic botany* 24: 95-103.

CALF News. 1973. Feed recycling showing promise. *CALF News* 11: 28, 29, 52.

Carlson, L. G. 1971. A total biochemical recycle process for cattle wastes. In *Proceedings of international symposium on livestock wastes*, pp. 89-91. Chicago: American Society of Agricultural Engineers publ. no. PROC-271.

Costigane, W. D., et al. 1972. *Methane production from anaerobic digestion of animal wastes*. Waterloo, Ontario: University of Waterloo.

Daft, F. S., et al. 1963. Role of coprophagy in utilization of B vitamins synthesized by intestinal bacteria. *Proceedings of Federation of American Societies for Experimental Biology* 22: 129-133.

Durham, R. M., et al. 1966. Coprophagy and the use of animal waste in livestock feeds. In *Proceedings of a symposium on livestock wastes*, pp. 89-91. Chicago: ASAE publ. no. SP-0366.

Dymond, G. C. 1949. The water hyacinth: A cinderella of the plant world. Appendix B in *Soil fertility and sewage* by Van Vuren, pp. 221-227 + pl. London: Faber and Faber.

Eden, A. 1940. Coprophagy in the rabbit. *Nature* 145: 36-37.

Halligan, J. E., and Sweazy, R. M. 1972. Thermochemical evaluation of animal waste conversion processes. Paper read at the 72nd national meeting of American Institute of Chemical Engineers, 21-24 May 1972, St. Louis, Mo.

Hamilton Standard Division of United Aircraft Corporation. 1973. *Draft development document for effluent limitations guidelines and standards of performance: feedlot industry*. Report on EPA contract no. 68-01-0595.

Hart, S. A., Taiganides, E. P., Eby, H. J. 1965. Waste disposal: pre-eminent challenge to agricultural engineers. *Agricultural engineering* 46: 220-221.

Loehr, R. C. 1973. *Pollution implications of animal wastes—a forward-oriented review*. Washington, D.C.: EPA publ. no. 13040-07/68.

McQuitty, J. B., and Barber, E. M. 1972. *Annotated bibliography of farm animal wastes*. Technical appraisal report EPS 3-WP-72-1. Ottawa: Environment Canada.

Meller, F. H. 1969. *Conversion of organic solid wastes into yeast*. 2nd ed. Washington, D.C.: U.S. Public Health Service.

Savery, C. W., and Cruzan, D. C. 1972. Methane recovery from chicken manure digestion. *Journal of Water Pollution Control Federation* 44: 2349-54.

Singh, Y. K., and Anthony, W. B. 1968. Yeast production in manure solubles. *Journal of animal science* 27: 1136.

Smith, L. W. 1971. Feeding value of animal wastes. In *Animal waste reuse—nutritive value and potential problems from feed additives: a review*, pp. 5-13, 42-56. Washington, D.C.: USDA publ. no. ARS-44-224.

Smith, L. W. 1973a. Nutritive evaluations of animal manures. In

Symposium: processing agricultural and municipal wastes, ed. G. E. Inglett, pp. 55-74. Westport, Conn.: Avi Publishing Company.

Smith, L. W. 1973b. Recycling animal wastes as a protein source. In *Alternative sources of protein for animal production*, pp. 146-173. Washington, D.C.: National Academy of Sciences.

Smith, L. W., Goering, H. K., Gordon, C. H. 1969. Influence of chemical treatments upon digestibility of ruminant feces. In *Proceedings of the conference on agricultural waste management*, pp. 88-97. Ithaca: Cornell University.

Smith, L. W., Goering, H. K., Gordon, C. H. 1970. In vitro digestibility of chemically-treated feces. *Journal of animal science* 31: 1205-09.

Smith, L. W., Goering, H. K., Gordon, C. H. 1971a. Nutritive evaluation of dairy cattle waste. In *Proceedings of Maryland nutrition conference*, pp. 1-6. College Park: University of Maryland Press.

Smith, L. W., Goering, H. K., Gordon, C. H. 1971b. Nutritive evaluation of untreated and chemically treated dairy cattle wastes. In *Proceedings of an international symposium on livestock wastes*, pp. 314-318. Chicago: ASAE publ. no. PROC-271.

Snapp, R. R., and Neumann, A. L. 1960. *Beef cattle*. New York: John Wiley and Sons, Inc.

Ward, G. M., and Beede, D. K. 1973. Digestibility of processed feedlot manure. *Feedstuffs* 45 (July 9): 25.

Whetstone, G. A., Parker, H. W., Wells, D. M. *Study of current and proposed practices in animal waste management*. U.S. Environmental Protection Agency Report 430/9-74-003. 84 pages plus four appendices. For sale by U.S. Government Printing Office. $4.70.

Zindel, H. C. 1972. DPW recycling facts updated. *Poultry digest* 31: 125-126.

VI
Low Cost Wastewater Treatment Facilities for Rural Areas

25. Low Cost Wastewater Treatment Facilities for Rural Areas

Philip H. Jones

Introduction

Rural Communities

The treatment of wastewater from a few hundred people has problems not encountered with the larger wastewater treatment plants in major metropolitan areas. Some of these problems are not even related to the low flow experienced in the smaller units but to the fact that small plants have little supervision or maintenance. This is generally true from those units designed to serve single residences or small groups of houses, right up to units serving the agricultural community that has developed into a small town or municipality.

The units treating single residences or small groups of houses are discussed elsewhere in this volume and perhaps represent only a special kind of the small wastewater-treatment plant. Clearly units, individual or community, which do not enjoy highly skilled or continuous maintenance must be, above all, highly dependable, because malfunctioning may continue for some time before it is noticed and corrected. During this period of malfunctioning, considerable environmental damage can be done.

In the case of the larger wastewater-treatment plants serving the metropolitan area, the sewage reaches the plant at a fairly constant rate, because of the variety of distances and velocities it travels from the source to the plant. In the case of the smaller unit, these distances are usually much smaller, and therefore the variability in rate of flow at the plant is much greater. Heavy surges of flow, which might be damped out in a city sewer system, will result in hydraulic or organic surcharges on the plant in the case of the smaller units. Such surges can on occasion reach as high as ten times the average flow rate and would, of course, strain the performance of the most sophisticated plant.

Another problem might be the release into the sewer system of a large volume of toxic or other substance which is difficult to treat. A dose that might pass undetected in a large plant because of dilution can prove serious in a small plant.

The important features of the small plant are as follows:

(a) It must be reliable without continuous skilled supervision;
(b) It must operate efficiently under a variety of flow conditions, including organic and hydraulic shock;
(c) It should not be unsightly;
(d) It should not generate large volumes of solids for subsequent disposal;
(e) It should not generate odors.

In the selection of a process to treat any wastewater a number of variables and conditions must be evaluated. In all cases the required degree of treatment must be maintained and the facilities designed to achieve that quality of effluent at the lowest possible cost. The following are a few of the considerations that must be made in process selection:

(a) Quality of effluent required

(i) Size and flow pattern of receiving stream or body of water. Clearly, a much higher quality of effluent would be required if the receiving stream were a seasonal creek than if it were a high-velocity, large-flow, year-round river. The tidal effect of an estuary would require different consideration than the calm of a stagnant lake

(ii) Susceptibility of receiving water to eutrophication. If a body of water is shallow, warm, and still during the summer months, it will be far more sensitive to the release of inorganic nutrients resulting in eutrophication. If the receiving body of water is a fast-flowing river or stream, eutrophication due to planktonic algae will be extremely unlikely, but the rooted variety of water plants may be a major nuisance. Under these conditions, it is quite possible that biologically available nitrogen may be more significant than phosphorus.

(iii) Subsequent and potential uses of receiving waters. If the subsequent uses involve potable water supplies, the public health will clearly be the significant consideration in determining the quality of effluent required. Alternatively, if the water is to be used for some industrial complex, the critical parameters of that industry must be considered in selecting a treatment process for the wastewater.

(iv) Availability and suitability of land for final effluent disposal. In rural areas where land is used largely for crops, it may be both feasible and practical for the final effluent to be disposed of as irrigation water for the crops. Under these circumstances, the quality of the effluent required would be grossly different from that of any of the previous methods of disposal.

(b) Availability and cost of land for treatment plant. Where land is available and inexpensive, it is frequently possible to select a wastewater treatment process which may require large areas of land but offer in return a relatively simple and inexpensive wastewater treatment process. Comparisons between high-rate, complex mechanical sewage treatment plants and sewage lagoons, as we will see later, suggest that the area of land required for a sewage treatment plant is inversely proportional to the degree of complexity of the process.

(c) Availability of skilled labor for maintenance and operation of the plant. The point which is all too often overlooked is the fact that labor skilled in the operation of complex sewage treatment plants may not be available in a rural area, and therefore complex treatment processes, although extremely efficient in theory, may fail completely because of inadequate or inefficient operation.

(d) Nature and quality of waste to be treated.

(i) Combined sewage (storm and sanitary).
One of the most common problems with small towns in a rural setting is that they have previously had a combined sewerage system built years before it was ever believed likely that sewage treatment would be necessary. Under these circumstances, when the flow is extremely high, the strength of the sewage is frequently low and vice versa.

(ii) Industrial wastes and the nature of industry.
Rarely if ever do we find a purely domestic waste, just as we rarely if ever find a town that does not have some kind of industry. In the rural environment small industries are frequently associated with the product of the land, such as dairies, cheese-making operations, canneries, and, of course, feedlots. Canneries carry on intensive operations for short periods of time during the seasons of the crops. Furthermore, a single cannery may be processing a variety of different crops, each of which may generate an entirely different waste. The seasonal nature of the waste and the seasonal volume of these wastes should be considered when selecting a process for this type of rural installation.

(e) Collection systems.

(i) Partly existing. Frequently a small town in a rural setting may have had a part of the sewer system built for a number of years which discharges the waste to the local river or lake. Under these circumstances, the extensions built to the plant site may to some extent control a number of parameters used in process selection.

(ii) Nonexistent. In some municipalities, it may be found that the septic tank or cesspool approach has been used in the past, and that in fact no sewers exist; in this case there is a great deal more flexibility in designing modern extensions to a location which would provide the optimum total system, i.e. plant and collector system.

Rural locations may be distinguished from urban locations by the density and nature of development. It is common for small agricultural settlements to develop into large communities without necessarily developing an urban identity.

Traditionally, a rural location was inhabited by farmers and the purveyors of goods and services required by farmers. Nowadays in many locations, particularly close to major urban centers, the population may include a high percentage of professionals living in the country and commuting to the city. They may be operating an active farm or leasing the farming facilities to a few local farmers.

Under these circumstances land values will undoubtedly be much higher than in the traditional case. In addition, the new rural dweller has probably left the city environment to seek the "fresh" country air and escape the pollution and bad odors of the modern city. In coming to the country, these urban refugees may be unprepared for the traditional country odors that prevail at the "muck" spreading season or those which are constantly present on a livestock farm. Thus the new rural resident may be a good deal more sensitive to and very much less tolerant of odors which the long-time country resident has learned to accept. Therefore, any odors resulting from a wastewater treatment facility may be of great importance, and the process selection must recognize this change in attitude of the rural resident.

Selected Process Alternatives

The land requirements for various types of wastewater treatment facilities are shown in Table 1. Since land is frequently both available and somewhat less expensive in rural areas, processes that require large expanses of land at low capital and operating cost might be the first ones seriously considered. A description of selected processes follows.

Table 1
Area Requirements for Biological Waste Treatment

Process	*Area Required per 1,000 Population*	
	Acres	*Hectares*
Irrigation	25-200	10-80
Intermittent Sand Filtration	1-4	0.5-2
Trickling Filters	0.005-0.05	0.002-0.02
Activated Sludge	0.001-0.005	0.0004-0.002
Stabilization Ponds	1-10	0.4-4

1. *Oxidation Ponds.* During the past twenty-five years, oxidation ponds have become popular because of the low capital cost and the relatively simple maintenance requirements. Although a variety of design procedures are recognized for oxidation ponds (Gloyna, 1971), they generally tend to be designed according to provincial or state regulations, such as 100 persons per acre of pond at an average depth of four feet. Ponds may be natural lagoon systems employing a facultative mode of operation; they may be totally aerobic, employing photosynthesis as a primary source of oxygen; or they be mechanically aerated. The last alternative would reduce the size of the pond necessary to accomplish the same degree of oxidation of wastes. The decision concerning type is usually related to the climate and the composition of the waste, whether it includes industrial waste or simple domestic wastes.

2. *Activated sludge systems.* Although the conventional activated sludge process (Arden and Lockett, 1914; Sawyer et al., 1939; McKinney and O'Brien, 1968) is not as commonly employed for small, isolated plants, a variety of modifications of this process are often used. The total-oxidation, the high-rate, and the contact-stabilization processes are examples of modified activated-sludge systems (Ulrich and Smith, 1951; Jones, 1965, 1968, 1970; Jenkins and Orhon, 1972) which are appropriate for the treatment of wastes from small communities.

In the extended-aeration process, raw sludge is screened, comminuted, and put into an aeration tank with a detention time of approximately twenty-four hours. Also added to the aeration tank is return activated sludge from the final settling tank. The main purpose of the total oxidation process is to eliminate or considerably reduce the quantity of solids which must be subsequently disposed of, and "total oxidation" means that all material that goes into the plant is totally oxidized. There is some question among researchers as to whether this actually takes place, some claiming that it is theoretically impossible and others claiming that it occurs in practice despite its theoretical impossibility.

The contact stabilization process has the advantage of not retaining for long within the plant the main flow of sewage. In this process, the bulk flow is admitted after comminution to a contact tank, where it is mixed with stabilized activated sludge. This contact period may be fifteen minutes to a half hour and is followed by final settling, then chlorination and release. Thus the bulk flow of the sewage in this process is detained in the sewage treatment plant for two and one half to three hours instead of the conventional ten to eleven hours or the twenty-four to twenty-six hours of the extended aeration process. Since the size of the units involved in the treatment process is dictated by the length of time the flow must be detained in these units, clearly the shorter the detention period, the smaller the overall installation and therefore the capital cost. When separated from the effluent in the final settling tank, the sludge is returned in the contact-stabilization process to a sludge reaeration or reactivation zone. Here concentrated sludge is aerated for a period of four to six hours, during which the accumulated organic material is oxidized and the cells multiply. With the contact stabilization process aerobic digestion is generally used to stabilize the activated sludge that is wasted from the system.

All these modifications of the activated-sludge process lend themselves very well to being housed in what are sometimes called package-activated sludge plants. The term "factory fabricated" might be more appropriate than "package", since the process and detention times should be selected specifically for the process, flow conditions, and sewage strength in any given location. In remote areas, however, where skilled construction labor may not be available, factory fabrication is a distinct advantage, for the units are rapidly installed on a prepared site but have been manufactured under the quality control one can exercise in a factory.

3. *Oxidation ditch.* The oxidation or Pasveer ditch (Ontario Water Resources Commission, 1964; Baars, 1962; Pasveer, 1963) is somewhat of a hybrid process between the aerated-lagoon process and the activated-sludge process. Originally conceived of as a low-cost, extended-aeration system, it comprises a closed-loop channel, three to four feet deep, filled with sewage and activated sludge. The mixed liquor contents of the ditch are maintained in motion at approximately one and a half to two feet per second by a large rotor or brush. Such a process may be operated either as a continuous or a batch

process. Where a continuous process is used, the waste is added continually and there is a continual overflow from one location on the ditch. The overflow would typically go to a final settling tank, where the underflow or activated sludge would be separated and returned to the ditch.

4. *Rotary biodiscs.* This process (Baehnke, 1966; Bartel and Noack, 1966; Welch, 1969) was developed in Germany and more recently came into use in the United Kingdom. The process currently in use includes a primary settling stage which generates raw sludge. The second step is a tank or series of tanks, where a series of closely spaced vertical discs are rotated while partially (about 45 percent) submerged. The rotation is at a rate of one to three revolutions per minute, and the process behaves very much like a biological trickling filter in which the discs serve as a physical surface for the growth of microbial films. By rotation the organisms are provided with alternating periods of oxygen and no oxygen. As the film builds up on the discs, the humus or floc periodically detaches and, as in the trickling filter process, passes to the final settling tank for separation and removal.

Modifications of the biodisc process are being developed to take advantage of power-saving devices and increased surface area. In Denmark a modification includes a large cage containing small plastic balls which is rotated while floating in a tank containing wastewater. The balls are only partially submerged in each rotation, but the surface area is increased by its spherical nature.

5. *Extended filtration.* This process, while not yet popular in North America, has enjoyed some limited success in the United Kingdom. It consists essentially of a biological filter comprising plastic media and built in the form of a tower. Raw sewage is passed through the filter after comminution and then to a final settling tank. A large portion of the solids settling out in the final settling tank plus a considerable fraction of the filter effluent are recycled through the plant. Thus the solids as well as the effluent are recycled and thus the term "extended filtration." The purpose is to achieve a very high degree of oxidation, and therefore stabilization, of the sludge solids.

6. *Other processes.* There are many other processes that are modification and hybridizations of the fundamental chemical and biological processes considered so far. These include the submerged-bed aeration system, which effectively combines the process of trickling filter, in which the biological floc is fixed, and the activated-sludge process, in which the air is added through bubbles by diffusers.

7. *Land application.* The ultimate disposal of wastes, whether they are the solids or the liquid effluents from the foregoing processes or are completely untreated, may very well be land application. This process, of course, is the original waste-disposal process devised by man and the animals and is now becoming more popular as the age of recycling is being thrust upon us. Where raw sewage is supplied to the land through irrigation, the type of crop growing on the land so fertilized must be controlled. It is important to know if the crop is to be eaten raw or if cooking will bring about some sterilization of pathogenic organisms which might be transmitted through the sewage-soil-plant system. A further hazard associated with the application of sludge to the land is the heavy metal content of sludges (Van Lonn, 1972). Many sludge products which have been traditionally used as fertilizers have recently been found to contain quite high concentrations of heavy metals, which, under certain circumstances, can be taken up and concentrated in the plant system. This process, therefore, has a danger of heavy metal poisoning through the food chain.

8. *Effluent polishing.* In some rural areas a fairly high degree of treatment may be required before the waste can be released to the receiving stream or body of water. Under these circumstances a variety of effluent-polishing devices have been developed and may be used. The release of the effluent to the land takes advantage of the natural processes of effluent polishing, but where this is not possible, upward-flow clarifiers, slow-sand filters, flocculation and filtration, and fine stainless-steel screening have been used with some success.

9. *Nutrient removal.* Since society has become aware of the eutrophication problem (Jones et al., 1969), the design of wastewater treatment plants has recently added nutrient removal as a further stage of treatment. Nutrient removal in the Great Lakes basin has tended to become synonymous with phosphorus removal, since a great deal of evidence has been accumulated to indicate that phosphorus is the critical nutrient in the eutrophication of the Great Lakes. Most processes have simply been modified by the addition of some chemical coagulation and sedimentation, but there are many processes available for the removal of nutrients, including adsorption (Bishop et al., 1967), electrodialysis, reverse osmosis (Merten and Nusbaum, 1968), as well as biological denitrification. One of the most effective ways of removing the two most significant nutrients, phosphorus and nitrogen, is land application, where this is at all possible. Some experiments are currently under way which apply a sixty-cycle current through an aluminum electrode and release soluble aluminum to combine with

the phosphate in the sewage and produce aluminum phosphate sludge.* These experiments deal with very small wastewater treatment processes, even down to the individual household size, largely because of the eutrophication problems caused by summer cottages in the holiday resort regions of Ontario and some parts of the United States. Summer cottages are often built in the Precambrian Shield area, where the bedrock is almost on the surface and the effect of a tile field is minimal. Given these circumstances and the fact that summer holiday families seem to require the use of automatic dishwashers and high-phosphate detergents, the problem of phosphorus eutrophication in the northern lakes of Ontario has in some instances become critical. Should it prove to be successful, the aluminum electric process will be extremely attractive because of its simplicity, lack of moving parts, and low maintenance requirements.

Ontario Experience

In Ontario there are some sixty-two sewage treatment plants and forty-eight waste stabilization ponds under the direct control of the Provincial Government. Figure 1 shows the distribution of size of the sewage treatment plants, and it will be noted that some 68 percent of the plants are under 2.0 MGD. Thus Ontario provides a good operating "laboratory" of smaller sewage-treatment plants serving essentially rural environments.

Table 2 shows some of the operating and cost parameters for ten selected plants. They are mainly less than 2.0 MGD and most are less than 1.0 MGD. Two larger plants were included in order to draw comparisons. The processes selected are mainly conventional activated-sludge and extended-aeration, but contact-stabilization and high-rate activated sludge were included once again to provide a comparative picture.

It can be seen from Table 2 and Figure 2 that the costs of operation are greatly dependent on the size of the plant and far less dependent on the process. This is an interesting observation because if it can be validated on a large-scale sample, the number of parameters used to decide on the wastewater treatment process for a given rural area can be reduced.

Further examination of Table 2 reveals that in the absence of a specific process to remove nutrients, the quantity of phosphorus re-

*Dr. Lorne Campbell, 1973: personal communication.

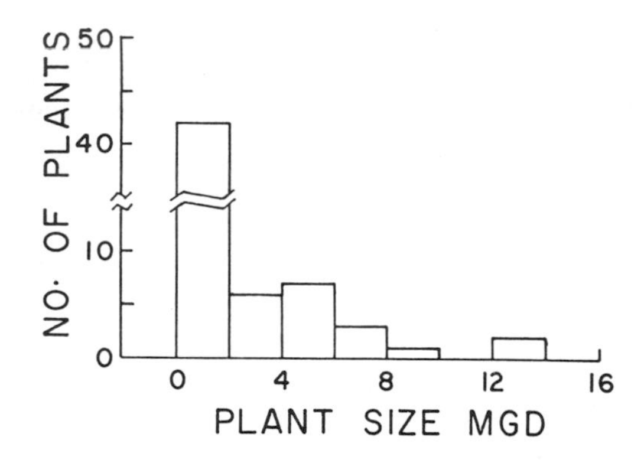

1. Size distribution of sewage treatment plants in Ontario.

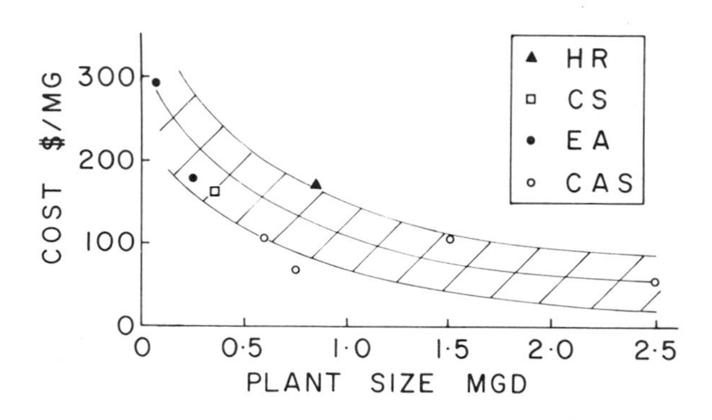

2. Costs of operation of ten selected wastewater treatment plants in Ontario (courtesy of the Ontario Ministry of the Environment).

HR = High rate	EA = Extended aeration
CS = Contact stabilization	CAS = Conventional activated sludge

moved in a small biological treatment plant appears to be independent of the process. Figures 3 and 4 suggest that although P removal may be independent of process, it may have some dependence upon influent concentration of P. This is entirely reasonable and consistent with much of the literature on this subject.

Of the 62 plants operated by the Ontario Ministry of the Environment, 12 are primary, 8 are extended-aeration, 2 are contact-stabilization, 37 are conventional activated-sludge, 1 is high-rate, and 1 is trickling-filter. An analysis of all the plants operated by OME is shown in Figures 5 and 6. These plots suggest that, in the cases studied,

Table 2
Operating Parameters of Ten Selected Wastewater Treatment Plants in Ontario

Town	*Design Flow MGD*	*Process*	*Load Ave. MGD*	*BOD rem %*	*SS rem %*	*% P red/rem*	*Cost Cap.*	*Date of Const.*	*Op. Cost $/MG*	*Op. Cost cent/lb BOD*
Moosonee	0.075	Ext.	0.036	79	89	30	—	1966	—	—
Elora	0.083	Ext.	0.11	75	90	44	361,285	1962	295.60	17
Westminster Twp.	0.25	Ext.	0.23	91	96	72	270,727	1959	179.79	15
Haileybury	0.35	C.S.	0.39	95	97	65	461,890	1967	165.00	9
Fergus	0.60	A/S	0.63	82	92	53	277,393	1958	106.50	9
Burlington[a] (Eliz Gdns)	0.75	A/S	1.1	84	91	41	382,773	1958	68.70	8
Meaford	0.86	High Rate	0.63	93	91	72	—	—	171.00	15
Georgetown[a]	1.5	A/S	1.7	74	92	44	871,677	1958	105.20	12
Burlington[a] (Drury Lane)	2.5	A/S	2.0	91	94	45	676,034	1960	51.60	3
Burlington[a] (Skyway)	6.0	Ext.	8.1	95	94	55	1,796,844	1962	43.10	2

[a]1972 Figures.

Courtesy of the Ontario Ministry of the Environment.

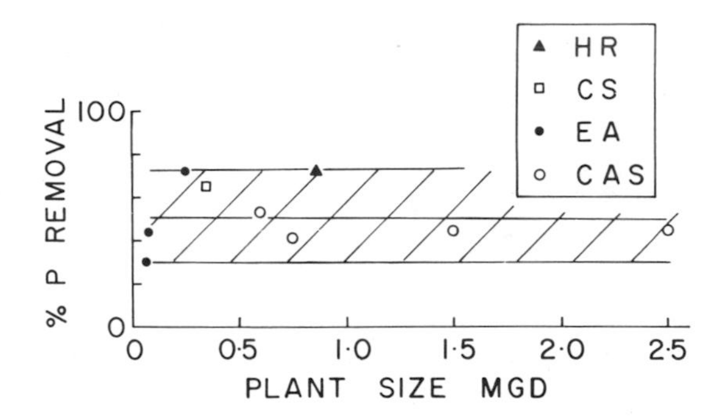

3. Phosphorus removal as a function of plant size in ten selected plants in Ontario (courtesy of the Ontario Ministry of the Environment). Note: No plants have P removal facilities per se. The P removal noted is incidental to the normal biological process.

HR = High rate

CS = Contact stabilization

EA = Extended aeration

CAS = Conventional activated sludge

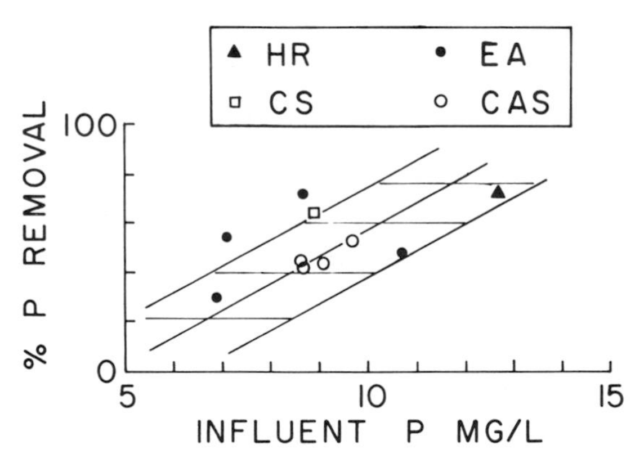

4. Phosphorus removal as a function of influent concentration of P in ten selected plants in Ontario (courtesy of the Ontario Ministry of the Environment). Note: No plants have P removal facilities per se. The P removal noted is incidental to the normal biological process.

HR = Righ Rate

CS = Contact stabilization

EA = Extended aeration

CAS = Conventional activated sludge

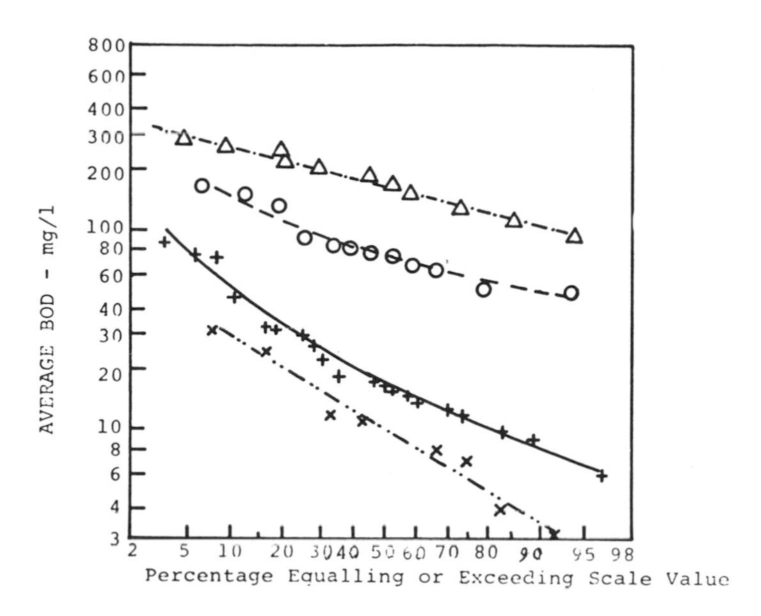

5. Average BOD in provincially operated wastewater treatment plants (courtesy of the Ontario Ministry of the Environment).

△ PLANT INFLUENT—all plants
○ PLANT EFFLUENT—primary treatment plants
X PLANT EFFLUENT—modified activated sludge plants
+ PLANT EFFLUENT—conventional activated sludge plants

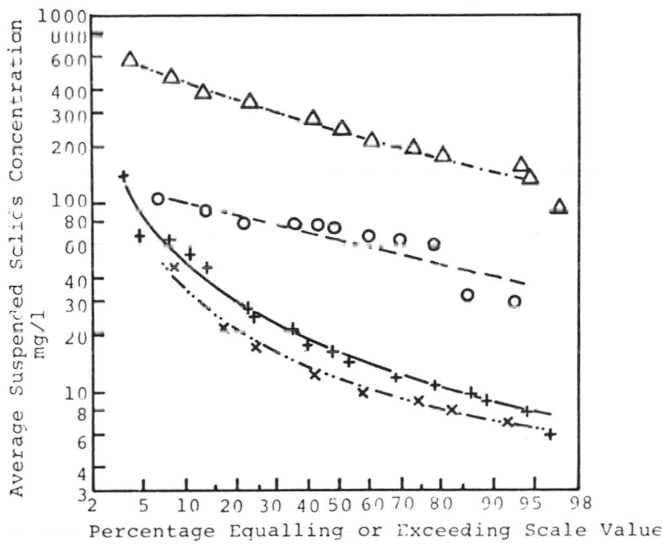

6. Average concentration of suspended solids in provincially operated wastewater treatment plants (courtesy of the Ontario Ministry of the Environment).

△ PLANT INFLUENT—all plants
○ PLANT EFFLUENT—primary treatment plants
X PLANT EFFLUENT—modified activated sludge plants
+ PLANT EFFLUENT—conventional activated sludge plants

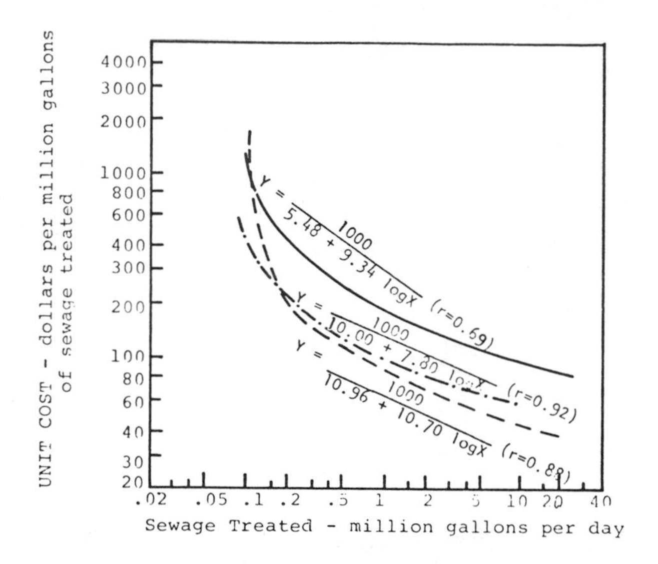

7. Unit operating costs in provincially operated wastewater treatment plants (courtesy of the Ontario Ministry of the Environment).

– – – – Primary Treatment Plants

–·–·–· Modified Activated Sludge Plants

———— Conventional Activated Sludge Plants

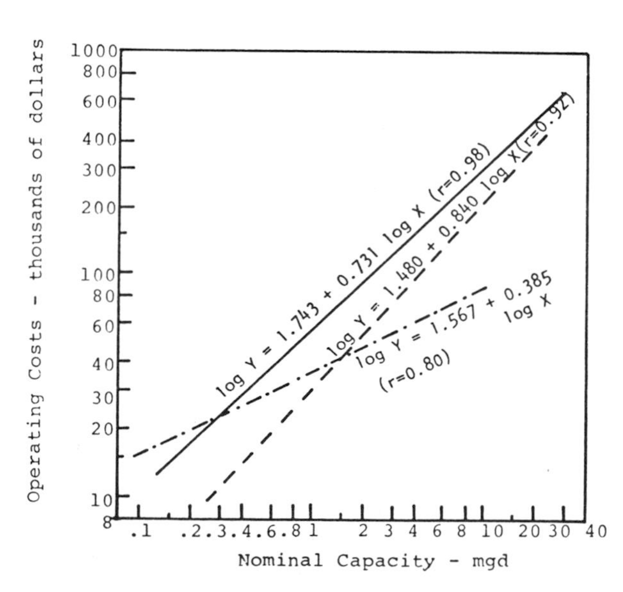

8. Total operating costs in provincially operated wastewater treatment plants (courtesy of the Ontario Ministry of the Environment).

– – – – Primary Treatment Plants

–·–·–· Modified Activated Sludge Plants

———— Conventional Activated Sludge Plants

modified activated sludge plants performed more effectively than the conventional activated sludge plants, and these in turn performed better than primary plants.

Figures 7 and 8 show the unit and total operating costs for these same three classes of treatment. The unit cost analysis suggests that at flows less than 200,000 gpd the modified activated-sludge process is a little more economical than the others.

Summary and Conclusions

Wastewater treatment facilities for rural areas should have special attention paid to certain features. Plants should exhibit reliability without skilled help, be efficient under variable quality and quantity of flow, blend into the countryside without unsightly features, generate little solids for further handling, and, finally, be relatively free from odors.

The design of such plants should consider the quality of effluent required in the light of characteristics of the receiving stream, its susceptibility to eutrophication, and the subsequent uses of said receiving stream. Design should also consider and provide for availability of land for accommodating the treatment unit and also for final disposal of effluent and/or sludge. The availability of skills for maintenance of the plant should also be a weighted consideration. Naturally, all designs consider the quality and quantity of wastes and the collection system, if any. A series of alternative processes which have application in rural waste-treatment situations follow.

Oxidation ponds in all their varieties have grown in popularity, because of their simplicity and low cost. The large area of land required is one major drawback of this process. A variety of activated sludge processes and modifications are considered, and some Ontario data presented suggests that process is less significant than flow. Alternative processes are the oxidation ditch, rotary biodiscs, extended filtration, and submerged bed-aeration system.

Land application may also be used for sewage, effluent, or sludge, but consideration must be given to the crops grown on this land and the possible public health effects either from pathogens or heavy metals.

Effluent polishing may be required, but this is perhaps less likely than nutrient removal. Eutrophication is proving to be a problem in the smaller, shallow warm lakes frequently found in rural Canada and the United States. There are a variety of simple possibilities involving

Table 3
Costs of Ten Selected Wastewater Treatment Plants—Ontario

Town	*Labor*	*Power*	*Chemicals*	*Sludge Haulage*	*Repairs & Maint.*	*Other*	*Total*
Georgetown 1.5 MGD A/S	49.00	12.50	4.60	20.20	6.60	12.30	105.20
Burlington (Drury Lane) 2.5 MGD A/S	29.00	11.45	1.00	4.35	1.35	4.45	51.60
Burlington (Eliz Gdns) 0.75 MGD A/S	34.30	11.10	1.30	8.15	1.80	12.05	68.70
Fergus 0.6 MGD A/S	55.00	9.10	5.80	16.20	2.90	17.50	106.50
Burlington Skyway 6 MGD Ext.	18.40	15.30	0.75	3.65	1.70	3.30	43.10
Elora 0.083 MGD Ext.	138.00	44.00	36.30[a]	—	22.50	54.80	295.60
Moosonee 0.075 MGD Ext.	—	—	—	—	—	—	—
Westminster Twp. 0.25 MGD Ext.	104.00	23.00	3.15	—	15.70	33.94	179.79
Haileybury 0.35 MGD C.S.	77.50						165.00
Meaford 0.86 MGD High Rate	82.00	40.40	3.00	—	13.10	32.50	171.00

[a]Poor effluent required addition of polyelectrolyte.

Courtesy of the Ontario Ministry of the Environment.

chemical addition and process modification which will remove substantial amounts of such nutrients as nitrogen and phosphorus.

The whole field of providing low-cost wastewater treatment facilities for rural areas, though not as scientifically glamorous as the large, complex treatment processes, represents probably the largest capital investment that society will be called upon to make. As such, this area of research should be encouraged, for the potential payoff is enormous.

References

Arden, E., and Lockett, W. T. 1914. Experiments on the oxidation of sewage without the aid of filters. *Journal of the Society of Chemical Industry* 33: 523–539.

Baars, J. K. 1962. *The use of oxidation ditches for treatment of sewage from small communities.* New York: World Health Organization bulletin no. 26.

Baehnke, B. 1966. Treatment: biological discs. *Technique sanitaire et municipale* 61: 362–365.

Bartel, H., and Noack, D. 1966. Advantage of using immersion trickling filters for the artificial biological cleaning of waste water in East Germany. *Wasserwirtsch-Wassertech* 16: 333–338.

Bishop, D. F., et al. 1967. Studies on activated carbon treatment. *Journal of Water Pollution Control Federation* 39: 217–229.

Gloyna, E. F. 1971. *Waste stabilization ponds.* World Health Organization Monograph Series no. 60. New York: WHO.

Jenkins, D., and Orhon, D. 1972. The mechanism and design of the contact stabilization activated sludge process. In *Proceedings of the Sixth International Conference on Water Pollution Research*, pp. B/5/10/1–B/5/10/10. Elmsford, New York: Pergamon Press.

Jones, P. H. 1965. Increased capacity for activated sludge plants at minimum cost. *Canadian Municipal Utilities* 103: 19–20.

Jones, P. H. 1968. Waste water treatment by contact stabilization at Penetanguishene Ontario. *Water and pollution control* 106: 34–35, 38–39, 43.

Jones, P. H. 1970. A mathematical model for contact stabilization - modification of the activated sludge process. In *Proceedings of the Fifth International Conference on Water Pollution Research*, pp. II–5/1 to II–5/7, ed. S. H. Jenkins. Elmsford, New York: Pergamon Press.

Jones, P. H., et al. 1969. *Report to the International Joint Commission on the Pollution of Lake Erie, Lake Ontario and the International Section of the St. Lawrence River.* 3 vols. International Lake Erie Water Pollution Board and the International Lake Ontario–St. Lawrence River Water Pollution Board.

McKinney, R. J., and O'Brien, W. J. 1968. Activated sludge—basic concepts. *Journal of Water Pollution Control Federation* 40: 1831–1843.

Merten, U., and Nusbaum, I. 1968. Organic removal by reverse osmosis. Paper read at 156th national meeting of the American Chemical Society, March 31–April 5, San Francisco.

Ontario Water Resources Commission. 1964. *Evaluation of the oxidation ditch as a means of waste water treatment in Ontario*. Toronto: OWRC, Research Division publ. no. 6.

Pasveer, A. 1963. Developments in activated sludge treatment in the Netherlands. In *Advances in biological waste treatment*, ed. W. W. Eckenfelder and B. J. McCabe, pp. 291–297. New York: Macmillan Company.

Sawyer, C. N., Nichols, M. S., Rohlich, G. A. 1939. Activated sludge oxidations. *Sewage works journal* 11: 51–67, 462, 595–606, 946–964.

Ullrich, A. H., and Smith, M. W. 1951. The biosorption process of sewage and waste treatment. *Sewage and industrial wastes* 23: 1248–1253.

Van Lonn, J. 1972. Where have all the metals gone? *Water and pollution control* 110: 20–21.

Welch, F. M. 1969. New approach to aerobic treatment of wastes. *Water and wastes engineering* 6: D/12–D/15.

26.
Wastewater Treatment for Small Communities

George Tchobanoglous

During the past ten years, concern over the design, construction, and operation of large regional systems has all but overshadowed interest in small treatment plants; yet it is interesting to note that over 75 million people in this country are served by small systems of various design. Recently, as the words "water quality control" have taken on new meaning, many of the problems associated with the design and operation of small treatment plants have become more painfully apparent. Therefore, the purposes of this paper are (1) to define some of the general problems associated with small plants, (2) to review alternative treatment processes that can be used for small communities, (3) to review some design considerations that are of major importance for small plants, and (4) to compare the alternative treatment processes from an economic standpoint.

Problems with Small Plants

The question might be raised, why discuss the problems associated with small plants before considering the various processes? The answer is to provide prospective for viewing the suitability of treatment alternatives relative to the real problems that have plagued most small plants. These problems can be divided into four general categories dealing with design, operation and maintenance, budget limitations, and communications.

Before discussing any of these problems it is important to define what is meant by a small plant. It is difficult to categorize treatment plants as small, medium, and large because any division must be based not only on flow but on the treatment function and the degree of treatment provided. For the purpose of this paper, treatment plants for a dry-weather flow of 5,000 gallons up to one million gallons per day, regardless of complexity, are considered to be small. The lower limit has been set to exclude septic-tank treatment systems from this discussion, as they have been amply covered elsewhere in this volume.

Design

The design of small treatment plants is an area that has not received the attention from engineers that it requires or deserves. Many times, large firms will pass up the design of small plants or assign them to a young, inexperienced engineer. In other cases, small firms with little or no experience have undertaken the design of these plants. Consequently, many of them have now proved to be inadequate, especially in meeting the more stringent discharge requirements established by various federal and state agencies. Clearly, the design of small plants that work is the responsibility of the engineer and not the operator or the contractor.

The successful operation of a treatment plant large or small is dependent on the adequacy of the design concepts as well as the physical implementation of these concepts. If the plant is poorly conceived and designed, no amount of operator skill can rectify the situation. Properly designed treatment plants should and can operate with a minimum amount of time spent on process control. In a recent article Warren R. Uhte (1970) noted that the following items must be incorporated into the design of an activated sludge treatment process: flexibility, reliability, freedom from extraneous problems, capacity, adequate control tools, problem anticipation, and a balanced design. These factors are especially important in the design of small treatment plants. Some critical factors related to the design of small plants are considered in the section dealing with special design considerations.

The disposal of sludge is currently another difficult problem faced by the operators of many small plants. Methods used in the past are no longer acceptable or adequate. For example, many of the smaller plants in the San Francisco Bay area that were once located in the lonely marshlands now find themselves in the midst of a growing residential and commercial community. Consequently, methods such as spreading raw sludge to dry before disposal are no longer tolerated. Because the disposal of sludge was not given adequate consideration in the original design, many operators now find themselves having to deal with this problem with little or no financial or technical help. In the future, it is hoped that sludge disposal will be one of the starting points in the design of small plants.

Operation and Maintenance

The ease with which small plants can be operated and maintained is of prime importance to the operator because he must live with the plant day in and day out. Here again it is the responsibility of the engineer to select methods of process operation and to specify equipment that will reduce the work associated with the operation and maintenance of small plants. Where possible, treatment processes should be designed so that they can be operated with a minimum of laboratory testing. The activated-sludge process is considered in this respect in the section dealing with special design considerations.

Under the tight operating budgets of most small plants, maintenance is a continuing headache for the operator. For this reason the small plant is no place to try an unknown piece of equipment or machinery. In selecting equipment, care should be taken to ensure that adequate service and parts centers are available in case of breakdown. Most importantly, by selecting proven equipment, the annual maintenance and repair bill can be reduced considerably and the overall operation improved.

Budget Limitations

For most large cities the legislative process calls for the city council to review the budget for the entire city as made up by the budget requests of the various departments comprising city government. Normally, sewer service charges are not adequate for needed expenditures, so an overrun is included in the tax rate for maintenance, operation, and capital improvements to the waste treatment system. In small cities, because every additional penny on the tax rate is felt throughout the community, most city councils are extremely sensitive to tax increases that do not produce a tangible asset to the community, such as a library or a city park. Capital improvements to the city's waste treatment system are not something that a voter can identify with, even in these times when pollution control is a topical item. Because of this, the small plant is, in effect, put on the welfare side of the budget and is the first in line when cuts are to be made in annual expenditures. Often the plant is put on emergency budgeting, and only when it gets into trouble with the local pollution control agency does the council consider its budget seriously. Small plants owned and operated by small special districts fare somewhat better when it comes to budgeting because they are single purpose agencies and the governing boards are concerned principally with water pollution control.

The budgetary limitations under which most small plants operate are reflected in both plant operation and maintenance staffing. Some of the problems related to operation and maintenance have been discussed previously. Treatment plant staffing is an area seriously affected by budgetary limitations. Usually it is impossible to staff small plants on a twenty-four hour basis. Thus the chance that something might happen to the process during the period when it is unmanned always exists. Limited staffing often prohibits on-the-job training programs because there are no free hours available for the purpose. Further, the agencies responsible for the operation of these facilities often provide little or no funds during the year for the periodic services of a consulting engineer. Normally, the plant is designed and constructed, the consulting engineer paid, and the operator left on his own to operate it as best he can. City engineers are normally either too busy with other matters or, when difficult operational problems arise, are not knowledgeable or experienced enough to assist in solving them. By the time the problem is worked out, the overall effectiveness of the plant on a yearly basis is considerably reduced.

Perhaps the most serious consequence of a limited budget is the salary differential that exists between large and small plants. Based on the results of a salary survey taken in the San Francisco Bay area it was found that superintendents and chief operators of plants with flows of 5 MGD or less received, on the average, a salary approximately 25 to 30 percent lower than operators of similar but larger plants. Although the operators of large plants have greater responsibility as a consequence of the larger employee complement and capital investment, it should be noted that the degree of technical expertise required for both large and small plants is the same. The different salary scales make it difficult if not impossible to recruit and keep a qualified staff in a small plant. In turn, this is reflected ultimately in the overall performance of the plant.

Communications

In the operation of small or large treatment plants a number of communication levels exist. For example, they exist between the operating staff and the chief operator and the superintendent, between the operating staff and the design engineer, and between the superintendent and the public works director. Of these, communications between the operating staff and the design engineer or consultant is a must for effective plant operation, and especially when a new plant is to be built. In planning for a new facility the staff should have the opportunity to express to the design engineer their ideas on layout, operation and maintenance, and equipment selection. The often heard cliché that the operator cannot articulate his ideas is unacceptable, for it probably means that the design engineer has not taken the time or made the effort to understand the operator's point of view. Often, communication difficulties can be overcome by holding discussions under more relaxed conditions, such as over a beer. In essence, what is required is involvement and understanding on the human level. It is worth the effort because it benefits everyone concerned and generally will result in improved morale and in the construction of a plant that will be easier to operate and maintain.

Alternative Treatment Processes

Over the years a variety of different processes have been developed for the treatment of domestic wastes. In fact, there are so many variations that it is impossible to list or discuss all of them. Rather, the approach to be followed in this presentation is to select and discuss the processes that have proven successful for application to small communities. It will be apparent that numerous combinations are possible with the systems presented. The material has been organized into four sections, dealing with (1) treatment function, (2) description of process flowsheets, (3) process and performance characteristics, and (4) process design data.

Treatment Function

Processes used for wastewater treatment can be grouped according to treatment function. The principal functions are: (1) removal of the suspended and floating material found in wastewater, (2) the stabilization of the organic matter and removal of the nonsettleable colloidal solids present in the wastewater, and (3) the removal of residual suspended material and specific constituents.

Suspended material is usually removed with physical unit operations such as plain sedimentation, whereas the removal of organic substances that are soluble or in the colloidal size range is accomplished most efficiently with biological unit processes. Residual solids and specific constituents are removed by a combination of physical, chemical, and biological processes. Although chemical treatment has been used, it is not recommended for small communities because of the uncertainties associated with the long-term availability of chemicals, especially for small users, and the added costs involved with sludge handling. Flowsheets and corresponding typical performance data for various processes designed to perform these functions are presented and discussed in the following sections.

Flowsheets

Flowsheets and typical application data for representative treatment processes for small communities are given in Figure 1 and Table 1, respectively. The flowsheets are discussed below.

Imhoff tank. The Imhoff tank consists of a two-story tank in which sedimentation is accomplished in the upper compartment and the digestion of solids is accomplished in the lower compartment. In terms of effluent quality the performance of such units is about the same as achieved using a separate primary sedimentation tank with external solids processing facilities. Sludge is removed periodically from the digestion compartment and spread to air dry, usually on sand beds.

Rotating biological-disk process. In this process a series of circular disks, constructed of various materials, are mounted on a central shaft. The disks are then partially submerged and rotated in a tank containing the wastewater to be treated. Treatment of the organic material is accomplished by bacteria that become attached to the disks. Biological solids that slough from the disks are removed by settling before the treated wastewater is discharged.

In its usual configuration this process would be preceded by a primary settling tank. Waste biological solids from the secondary settling tank can be thickened in the primary settling tank or in a separate thickener. Solids are usually digested to reduce their organic contents so that they may be dewatered or air dried more easily. The digestion process can be either aerobic or anaerobic depending on local conditions. Although easier to operate, aerobic digestion has the disadvantage of requiring a continuous power input. Design criteria for both processes are given at the end of this section.

Trickling-filter processes. In the tricking-filter processes the liquid to be treated is applied over solid media to which microorganisms have become attached. Rock, slag, plastic, and redwood slats have been used for this purpose. Conventional filters filled with rock or slag vary in depth from 4 to 10 feet, whereas plastic media and redwood slats are employed in tower filters that vary in height from 20 to 40 feet. The process flowsheet most often used is essentially the same as that described above for the rotating biological-disk process (see Figure 1).

Activated-sludge processes. The five processes included in this grouping are similar, with the exception that the first two are assumed to be used with external digestion (see Figure 1). In the last three it is assumed that the biological solids produced during treatment are di-

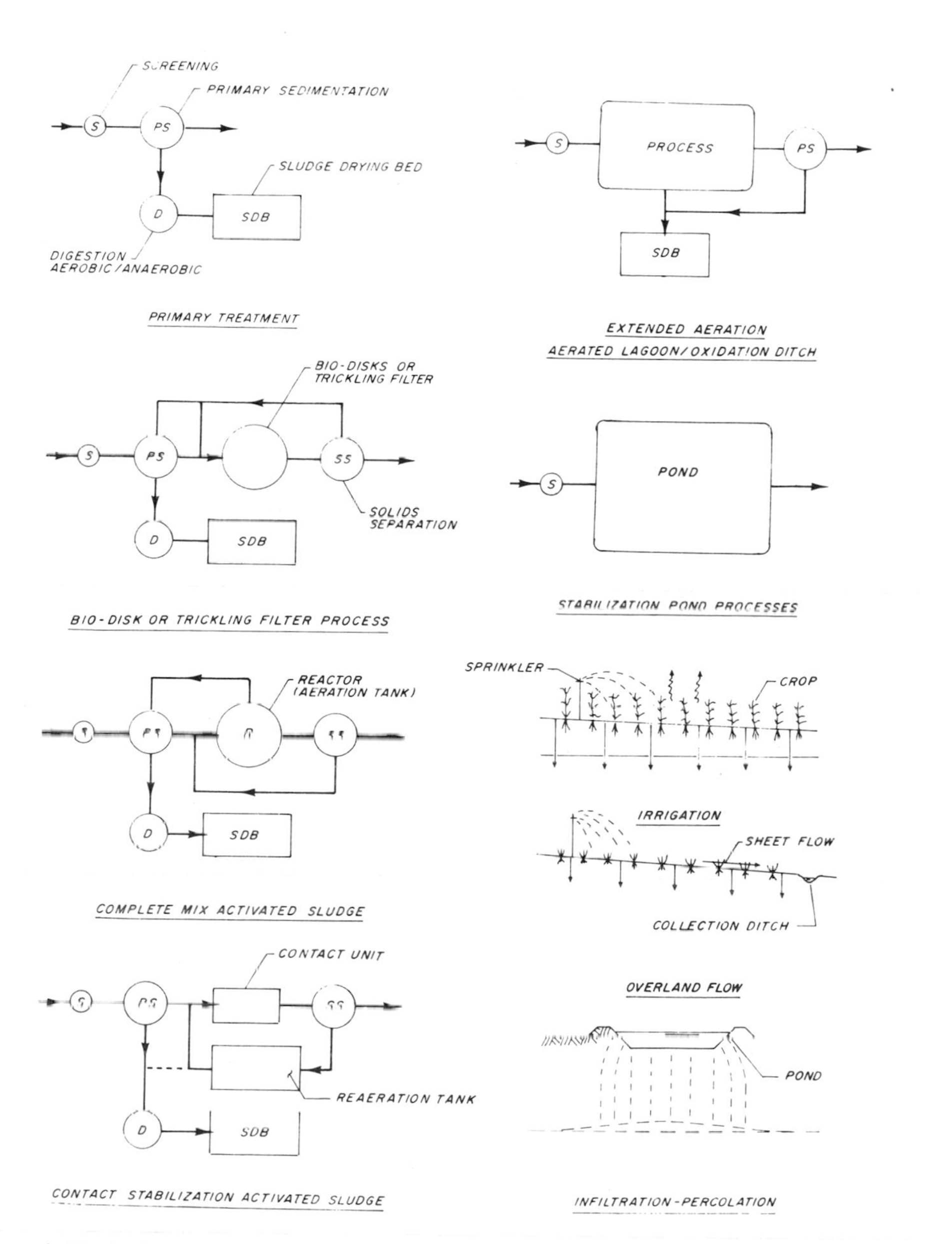

1. Typical process flowsheets (adapted in part from Pound and Crites, 1973).

Table 1
Treatment Processes for Small Communities

Process	*Application*
Imhoff tank	Very small communities, housing tracks, and small institutional facilities where primary treatment is acceptable.
Rotating biological disks	Domestic wastes; small communities; may be covered for odor control; where growth of filamentous organisms may be a problem.
Trickling-filter processes Conventional (low rate) Conventional (high rate) Tower filters	Domestic wastes; low application rates are necessary to achieve high removal efficiencies; nitrification can be achieved with very low loading rates; can be used where growth of filamentous organisms may be a problem.
Activated-sludge processes	
Complete mix	General application; resistant to shock loads; package plants.
Contact stabilization	Treatment of wastes where BOD is colloidal or suspended; package plants; recreational areas.
Extended aeration	Small communities; package plants; housing developments recreational areas; industrial wastes
Aerated lagoon (with settling) Oxidation ditch (with settling)	General application; lower efficiencies during cold weather because of temperature sensitivity.
Stabilization-pond processes High-rate aerobic Facultative (alyal surface layer) Facultative (aerated surface layer) Anaerobic	Domestic wastes; suitable for use where large land areas are available and effluent quality need not be constant; effluent quality can be improved by adding a series of settling ponds, settling facilities, or mixed media filtration; anaerobic ponds usually followed by aerobic or facultative ponds for improvement of effluent quality.
Land disposal Irrigation Overland flow Infiltration-percolation	Domestic wastes; suitable for use where very large land areas are available; pretreatment of wastes usually required; overall effluent quality similar to that obtained from conventional plus advanced wastewater treatment; sale of crops can significantly reduce annual operating costs; choice of method depends on solid and other local characteristics.

gested internally and discharged periodically to drying beds or sludge lagoons (see Figure 1).

Although the last three processes are often used without secondary settling, it is not recommended; the quality of the effluent will be variable because of solids carryover and will generally even fail to meet the discharge requirements established by state pollution control agencies. With the addition of secondary settling and solids return, the latter three processes are fundamentally the same as the first two with the exception of solids handling. If used without secondary settling, these processes can be followed by aerobic stabilization ponds for the removal of solids. Such ponds are often called "maturation ponds" (Caldwell et al., 1973) and are lightly loaded (5 to 15 lb. BOD/acre/day) to achieve the required effluent quality. Such ponds are also used to polish or upgrade the effluent from conventional activated sludge and tricking filter processes.

Stabilization-pond processes. The types of ponds or lagoons most often used may be classified with respect to biological activity as high-rate aerobic, facultative, and anaerobic (see Table 1). In high-rate aerobic ponds, approximately 80 to 85 percent of the organic material in the incoming wastewater is converted to algae (Caldwell et al., 1973). The problem with this process is the removal of algae. Although maturation ponds have been used in some locations, this problem has limited their widespread use.

Facultative ponds, by far the most common, have been used extensively throughout the United States. Stabilization is brought about by both aerobic oxidation and anaerobic digestion (Metcalf and Eddy, 1972). Two different approaches have been used to maintain an aerated upper layer. In the first, oxygen in the upper layer is maintained by the presence of algae. In the second, oxygen is maintained with surface aerators or plastic tube aerators located on the bottom of the pond. The second method is preferred in cold-climate locations where the continuous maintenance of algae in the surface layers may be a problem. Because the effluent quality from facultative lagoons is variable, facultative ponds are often followed in series by maturation ponds.

Ponds in which the organic loading is so high that anaerobic conditions develop throughout are called anaerobic ponds. They have been widely used for partially reducing the organic content of various industrial wastes. When used for domestic waste treatment, anaerobic ponds are usually followed by facultative and/or aerobic ponds to reduce the BOD and suspended solids in the effluent (Caldwell et al., 1973).

Land-disposal processes. The disposal of wastewater on land can be accomplished by irrigation, overland flow, and infiltration-percolation. Irrigation involves the controlled discharge of (processed) wastewater to the land to support plant growth. Wastewater can be applied by sprinkler, by flooding, or by ridge and furrows. In this process most of the wastewater is lost by evapotranspiration and deep percolation (Pound and Crites, 1973).

Overland flow involves the controlled discharge of (processed) wastewater onto graded land (2 to 4 percent), where it flows in a thin layer down the grade and appears as runoff. The land, which should be relatively impermeable, is generally planted with a grass cover crop to provide a habitat for microorganisms, to serve as a living filter, and to prevent erosion (Pound and Crites, 1973). To date, Reed Canary grass has been most commonly used.

Infiltration-percolation involves the application of (processed) wastewater to spreading basins, where it is allowed to percolate down to the groundwater. Because application rates are normally measured in feet, the underlying soils must be highly permeable. To maintain an adequate infiltration capacity the basins are closed on an intermittent basis (Pound and Crites, 1973).

In discussing each of the land-disposal processes it will be noted that the word "processed" in front of "wastewater" has been set in parentheses. The reason for this is that the need for pretreatment, especially with the first two processes, has not been established firmly (Pound and Crites, 1973). As a minimum, screening would be required to protect equipment. With proper wastewater pretreatment, land application can be considered as an alternative to conventional and advanced wastewater treatment (Pound and Crites, 1973).

Process and performance characteristics

Process and performance characteristics for the various processes discussed previously are considered in this section.

Operational characteristics. Important operational characteristics for the various processes have been summarized and are presented in Table 2. As shown, most of these factors are significant from the standpoint of operation and plant performance. Because many of these characteristics are often not considered in the preparation of cost estimates, inappropriate processes have been selected. To avoid such mistakes, the characteristics should be reviewed carefully in light of staffing and budgetary limitations.

Table 2
Operational Characteristics of Various Treatment Processes

Item	*Process*					
	Rotating Disk	*Trickling Filters*	*Act. Sludge External Digestion*	*Act. Sludge Internal Digestion*	*Facultative Ponds*	*Land Disposal*
Process characteristics						
Reliability with respect to						
Basic process	Good	Good	Good	Very Good	Good	Excellent
Influent flow variations	Fair	Fair	Fair	Good	Good	Good
Influent load variations	Fair	Fair	Fair	Good	Good	Good
Presence of industrial waste	Good	Good	Good	Good	Good	Good
Industrial shock loadings	Fair	Fair	Fair	Good	Fair	Good
Low temperatures (20°C)	Sensitive	Sensitive	Good	Sensitive	Very Sensitive	Good
Expandability to meet						
Increased plant loadings	Good, must add additional disk module	Limited	Fair to good if designed conservatively	Good, ultimately more volume will be required	Fair, additional ponds required	Good
More stringent discharge requirements with respect to						
Suspended solids	Good, add filtration or polishing ponds	Good, add filtration or polishing ponds	Good, add filtration or polishing ponds	Good, add filtration or polishing ponds	Add additional ponds and filtration	Additional pretreatment
BOD	Improved by filtration	Improved by filtration	Improved by filtration	Improved by filtration	Improved by filtration	Lower application rates
Nitrogen	Good, denitrification must be added	Good, denitrification must be added	Good, nitrification-dentrification must be added	Good, denitrification must be added	Fair	—
Operational complexity	Some	Some	Moderately complex	Some	Simple	Simple
Ease of operation and maintenance	Very good	Very good	Fair	Excellent	Good	Excellent
Power requirements	Low	Relatively high	High	Relatively high	Low	Moderate
Waste products	Sludges	Sludges	Sludges	Sludges	Sludges	—
Potential environmental impacts	Odors	Odors	—	—	Odors	—
Site considerations						
Land area requirements	Moderate plus buffer zone	Moderate plus buffer zone	Moderate plus buffer zone	Large plus buffer zone	Large plus buffer zone	Large
Topography	Relatively level	Relatively level	Relatively level	Relatively level	Relatively level	Not critical

Performance data. Of major importance in the selection of any process is knowledge concerning expected performance. Estimated performance data for various wastewater treatment processes are reported in Table 3. These data were derived from a variety of sources and represent the effluent quality that can be achieved when the processes are properly designed and operated.

As will be noted, effluent quality for the rotating biological-disk, trickling-filter, and activated-sludge processes will be similar under ideal conditions. The choice usually depends on local conditions. It should be noted that performance differences reported in the literature between and among these processes were often caused by inadequately designed secondary settling tanks and therefore are of little value. Appropriate design criteria are given below.

Design Data

Typical design data have been summarized in Tables 4 through 8 for each of the processes listed in Table 2 with the exception of the Imhoff tank. Data on aeration equipment and digestion processes have also been included and are reported in Tables 9 through 11.

It is intended that these data be used as a guide in the preliminary sizing of processes and in checking proposed design criteria. They are not meant to be used as a substitute for a complete and detailed design analysis. Process design details may be found in several references (Caldwell and Uhte, 1973; Eckenfelder, 1970; McKinney, 1970; Metcalf and Eddy, Inc., 1972; Salvato, 1972; Austin Center for Research, 1971). A few special design considerations are discussed in the following section.

Treatment process data. Data for the rotating biological-disk, trickling-filter, activated-sludge, stabilization-pond, and land-disposal processes are reported in Tables 4 through 8 respectively. The range of values given for each design variable in these tables was selected so as to be consistent with the effluent quality given in Table 3. Although higher values will be found in the literature, the reported values are recommended.

Aeration equipment. Because aeration equipment is used in so many of the treatment processes, it was felt that some design information and data should be included. The types of aeration equipment commonly used are shown schematically in Figure 2. General information about the various types of aeration equipment may be found in Table 9. Although pure oxygen is also used, cost considerations normally limit its use in small plants.

Table 3
Estimated Performance Data for Alternative Wastewater Treatment Processes[a]

Process	*Constituent*						*Waste for Ultimate Disposal*
	SS	*BOD*	*COD*	*N*	NH_3	*P*	
	Incoming Wastewater						
	225	*200*	*450*	*40*	*25*	*10*	
	Effluent from Treatment Process						
Imhoff tank	80	120	350	35	25	9	Sludge
Rotating biological disks	25	18	100	25	3	7	Sludge
Trickling-filter processes							
Conventional (low rate)	25	18	100	25	1	7	Sludge
Conventional (high rate)	30	20	100	30	25	7	Sludge
Tower filter	30	20	100	30	25	7	Sludge
Activated-sludge process							
Complete mix	20	15	90	25	20	7	Sludge
Contact stabilization	20	15	90	25	20	7	Sludge
Extended aeration	20	15	90	30	2	7	Sludge
Aerated lagoon (with settling)	20	15	90	30	2	7	Sludge
Oxidation ditch (with settling)	20	15	90	30	2	7	Sludge
Stabilization pond processes							
High-Rate aerobic	170	60	200	30	1	9	
Facultative (algal surface layer)	120	40	160	20	1	4	Sludge[c]
Facultative (aerated surface layer)	80	25	140	20	1	4	Sludge[c]
Anaerobic[b]	100	40	140	20	1	4	Sludge[c]
Land-disposal processes[d]							
Irrigation	2	4	80	6	1	0.5	
Overland flow	4	6	90	10	4	3	
Infiltration-percolation	2	4	50	10	1	0.5	

[a]Under ideal conditions.
[b]Usually followed by aerobic or facultative ponds.
[c]Sludge accumulated in ponds.
[d]Following pretreatment.

Table 4
Typical Design Data for Rotating Biological Disk Process[a]

Item	*Range*
Hydraulic loading rate, gpd/sq. ft.	0.5-2
Organic loading rate[b], lb. BOD/day/1000 sq. ft.	2-6
Detention time, min.	50-70
Temperature coefficient, θ (uncovered)	1.02-1.04
Peripheral speed of shafts, fpm	4-10
Number of disks per shaft	40-60
No. of shaft-disk assemblies in series	4-5
Secondary settling tank, lb/sq. ft./hr at peak flow	1.25-1.50

[a]Adapted in part from Autotrol Corporation, 1971; George A. Hormel and Company, 1972; Lager and Smith, 1973.

[b]Loading range to produce a high quality effluent after settling (see Table 3) at temperatures above 60 F.

Typical performance data are summarized in Table 10. In specifying aeration equipment, caution should be exercised in accepting manufacturer's claims concerning field performance, especially in untried situations. As a rule of thumb, 2.0 lb O_2/hr/hp should be used for estimating purposes when using floating surface aerators. Also, under continuous service the cost of aeration will be about $72/hp/yr when the cost of power is $0.01/KWH.

Digestion process data. Typical design data and related information for both anaerobic and aerobic digestion processes have been gathered and are reported in Table 11. With either process the volatile solids reduction will generally be about 40 or 45 percent. In small plants the usual practice is to dewater the sludge by allowing it to air dry on sand beds or shallow lagoons. Although vacuum filters or centrifuges have been used to dewater the sludge, the cost associated with their operation and maintenance normally precludes doing so in small plants.

Special Design Considerations

Although attention should be paid to all aspects of design, there are a few areas that are critical from the standpoint of plant design and operation. They are related to (1) the characteristics of the incoming wastewater, (2) the design of secondary settling tanks, (3) plant control, (4) maintenance, and (5) power requirements.

Table 5
Typical Design Data and Information for Trickling Filter Processes

	Conventional Filters		
Item	*Low Rate*	*High Rate*	*Tower Filter*
Hydraulic loading,[a, b] gpd/sq. ft.	30–60	100–200	200–600
Organic loading,[b] lb. BOD/day/1000 cu. ft.	6–12	20–60	60–140
Temperature coefficient, θ	1.02–1.06	1.02–1.04	1.02–1.04
Depth, ft.	6–10	4–10	20–40
Recirculation ration, R/Q	None	Up to 4/1	Up to 4/1
Packing material	Rock, slag	Rock, slag	Plastic or redwood slats
Dosing interval	Not more than 5 min.	Continuous	Continuous
Sloughing	Intermittent	Continuous	Continuous
Nitrification	Usually nitrified	Nitrification at low loadings	Nitrification at low loadings

[a]Hydraulic loading rates based on plan flow.
[b]Loading range to produce high quality effluent after settling (see Table 3).

Table 6
Typical Design Data for Activated Sludge Processes[a]

	Process				
Item	*Complete Mix*	*Contact Stabilization*	*Extended Aeration*	*Aerated Lagoons*	*Oxidation Ditch*
Mean cell residence time, θ_c, 1/day	6–12	6–12	20–30	10–30	20–30
Food-to-microorganism ratio, F/M lb BOD/lb MLVSS/day	0.2–0.4	(0.2–0.6)[b]	0.05–0.15	0.05–0.2	0.03–0.10
Volumetic loading lb BOD/1000 cu. ft./day	50–100	30–80	10–25	60–70	10–20
Temperature coefficient, θ	1.0–1.02	1.0–1.02	1.06–1.09	1.06–1.09	1.06–1.09
Mixed liquor suspended solids, MLSS, mg/L	2000–5000	1000–3000 (4000–10000)[c]	3000–6000	2000–3000	3000–8000
Volatile fraction of MLSS	0.7–0.9	0.6–0.9	0.6–0.8	0.6–0.8	0.6–0.8
Hydraulic detention time, θ_h, hr.	2–6	0.3–0.5 (3–6)[c]	18–36	0.5–6 (days)	0.5–4 (days)
Recycle ration, R/Q	0.25–1.0	0.25–1.0	0.5–1.5	0.25–0.75	0.25–0.75

[a]Adapted from Metcalf and Eddy, Inc., 1972.
[b]Contact unit.
[c]Solids stabilization unit.

Table 7
Typical Design Data and Information Stabilization Pond Processes[a]

Item	High Rate	Facultative (alyal surface layer)	Facultative (aerated surface)	Anaerobic
	Process			
Organic loading, lb BOD/acre/day	60–120	15–60	30–100	400–800
Power requirements hp/1000 cu. ft.	0.05–0.2 (mixing)	None	0.05–0.7 (aeration)	None
Detention time, days	4–10	7–30	7–20	20–50
Depth, ft.	1–1.5	3–6	3–10	8–15
Pond size, acres	10	10	10	2
Number of ponds	2–6	2–6	2–6	2–6
Operation	Series/parallel	Series/parallel	Series/parallel	Series/parallel
Pond Configuration				
Shape	Not Important	Not Important	Not Important	Not Important
Inlet	Multiple-entry arranged to obtain maximum dispersion of incoming wastewater	Multiple-entry near bottom in center	Multiple-entry near bottom in center	Multiple-entry near bottom in center
Outlet	Multiple or single exit	Multiple-exit designed to reduce algae carryover	Multiple or single exit designed to reduce algae carryover	Multiple or single exit

[a]Adapted from Caldwell et al., 1973; Lager and Smith, 1973; Metcalf and Eddy, Inc., 1972.

Table 8
Typical Design Data and Operational Characteristics for Land Disposal Processes[a]

	Process		
Item	*Irrigation*	*Overland Flow*	*Infiltration-Percolation*
Liquid loading rate, in/wk	0.5–4	2–5.5	4–120
Annual application, ft.	2–8	8–24	18–500
Land required for 1-mgd flow, acres	140–560 (plus buffer zone)	46–140 (plus buffer zone)	2–62 (plus buffer zone)
Application techniques	Spray or surface	Usually spray	Usually surface
Soils	Moderately permeable with good productivity when irrigated	Slowly permeable such as clay loams and clay	Very permeable soils such as sand, loamy sands, and sandy loams
Probability of influencing ground water quality	Moderate	Slight	Certain
Needed depth to ground water, ft.	5	Undetermined	15
Wastewater lost to	Predominantly by evaporation and deep percolation	Surface discharge dominates over evaporation and percolation	Percolation to ground water
Use in cold climates	Fair, conflicting data	Insufficient data	Excellent

[a]From Pound and Crites, 1973.

Table 9
Types of Aeration Equipment[a]

Type	*Application*	*Depth at which commonly used, ft*	*Advantages*	*Disadvantages*
Floating mechanical aerator	Aerated lagoons, facultative ponds.	10–15	Good mixing and aeration capabilities; easily removed for maintenance.	Ice problems during freezing weather; ragging problem without clogless impeller.
Rotor aeration unit	Oxidation ditch aerated lagoons.	3–10	Probably unaffected by ice; not affected by sludge deposits.	Possible ragging problem.
Plastic tubing diffuser	Aerated lagoons; facultative lagoons.	3–10	Not affected by floating debris or ice; no ragging problem.	Calcium carbonate build-up blocks air diffusion holes; is affected by sludge deposits.
Air guns	Aerated lagoons; deep lagoons; lakes.	12–20	Not affected by ice; good mixing.	Calcium carbonate build-up blocks air holes; potential ragging problem; affected by sludge deposits.
INKA system	Aerated lagoons; activated sludge reactors.	8–15	Not affected by ice; good mixing.	Potential ragging problem.
Helical diffuser	Aerated lagoons.	8–15	Not affected by ice; relatively good mixing.	Potential ragging problem; affected by sludge deposits.
Simplex cone	Activated sludge.		Good mixing and aeration capabilities.	
Dissolved air aeration	Activated sludge reactors of various types.		Good aeration and mixing capabilities.	

[a]Adapted from Lager and Smith, 1973.

Table 10
Typical Performance Data for Different Aeration Equipment[a]

Type	*Value Range*	*Remarks*
Floating surface aerator, lb. O_2/hp/hr	1.8–4.5	
Floating rotor aerator, lb. O_2/hp/hr	3.5–4.2	Available in various lengths
Plastic tube aerator		
Transfer rate, lb. O_2/hr/100 ft	0.2–0.7	
Air supplied, scfm/100 ft.	1–2	Horsepower will depend on system piping configuration
Air gun aerator		
Transfer rate, lb. O_2/hr/unit	0.8–1.6	
Air supplied, scfm/unit	3–18	
INKA aerator		
Transfer rate, lb. O_2/hr/1000 cu. ft.	10–100	log-log relationship
Air supplied, scfm/1000 cu. ft.	20–250	log-log relationship
Helical aerator		
Transfer rate, lb. O_2/hr/unit	1.2–4.2	
Air supplied, scfm/unit	8–30	
Simplex surface aerator		
Transfer rate, lb. O_2/hp/hr	3.0–4.2	
Rotation speed, rpm	30–45	
Dissolved air aeration		Saran tub diffusers
Transfer rate, lb. O_2/hr/unit	0.3–1.0	log-log relationship
Air supplied, scfm/unit	2–10	log-log relationship
Rotor aerator		Available in various lengths up to about 18 ft. for use in oxidation ditches
Transfer rate, lb. O_2/hp/hr	3.0–5.0	
Transfer rate, lbs. O_2/hr/ft of length	1.0–3.0	

[a]Derived, in part, from Eckenfelder, 1970; Lager and Smith, 1973; Metcalf and Eddy, Inc., 1972; Federal Highway Administration, 1972.

Wastewater Characteristics

A common assumption made by most engineers in the design of biological treatment systems is that wastewater is a balanced medium for the growth of bacteria. Although it is true that most wastewaters will support biological growth it is the type of organism or organisms that is important.

Often, small as well as large activated sludge systems have been plagued with the growth of filamentous microorganisms and the problems associated with their growth (Tchobanoglous et al., 1973). It is proposed that many of these problems may be due to trace element deficiencies that are related to the growth rate at which the system is operated. Further, if hydrogen sulfide is generated in the incoming

Table 11
Typical Design Data and Information for Biological Sludge Digestion Processes[a]

Item	Value		Remarks
	Anaerobic Digestion[b]		
	Range	Type	
Flow regime			Complete mix
Mixing equipment			Gas recirculation, mechanical mixers, and combination systems have been used successfully
Mean cell residence time, θ_c, days			
Operating temperature °F			
65	23–33	28	
75	16–24	20	
25	12–18	14	
95	8–12	10	
105	8–12	10	
Hydraulic residence time, θ_h, days			Same as mean cell residence time given above
	Aerobic Digestion		
Flow regime			Complete mix
Mixing equipment			Usually same as aeration equipment
Mean cell residence time, θ_c, days at 20° C			
Activated sludge only	15-20	16	
Activated sludge from plant without primary settling	16-22	20	
Primary plus activated or trickling filter sludge	18-26	24	
Temperature coefficient, θ	1.06-1.10	1.08	For adjusting volume requirements below 20°C
Solids loading, lb. vol. solids/cu. ft./day	.01-.02	0.15	
Oxygen requirement, lb/lb. vol. solids destroyed	1.6-2.2	2.0	Volatile solids, reduction typically varies from 36 to 45 percent
Energy requirements for mixing			
Mechanical aerator, hp/1000 cu. ft.	0.5-1.5	1.0	Depends on basin geometry
Air mixing, scfm/1000 cu. ft.	20-40	30	
Dissolved oxygen level in liquid, mall	1-3	2	

[a]Adapted from Metcalf and Eddy, Inc., 1972.
[b]Single stage digester of first stage of two stage digestion process.

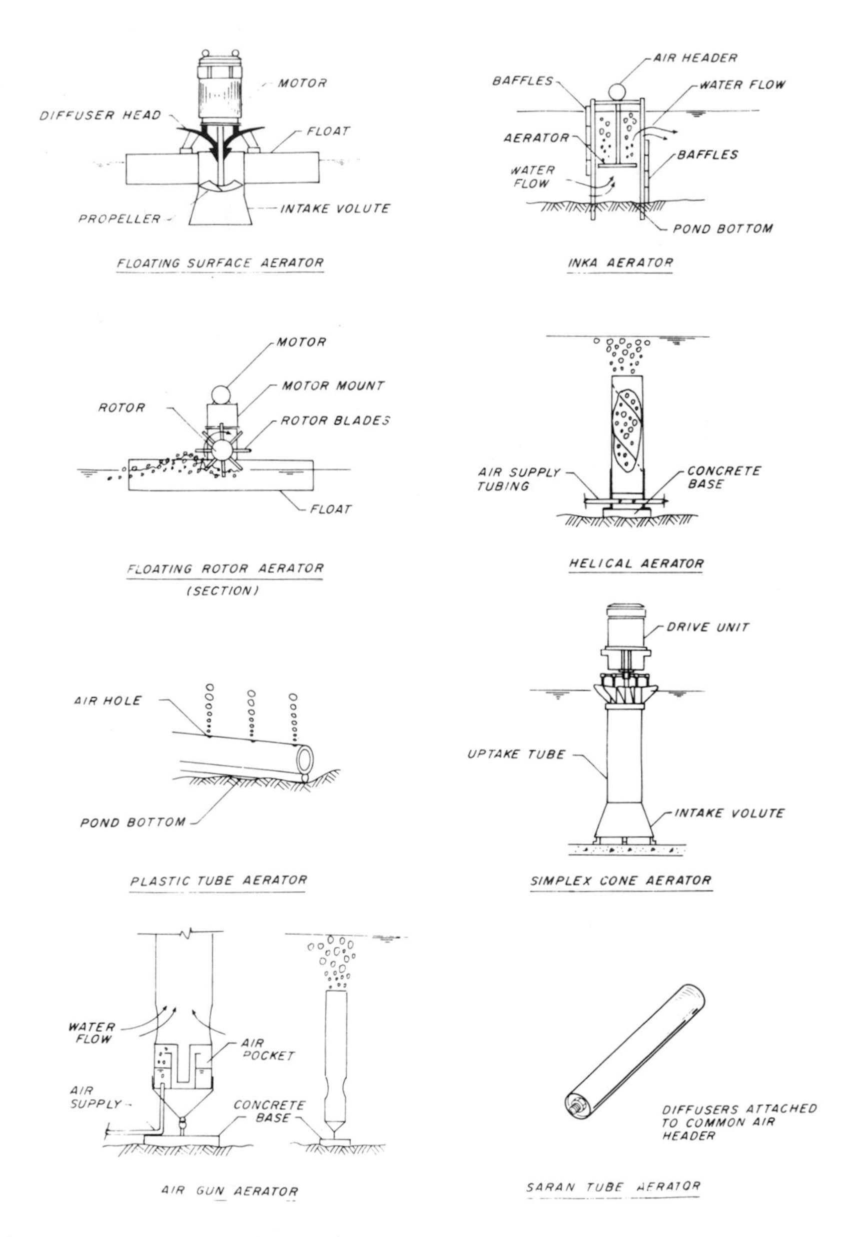

2. Schematic view of various types of aerators (adapted from Metcalf and Eddy, Inc., 1972, and Lager and Smith, 1973).

sewer, problems caused by trace element deficiencies will be aggravated. These topics are considered in more detail below.

Trace element requirements. Numerous studies have found that sixteen elements and three vitamins are necessary for the growth of one or more organisms (Nicholas, 1963). The elements are N, P, K, Ca, Mg, Na, S, Fe, Mn, Cu, Zn, Mo, B, C1, Co, and V. Although other elements have been found in the ash of microorganisms, their necessity for growth has yet to be proven. The required concentration of each of these elements and vitamins is unknown at the present time and, in fact, may never be known. Some general indications can be obtained by considering the composition of E coli reported in Table 12. The data for iron from this table can be related to a typical biological treatment situation, as follows. If the incoming BOD_5 to the aeration basin of an activated-sludge plant is equal to 150 mg/l, and 50 percent of the BOD_5 is converted to cells, then 0.15 mg/l of iron will be required ($0.5 \times 150 \times 0.002$). The actual concentration of iron required in the liquid phase will, of course, depend on local conditions. Quantities for the other elements could be derived similarly if their requirements were known.

Before leaving this subject it is important to note that the trace element requirements for filamentous microorganisms are significantly less than those for E coli. In fact, one of the outstanding characteristics of filamentous organisms is their ability to live on practically nothing. The significance of this observation is considered in the following discussion.

Growth rate and trace-element requirements. The relationship of growth rate to trace-element requirements can be understood more clearly by referring to Figure 3, in which the observed growth rate, Y_{obs}, is plotted against the mean cell residence time, θ_c. For example, the observed growth rate at θ_c value of 4 days is about 4 times greater than the value at 20 days. Now, suppose that the incoming wastewater has just sufficient nutrients to support the growth of "conventional organisms" at a θ_c value of 20 days. If a new system were now designed with a θ_c of 4 days, it is clear that sufficient nutrients would not be present to support the growth of "conventional organisms." What often happens under these circumstances is the development of filamentous organisms which are able to compete effectively with the conventional organisms because of their reduced nutrient requirements.

A similar condition can often occur when a new industry which discharges organic wastes is brought into the system. Suppose, for example, that there are sufficient nutrients in the domestic wastewater to support the growth of conventional organisms at a θ_c value

Table 12
Approximate Elementary Composition of the Microbial Cell (Data for Escherichia Coli)[a]

Element	*Percentage by weight*
Carbon	50
Oxygen	20
Nitrogen	14
Hydrogen	8
Phosphorus	3
Sulfur	1
Potassium	1
Sodium	1
Calcium	0.5
Magnesium	0.5
Chlorine	0.5
Iron	0.2
All others	0.3

[a]From Luria, 1960.

of 4 days. Now, if the organic content in the wastewater is doubled without a corresponding increase in the concentration of trace elements, the resulting waste characteristics will favor the growth of filamentous microorganisms. Following this line of reasoning, it can be shown that because of its low solids production, an extended aeration-activated sludge system could operate successfully where a conventional system would not.

Trace-element precipitation. Trace element deficiencies may be aggravated further if hydrogen sulfide is generated in the incoming sewer. Because this is a common occurrence in many small wastewater collection systems such a deficiency is a very real possibility (Thistlethwayte, 1972). Using iron as an example, its precipitation as ferrous sulfide is given by the following reaction (Hildegrand and Powell, 1964).

$$FeS(s) + 2H^+ \rightleftharpoons H_2S\,(g) + Fe^{++} \tag{1}$$

In addition to iron, zinc, copper, cobalt, and vanadium will also be precipitated as sulfides at the pH values normally encountered in domestic wastewater.

The presence of hydrogen sulfide is also troublesome, for it may inactivate the vitamins found in wastewater which are necessary for balanced microorganism growth.

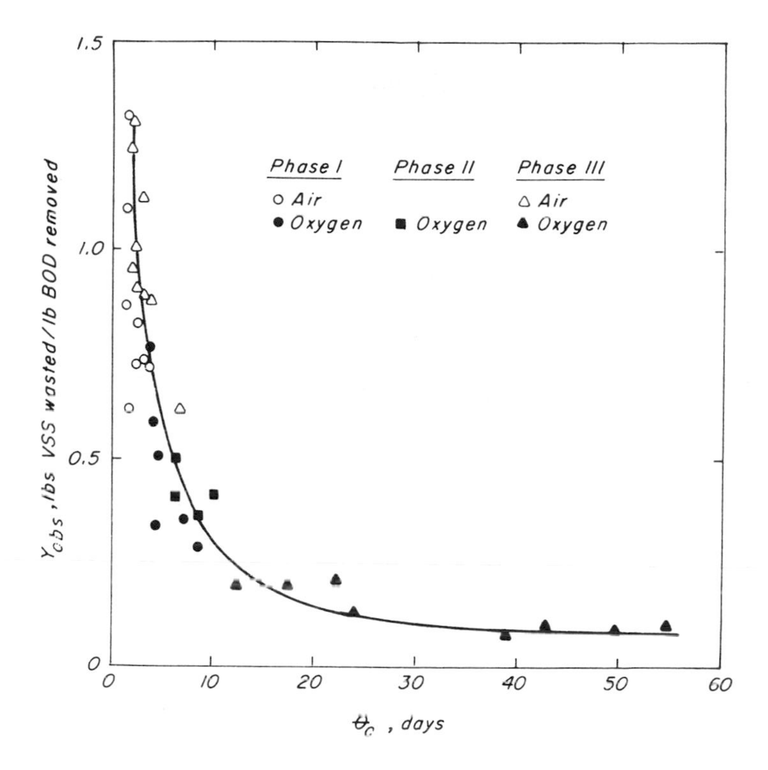

3. Observed cell yield versus mean cell residence time (from Sherrard and Schroeder, 1972).

Design implications. The implications of trace element or nutrient deficiencies in the design of small systems are significant. For example, if the production of hydrogen sulfide occurs in the sewer, it may be necessary to add trace elements when conventional biological treatment is used. On the other hand, land treatment after primary treatment would be suitable for such a waste, since most soils will contain sufficient quantities of required trace elements. In a situation where trace elements are low in the local water supply and in the wastewater, a low-growth system such as extended aeration may be necessary if the activated sludge process is to be employed.

Alternatively, if filamentous organisms are going to develop, it may be appropriate to select a system that will work with these organisms. It should be noted that from a treatment standpoint the filtered BOD_5 of the effluent will be as low as that obtained from a conventional plant (typically in the range from 2 to 8 mg/l). As an example, the use of biodisks may be feasible in such a situation. No matter what process is used, the key issue is to identify whether or not the conditions that might promote the growth of filamentous organisms will occur and to allow for this in the design.

Design of Secondary Settling Tanks

Perhaps the most critical link in the operation of small activated sludge plants is the secondary settling tank(s). The reason is that the process will almost certainly not perform properly unless the mixed liquor solids can be separated and returned to the tank. Although this seems fundamental, the settling tanks for a large number of small activated sludge plants are designed improperly with respect to loading rates and solids removal and return facilities. In the past, overflow rates were often based on average design flows. But to avoid the loss of solids that occurs when the design flow is exceeded, the effluent overflow rates must be based on peak dry weather flow conditions. Overflow rates used in design must also take into account the concentration of mixed liquor solids, the recycle rate, and the solids loading rate. At mixed liquor solids concentrations above about 2,000 mg/l recommended solids loadings for small plants vary from 1.0 to 1.25 lb/sq ft/hr. Below 2,000 mg/l the overflow rate should be limited to values consistent with satisfactory operating experience (800–1,000 gal/sq ft/day). Because of the high peak to average flow ratios for small plants, use of the above loading rates requires that special attention be given the design of the solids removal facilities if problems with rising sludge are to be avoided. In extremely small plants, positive means such as screw conveyors should be used to return the sludge.

Plant control

Process control is especially important in small plants because of manpower limitations. The key word in designing the physical facilities from a process control standpoint should be simplicity. Also, because many small plants are designed to service recreational areas where the load varies seasonally, the operator should be able to adjust plant operations to different modes without the need to have special equipment or facilities. Some aspects of simplicity in the control of the activated-sludge process are considered below.

When using the activated-sludge process, control is usually accomplished by adjusting the food to microorganism, F/M, or by controlling the mean cell residence time, θ_c (Burchett & Tchobanoglous, 1972; Metcalf and Eddy, Inc., 1972). The difficulty with the first method is that the operator must have available some means of measuring the organic content of waste, whether its biochemical oxygen demand (BOD), total organic carbon (TOC), or chemical oxygen demand (COD), and means for measuring the concentration of the organism in the aeration tank. Both of these requirements are often unrealistic at small treatment plants.

In the second method, control is achieved by wasting a given amount of sludge on a daily basis. The mean cell residence time, θ_c , in days can be defined as:

$$\theta_c = \frac{X}{X_w} \tag{2}$$

where X = mass of organisms in the aeration tank, lb

X_w = mass of organisms lost from the system each day including those solids purposely wasted and those lost in the effluent, lb/day

The mass of organisms in the aeration tank is determined by multiplying the tank volume, V, expressed in million gallons, by the cell concentration, X, in milligrams per liter by 8.34. The factor 8.34 is used to convert mg/l per million gallons to lb per million gallons.

In practice, sludge can be wasted from either the aeration tank or the return line from the clarifier. If wasting is from the aeration tank and the solids lost in the effluent are assumed to be negligible, then:

$$\theta_c = \frac{xV}{xQ} \tag{3}$$

where x = cell concentration in aeration tank, mg/l
V = volume of aeration tank, mg
Q_w = waste flow rate, mgd

Thus, for a given value of θ_c the wasting rate is:

$$Q_w = \frac{V}{\theta_c} \tag{4}$$

The advantage of this method of wasting is that the operator need not make any tests to control the process. He simply sets the waste pump to pump a fraction of the aeration tank volume in a day. Additional details on this method of process control including the design of physical facilities for its implementation may be found in Burchett and Tchobanoglous (1972).

If wasting is accomplished from the sludge return line and it is assumed that the solids lost in the effluent are negligible, then the wasting rate can be approximated as follows:

$$Q_w = \frac{xV}{x_r \theta_c} \tag{5}$$

where x_r = sludge concentration in return line

Because the ratio, x/x_r, and not the actual concentrations need to be known to use equation (5), appropriate data can be derived using a small manual or electrically operated centrifuge. By measuring the sediment height in the tubes after centrifugation for samples taken from the aeration tank and the return line, the ratio x/x_r can be determined readily in the field and the wasting rate can be determined as both V and θ_c are known.

By adopting the mean cell residence time method of control, the operation of small activated sludge plants can be simplified greatly. Similar techniques can and must be applied to the control of the other processes considered in the previous section.

Maintenance

One of the most difficult things to provide in a small plant is reserve tank capacity to allow for adequate maintenance and repairs. This can be overcome partially by providing in duplicate those critical facilities that are routine maintenance items. For example, if air diffusers are used, they should be provided in duplicate and designed so that repairs can be made without shutting down or bypassing the process. Although this will increase the cost, it is necessary if the small treatment facility is to remain in continuous operation.

Power Requirements

As the availability of power becomes limited, it is essential that process power requirements be considered carefully. Specifically, each process should be evaluated to determine what will happen if power is temporarily interrupted, purposely shut off for a given time interval such as 4 to 6 hours, or totally lost for days at a time. The impact of these effects can then be compared among the alternatives and trade-offs between initial capital costs and daily power requirements can be investigated. Such an analysis must be performed before selecting any process.

Economic Evaluation

The economic evaluation of alternative processes is considered and illustrated in this section. The material to be presented has been divided into three sections dealing with economic criteria, costs for the alternative processes, and a summary discussion.

Economic Criteria

To assess properly the economic feasibility of alternative processes it is necessary to prepare detailed cost estimates. Before such estimates can be prepared, however, economic criteria must be selected so that equivalent costs are compared. For example, a true evaluation of alternative processes must be based on total annual cost. The reason is that all costs, including operation, maintenance, supervision, and depreciation and interest on borrowed capital, are included in the computation of annual cost.

Annual interest and depreciation, commonly referred to as "fixed costs," are usually computed using the capital recovery method. The recommended recovery period for small plants varies from 10 to 20 years. Such short periods are used because of the uncertainties associated with effluent disposal and reuse relative to regulation and regional plans that may be adopted in the future. In November 1973 the interest rate charged on borrowed money varied from 7 to 9 percent. It should be noted that when interest rates are high, questions of time-capacity expansion are very important and should also be considered in any economic analysis (Rachford et al., 1969).

Because costs are changing so rapidly both nationally and locally, it is extremely important that any cost evaluation be referenced to some cost index. One of the most common is the Engineering News Record Construction Cost (ENRCC) index. Other important indexes include the Environmental Protection Agency Sewer Cost and Treatment Plant Indexes. Where possible, index values should also be adjusted to reflect current local costs.

Costs

Capital, operation and maintenance, and annual and unit costs are considered in this section.

Capital costs. Expenditures for (1) the purchase of land, (2) earthwork, (3) the purchase of equipment, and (4) the construction of plant and related facilities comprise the major capital costs associated with the construction of a treatment plant.

Ideally, these costs should be determined for each alternative process to be considered by obtaining quotes from local equipment suppliers and contractors. Unfortunately, this is often not possible because of time and budgetary considerations. When such is the case, the only recourse is to use generalized cost data, which are available from a variety of sources. The federal government through the Environmental Protection Agency has issued a number of reports dealing with the cost of treatment plants.

Because it is beyond the scope of this paper to present detailed estimating data on capital cost, generalized data on the capital costs of the various processes considered previously are presented in Table 13. As shown, the individual trickling-filter processes have been grouped under one heading. This has been done because the cost of such facilities will, within limits, be about the same. Similarly, where appropriate, the activated-sludge, stabilization-pond, and land-disposal processes have been grouped. Although the range in the cost data is derived principally from local and geographical conditions, process differences will have some effect. It should also be noted that land costs have not been included.

Because it was intended that the costs given in Table 13 would be used for estimating purposes, capacity factors are also given so that the reported costs can be scaled down for plants with a smaller capacity than 1.0 MGD. Scaled costs are obtained using the following expression:

$$C_s = C_1 \, (Q_1 / Q_s)^{\alpha} \tag{6}$$

where C_s = cost of smaller plant, dollars

C_1 = cost of larger plant, dollars

Q_1 = flowrate of larger plant, MGD

Q_s = flowrate of smaller plant, MGD

α = scale factor

The scale factor accounts for the observation that many unit costs

Table 13
Estimated Capital Costs for Alternative Treatment Processes with a Design Flow of 1.0 MGD[a]

Process	*Cost Range dollars*[b, c, d]	*Scale Factor,* α[d, e]
Imhoff tank	340,000–420,000	0.54
Rotating biological disks	700,000–1,000,000	0.62
Trickling-filter processes	700,000–1,000,000	0.62
Activated-sludge processes		
With external digestion	900,000–1,100,000	0.62
With internal digestion[f]	400,000–600,000	0.67
Stabilization-pond processes	200,000–300,000	0.60
Land-disposal processes[g]		
Basic system	300,000–360,000	0.58
With primary treatment	880,000–1,040,000	
With secondary treatment	700,000–1,460,000	
Land disposal[h]		
Basic system	180,000–220,000	0.58
With primary treatment	760,000–900,000	
With secondary treatment	580,000–1,320,000	

[a]Components comprising process are shown in Figure 1.
[b]Excludes land costs and the costs involved in transporting wastewater to treatment site.
[c]Based on an ENRCC index of 1900.
[d]Includes contractors' profit and engineering contingency allowance.
[e]Used to scale costs for smaller sized plants (see text).
[f]Extended aeration, aerated lagoons, oxidation ditches.
[g]Irrigation and overland flow.
[h]Infiltration-percolation.

decrease as the size of the project increases (Rachford et al., 1969). Typically, scale factors for wastewater treatment plants reported in the literature are considerable higher (0.67–0.74) than those reported in Table 13 (Rachford et al., 1969). The reason for this difference is that the higher factors commonly used have been derived by evaluating cost data for plants with flows varying from 1.0 to 100 MGD. Based on a detailed review of the literature and other cost data for small facilities, it is felt that the factors given in Table 13 are appropriate for adjusting costs for flows in the range from 0.05 to 1.0 MGD. The use of such factors is illustrated in the following section.

Operation and maintenance costs. Operation and maintenance costs for most treatment processes usually fall into the following categories: labor, power, chemicals, parts, and supplies. The labor cost is derived by considering the time spent on the following tasks: supervision and report preparation, clerical, laboratory, yard, operation, and maintenance (EPA, 1973). Estimated man-hour requirements for these tasks for various processes with a plant design capacity of 1.0 MGD are reported in Table 14. The reported data are intended to serve as a guide and should be revised to reflect local operating conditions. Staffing requirements are considered more fully in references (EPA, 1973; Patterson and Banker, 1971). Using the total man-hour figures from Table 14 and cost data derived from a variety of sources, annual operation and maintenance costs have been derived for the various processes (see Table 15). The labor costs in Table 15 were computed by assuming 1,500 productive hours per man per year and an average cost of $15,000 per man per year. It was assumed that the $15,000 figure includes fringe benefits.

Using the cost data from Table 15 as a basis, annual operation and maintenance costs for smaller plants have been derived and are presented in Table 16. The scale factors used to derive costs for the smaller capacity plants are also given in Table 16. As will be noted, these factors vary with the size of the plant. This adjustment is necessary to account for the fact that below some lower limit the cost of labor does not change significantly as the size of the plant decreases. The scale adjustment curve used in these computations is given in Figure 4.

Annual and unit costs. The total annual cost of the various processes can be derived by combining the annual capital recovery and operation and maintenance costs. The results of such a computation for a 1.0 MGD plant are given in Table 17. The capital recovery costs were obtained by multiplying the estimated initial capital costs by the capital recovery factor for a return period of fifteen years and an interest of 7 percent.

The unit treatment cost per 1,000 gallons has been computed and is also reported in Table 17. Comparing the computed value for the 1.0 MGD activated sludge process with external digestion (50.5¢ / 1,000 gal) to typical values reported in the literature for much larger plants (20¢–30¢ /1,000 gal) the effect of scale is apparent. The unit difference is even more dramatic as the plant size decreases below 1.0 MGD.

Discussion

The purpose of this section was to present and illustrate some of the

Table 14
Estimated Annual Man-Hour Requirements for Alternative Treatment Processes with a Design Flow of 1.0 MGD[a]

Process	Annual Man Hours						
	Supervisory[b]	*Clerical*	*Laboratory*	*Yard*	*Operation*	*Maintenance*	*Total*
Imhoff tank	208	52	208	208	416	208	1300
Rotating biological disks	624	208	416	416	2080	1664	5408
Trickling-filter processes							
Conventional (low rate)	624	208	416	416	2080	1664	5408
Conventional (high rate)	624	208	416	416	2080	1664	5408
Tower filters	624	208	416	416	2080	1664	5408
Activated-sludge processes							
Complete mix	624	208	520	416	2912	2080	6760
Contact stabilization	624	208	520	416	2496	2080	6342
Extended aeration	624	208	416	416	1248	1248	4160
Aerated lagoon (with settling)	624	208	416	416	1248	1248	4160
Oxidation ditch (with settling)	624	208	416	416	1248	1248	4160
Stabilization pond processes							
High-rate aerobic	416	104	208	416	416	520	2080
Facultative (algal surface layer)	416	104	208	416	416	416	1976
Facultative (aerated surface layer)	416	104	208	416	416	520	2080
Anaerobic	416	104	208	416	416	416	1976
Land-disposal processes[c]							
Irrigation	416	104	416	208	1040	1248	3432
Overland flow	416	104	416	208	832	1040	3016
Infiltration-percolation	416	104	416	208	520	416	2080

[a]To provide an effluent quality such as shown in Table 3.
[b]Includes preparation of reports.
[c]Labor requirements for pretreatment not included.

Table 15
Estimated Annual Operation and Maintenance Costs for Alternative Treatment Processes with a Design Flow of 1.0 MGD

	Cost, Dollars				
Process	*Labor*[a]	*Power*[b]	*Chemicals*[c]	*Parts/ Supplies*[c]	*Total*
Imhoff tank	13,000	500	1800	200	15,550
Rotating-biological disks	54,000	1200	1400	1000	57,680
Trickling-filter processes					
Conventional (low rate)	54,080	2000	1400	1000	58,480
Conventional (high rate)	54,080	2000	1400	1000	58,480
Tower filters	54,080	3000	1400	1000	59,480
Activated-sludge processes					
Complete mix	67,600	7000	1200	1200	77,000
Contact stabilization	63,420	6000	1200	1200	71,820
Extended aeration	41,600	5000	1200	1000	48,800
Aerated lagoon (with settling)	41,600	5000	1200	1000	48,800
Oxidation ditch (with settling)	41,600	5000	1200	1000	48,800
Stabilization-pond processes					
High-rate aerobic	20,800	1500	1800	800	24,900
Facultative (algal surface layer)	19,760	500	1800	400	22,460
Facultative (aerated surface layer)	20,800	1500	1800	800	24,900
Anaerobic	19,760	500	1800	400	22,460
Land-disposal processes[d]					
Irrigation	34,320	6500	1800	1000	43,620
Overland flow	30,160	6500	1800	1000	39,460
Infiltration-Percolation	20,800	1500	1800	1000	25,100

[a]Computed from data in Table 14 assuming 1500 productive hours per man per year and a cost of $15,000 per man per year (includes fringe benefits).

[b]Estimated at $0.02/KWH.

[c]Based on an ENRCC Index of 1900.

[d]Pretreatment costs not included.

Table 16
Estimated Annual Operation and Maintenance Costs in Dollars for Alternative Treatment Processes for Plants of Various Capacity[a]

	Flow, MGD (Scale Factor)					
Process	*0.05 (0.45)*	*0.1 (0.50)*	*0.25 (0.55)*	*0.5 (0.60)*	*0.75 (0.60)*	*1.0[a]*
Imhoff tank	4,000	4,900	7,200	10,400	12,900	15,500
Rotating-biological disks	15,000	18,240	27,000	38,500	48,100	57,680
Trickling-filter processes						
Conventional (low rate)	15,200	18,500	27,300	39,000	48,800	58,480
Conventional (high rate)	15,200	18,500	27,300	39,000	48,800	58,480
Tower filters	15,500	18,800	27,300	39,600	49,600	59,480
Activated-sludge processes						
Complete mix	20,000	24,400	33,000	51,300	64,200	77,000
Contact stabilization	18,700	22,700	33,500	47,900	59,900	71,820
Extended aeration	12,700	15,500	22,800	32,100	40,700	48,800
Aerated lagoon (with settling)	12,700	15,500	22,800	32,100	40,700	48,800
Oxidation ditch (with settling)	12,700	15,500	22,800	32,100	40,700	48,800
Stabilization-pond processes						
High-rate aerobic	6,500	7,900	11,600	16,600	20,800	24,900
Facultative (algal surface layer)	5,800	7,100	10,500	15,000	18,800	22,460
Facultative (aerated surface layer)	6,500	7,900	11,600	16,600	20,800	24,900
Anaerobic	5,800	7,100	10,500	15,000	18,800	22,460
Land-disposal processes[b]						
Irrigation	11,400	13,800	20,400	29,000	36,400	43,620
Overland flow	10,300	12,500	18,500	26,000	32,900	39,460
Infiltration-percolation	6,500	7,900	11,700	16,500	20,900	25,100

[a]See Table 15 for basic assumptions.
[b]Pretreatment costs not included.

Table 17
Estimated Total Annual and Unit Costs for Alternative Treatment Processes with a Design Flow of 1.0 MGD

Process	*Initial Capital Cost Dollars*[a, b]	*Annual Cost, Dollars*[b]			*Unit Cost Cents/ 1000 gal*[b]
		Capital[c]	*O&M*[d]	*Total*	
Imhoff tank	380,000	41,720	15,550	57,270	15.7
Rotating-biological disks	800,000	87,832	57,680	145,512	39.9
Trickling-filter processes	900,000	98,811	58,480	157,291	43.1
Activated-sludge processes					
With external digestion	1,000,000	109,790	74,410	184,200	50.5
With internal digestion	500,000	54,895	48,800	103,695	28.4
Stabilization-pond processes	250,000	27,447	23,680	51,127	14.0
Land-disposal processes					
Basic system	340,000	37,328	41,540	78,869	21.6
With primary treatment	940,000	103,302	81,540	184,742	50.6
With secondary treatment	1,240,000	136,139	115,950	252,089	69.1
Land-disposal processes					
Basic system	200,000	21,958	25,100	47,058	12.9
With primary treatment	800,000	87,832	65,100	152,932	41.9
With secondary treatment	1,000,000	109,790	99,510	209,300	57.3

[a]Estimated from Table 13.

[b]Based on an ENRCC index of 1900.

[c]Capital recovery factor = 0.10979 (15 years at 7 percent).

[d]Average values from Table 16.

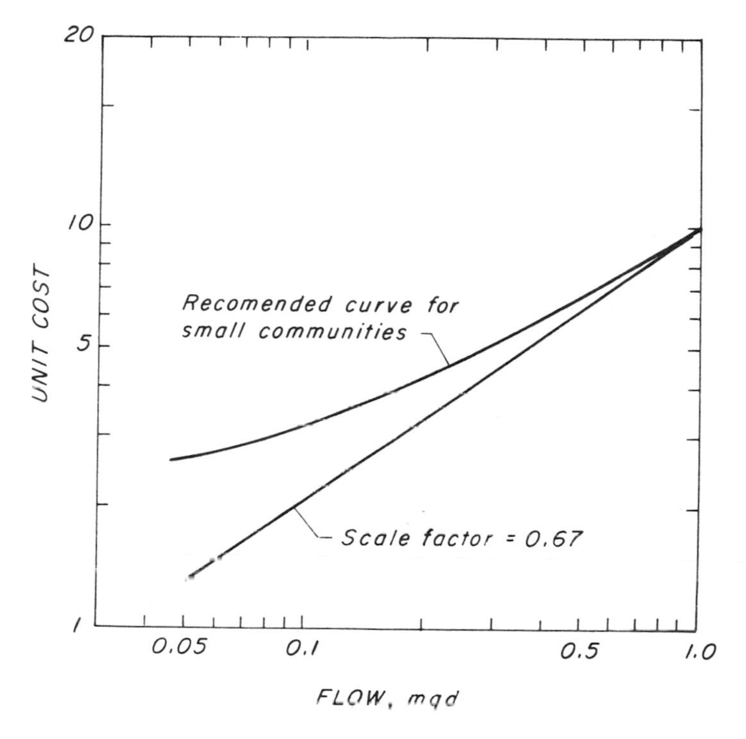

4. Unit operation and maintenance cost versus flow rate for small plants.

aspects involved in the economic evaluation of treatment processes. It is intended to serve as a guide for the collection and analysis of cost data for any process. To this end, what is important is the methodology and not the specific numbers.

Summary

Over the past twenty years a variety of processes have been developed that are suitable for the treatment of wastewater from small communities. Because local conditions are so variable, it is clear that there is no "best" process that can be applied universally. What is required is consideration of a variety of alternative processes. Because of the budget limitations faced by most small cities, special attention must be given to operation and maintenance requirements. Selection of the best process should then be based on an analysis of the characteristics of the wastewater, effluent requirements, local budgetary limitations, costs, and assessment of local impacts.

References

Autotrol Corporation, Bio-Systems Division. 1971. *Application of rotating disk process to municipal wastewater treatment*. Washington, D.C.: Office of Research and Monitoring, EPA.

Burchett, M. E., and Tchobanoglous, G. 1972. *Physical facilities for controlling the activated sludge process by mean cell residence time*. Paper read at the 45th Annual Water Pollution Control Federation Conference, October 1972, Atlanta, Georgia.

Burd, R. S. 1968. *A study of sludge handling and disposal*. Federal Water Pollution Control Administration, U.S. Department of the Interior Publication no. WP-20-4.

Caldwell, D. H., Parker, D. S., Uhte, W. R. 1973. *Upgrading lagoons*. Washington, D.C.: EPA.

Center for Research in Water Resources, Austin, Texas. 1971. *Design guides for biological wastewater treatment processes*. Water Pollution Control Research Series no. 11010 ESQ 08/71. Washington, D.C.: EPA.

Eckenfelder, W. W., Jr. 1970. *Water quality engineering for the practicing engineer*. New York: Barnes and Noble.

EPA. 1973. *Estimating staffing for municipal wastewater treatment facilities*. Washington, D.C.: Operation and Maintenance Program, Office of Water Program Operations, EPA.

Federal Highway Administration. 1972. *Oxidation ditch sewage waste treatment process*. Waste Disposal Series, vol. 6. Washington, D.C.: FHA, U.S. Dept. of Transportation.

George A. Hormel and Company, Environmental Pollution Control Division. 1972. *Rotating biological surface system of waste water treatment*. Austin, Minnesota: George A. Hormel and Company brochure no. 373.

Hildegrand, J. H., and Powell, R. E. 1964. *Principles of chemistry*. 7th ed. New York: MacMillan Company.

Lager, J. A., and Smith, W. G. 1973. *Urban stormwater management: an assessment*. Washington, D.C.: Office of Research and Development, EPA.

Luria, S. E. 1960. Approximate elementary composition of the microbial cell. In *The Bacteria*, vol. 1, ed. I. C. Gunsalus and R. Y. Stanier, pp. 14-16. New York: Academic Press.

McKinney, R. E., ed. 1970. *2nd International Symposium for Waste Treatment Lagoons*. Lawrence: University of Kansas Nuclear Reactor Center.

Metcalf, L., and Eddy, H. P. 1935. *Disposal of sewage. American sewerage practice*, vol. 3. 3rd ed. New York: McGraw-Hill.

Metcalf and Eddy, Inc. 1972. *Wastewater engineering: collection, treatment, disposal*. New York: McGraw-Hill.

Michel, R. L. 1970. Costs and manpower for municipal wastewater treatment plant operation and maintenance, 1965-1968. *Journal of Water Pollution Control Federation* 42: 1883-1910.

Nicholas, D. J. D. 1963. Inorganic nutrient nutrition of microorganisms. In *Plant physiology* II, ed. F. C. Steward, pp. 363-447. New York: Academic Press.

Patterson, W. L., and Banker, R. F. 1971. *Estimating costs and manpower requirements for conventional wastewater treatment facilities*. Washington, D.C.: Office of Research and Monitoring, EPA.

Pound, C. E., and Crites, R. W. 1973. *Wastewater treatment and reuse by land application*, vols. 1 and 2. Washington, D.C.: Office of Research and Development, EPA.

Rachford, T. M., Scarato, R. F., Tchobanoglous, G. 1969. Time-capacity expansion of waste treatment systems. *Journal of the Sanitary Engineering Division, American Society of Civil Engineers* 95: 1063–77.

Salvato, J. A., Jr. 1972. *Environmental engineering and sanitation*. 2nd ed. New York: Wiley-Interscience.

Sherrard, J. H., and Schroeder, E. D. 1972. Variation of cell field and growth rate in the completely mixed activated sludge process. Paper presented at the 27th Industrial Waste Conference, Purdue University, May 1972, Lafayette, Indiana.

Smith, R., and Eilers, R. G. 1970. *Cost to the consumer for collection and treatment of wastewater*. Washington, D.C.: Office of Research and Monitoring, EPA.

Tchobanoglous, G., Wanderer, W. C., Jr., Glide, L. G., Jr., and McDermott, G. N. 1973. Filamentous bulking, trace metal deficiencies and sulfide interrelationships. Paper presented at 46th Annual Water Pollution Control Federation Conference, October 1973, Cleveland, Ohio.

Thistlethwayte, D. K. B., ed. 1972. *Control of sulfides in sewage systems*. Ann Arbor: Ann Arbor Science Publishers, Inc.

Uhte, W. R. 1970. What design engineers owe the operator. *Bulletin Canadian Water Pollution Control Federation* 8: 4–5.

27. The Clivus-Multrum System: Composting of Toilet Waste, Food Waste, and Sludge within the Household

Carl R. Lindström

The Spaceship Earth Concept

Comparing the earth and its life support systems to a spaceship is a vivid way of illustrating the finite nature of our ecosystem. As described by Lynton Caldwell this model

> illustrates relationships between man and his environment that are basic to his welfare and survival. Ecological facts that man prefers to evade on earth are universally acknowledged for the spaceship. For example, no one doubts that there is a limit to the number of passengers that the ship can accommodate, and the need for reserve capacity to meet unforeseeable contingencies is not questioned. *It is obvious that the spaceship cannot indefinitely transform its nutrients into waste.*
>
> If extruded from the ship as waste, energy sources are irretrievably lost; if accumulated as waste, viability of the ship is ultimately destroyed from within. There is no escape from the necessity of recycling waste materials. For the duration of the voyage, the ship must remain in ecological balance. Disruption of any of its systems may mean disaster for the mission and the crew. Systems maintenance is, therefore, one of the essential components of a program of space exploration.

One of the main reasons for water pollution from the household is the fact that we use water as a transportation medium for dissolvable materials such as toilet wastes, waste food, detergents, etc. Consequently, the logical and most effective way to decrease water pollution from households is to develop and use technologies that eliminate water transports, if possible. It is important to keep in mind that practically all material coming into a household sooner or later leaves as gas, liquid, or solid material. Therefore, we must regard the total problem and not "solve" one problem by creating another or different problem.

Process Combinations

The present division between solid and liquid waste (Figures 1 and 2) is not necessary or, for that matter, even natural. If, in fact, the *organic* material, including soiled paper, that now is disposed of as solids is mixed with toilet wastes, the conditions for composting processes are almost ideal. Heat production and moisture in the material are favorable, as well as the balance between carbon, nitrogen, and phosphorus. This also has great importance for the end product and its value as a fertilizer. Unlike ordinary chemical fertilizers, this end product also contains trace metals like sodium, iron, magnesium, zinc, etc., which are of great importance for balanced soil conditions. The end product is a humus soil material with an odor similar to that of composted leaves. The average amount corresponds to a few buckets per person per year. The important thing is that it appears possible to process the organic waste, including the toilet deposits, safely and hygienically within the household into a harmless and even useful end product. Tables 1 and 2 show the analyses of the end product from a compost reactor that has been in operation for eight years. If the NH_3 but not the NO_2 or NO_3 leaves, the unspecified mineral constituents would be 63 - (10.79 - 1.94) = 54.2%. This substance is most likely silicates from the original "earth-bed."

If one studies only the "main nutrients," which are those containing K, N, P, Mg, Ca, he finds that the sum of these as oxides and NH_3 is 9.11% of the total sample or 14% of the mineral substance = ashes and ammonia. Table 3 shows the proportion of these different nutrients in soil from the composter and two chemical fertilizers (Y-lannos from Finland).

When comparing the contents of toilet wastes and of other domestic wastewaters (Table 4), one finds that the toilet waste ("black water") contributes about 90% of the nitrogen content in "normal" sewage water, about half of the phosphorus, about 65% of the COD, and also the major portion of the thermostable bacteria. The wastewater from laundry, bath, and sink ("grey water") contributes mainly BOD, phosphorus, and bacteria. The relatively small amount of nitrogen in grey water gives it a special BOD characteristic. The BOD_1 is almost 40% of the ultimate oxygen demand (UOD), and BOD_5 has nearly completed the breakdown (see Figure 3). The corresponding oxidation process for the toilet water is considerably slower. The oxygen of the BOD_1 is only about 10% of the UOD. This must be considered when choosing methods of treatment.

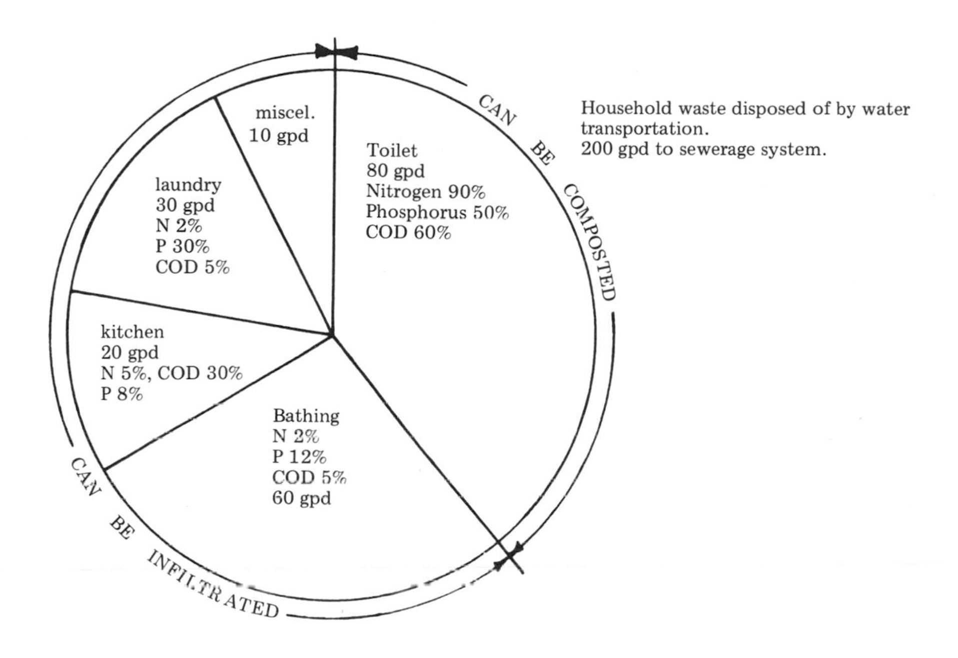

Household waste disposed of by water transportation.
200 gpd to sewerage system.

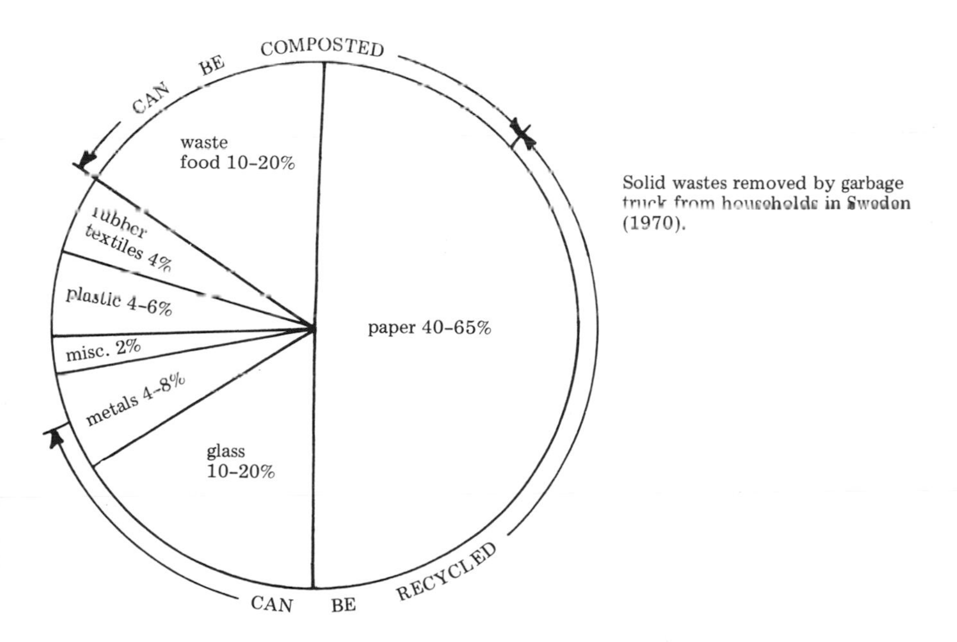

Solid wastes removed by garbage truck from households in Sweden (1970).

1. Conventional disposal of household waste.

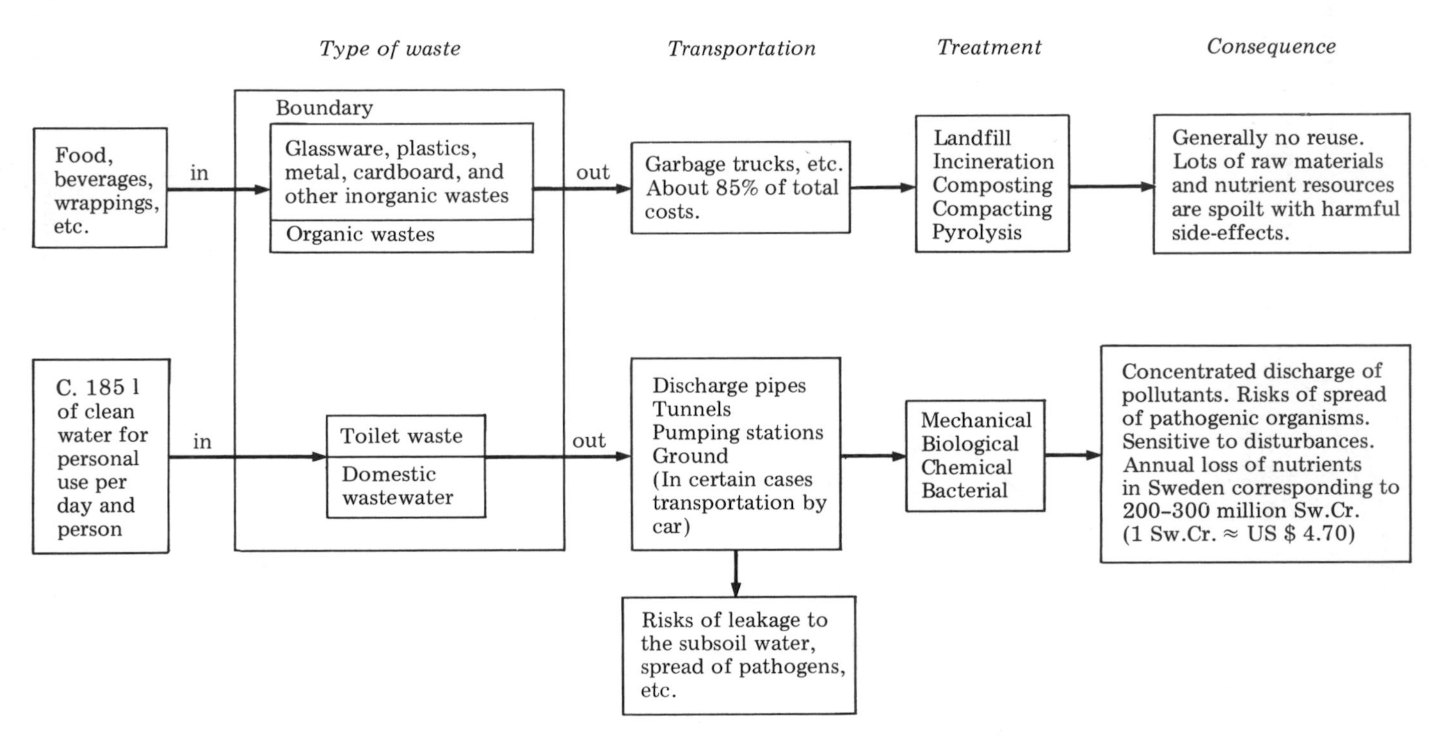

2. Conventional waste treatment methods.

Table 1[a]
Composition of Contents of Compost Reactor Receiving Human Organic Waste for Eight Years

pH = 7.6
Ashes = 63.0%
"Humus" = 23.7%
Rest 13.3% (essentially water)
Conductivity 92
Volume weight 735 gms/liter

	Mg/liter	*Percentage*
Ca^{2+}	14000	1.9
K^+	8200	1.12
B	4.2	
Cu	32	0.044
Mn	145	0.020
Mg	960	0.13
P total	3700	0.50
N total		0.84
N in NH_3	1150	0.16
N in NO_2	3200	0.44
N in NO_3	980	0.13
Cl	350	0.05
SO_4^{2-}	650	0.09
Al^{+3} dissolvable in HCL		0.87

[a]Analyses made by Markkarte Ringstjänst, Helsinki, Finland.

Table 2
Corresponding Oxides and NH_3 Values of the Main Constituents given in Table 1

CaO	2.66
K_2O	1.35
MgO	0.22
N_2O_5	0.60
N_2O_3	1.19
Al_2O_3	1.68
P_2O_5	1.15
NH_3	1.94
Total	10.79

Table 3
Comparison of the Compositions of Composted Home Wastes and Two Chemical Fertilizers

		Chemical Fertilizer Analysis	
	Domestic Waste Multrum Compost	*Väk. Oulu Y-lannos*	*Normal Super-Y-lannos*
K_2O	19.6%	18.0%	15.0%
P_2O_5	16.7%	15.0%	20.0%
Total N	12.2%	13.0%	15.0%
NO_3-N	8.0%	5.2%	4.3%

With chemical flocculation and precipitation by $FeCl_3$ an average reduction of about 95% of the phosphorus, 50% of the nitrogen, and 60% of the COD in the grey water has been measured in an operating plant at the Royal Institute of Technology in Stockholm. The combined effect of composting toilet wastes and $FeCl_3$ treatment of the grey water would bring about an overall reduction of about 95% of the nitrogen and phosphorus and 85 to 90% COD reduction. This means an average output of 6–10 gpd COD, 0.5 gpd N, and 0.015 gpd P. According to a report from the National Swedish Environment Protection Board dated August 1973 on infiltration of sewage water, the main impact on the subsoil waters is an increased concentration of nitrates, chlorides, and sulphates. As measured about nine feet below the infiltration pipes, filtration through the soil was reported to have removed about 90% or more of BOD and almost all of the phosphorus and bacteria, the main constituents of grey water.

The effectiveness of this decomposition through natural soil processes suggests a different segregation of household wastes. Figure 4 shows a combination of wastes and processes with considerably less impact on the recipient media than the traditional methods. The grey water from the bath, sink, and laundry gets most of its salts from detergents and most of its BOD from the dishwater. Depending on the density of population, soil conditions, and hydrological circumstances, different levels of pretreatment may be necessary. In urban areas with access to service, a chemical precipitation technique may be suitable. Figure 5 shows an example of a system where the sludge is automatically transferred to the compost reactor each day. Unless the sludge is dewatered, this process will end with an energy demand of about 0.8 KWH per person per day for evaporation (Table 5). This com-

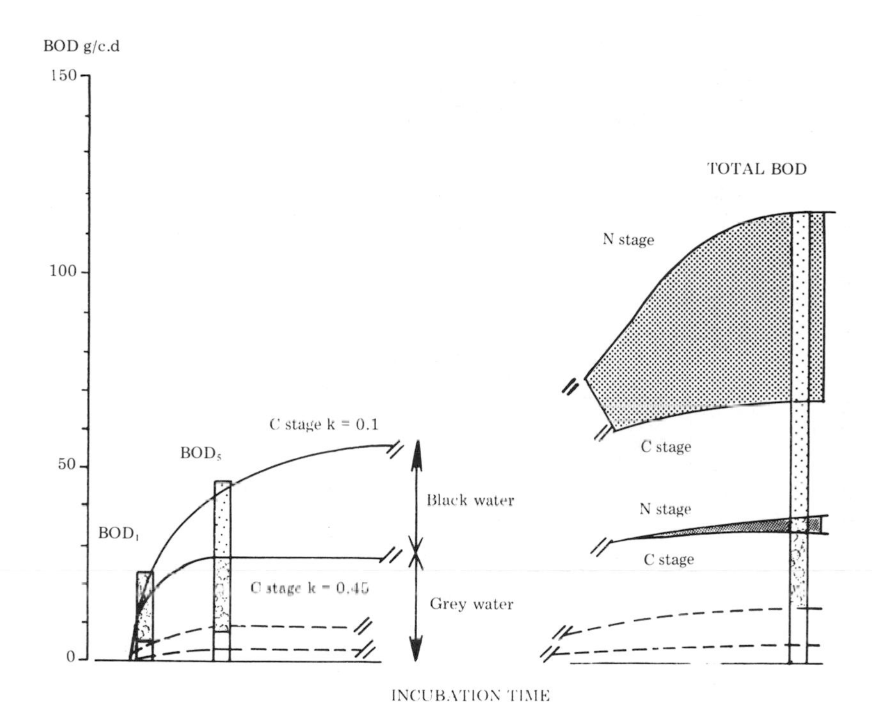

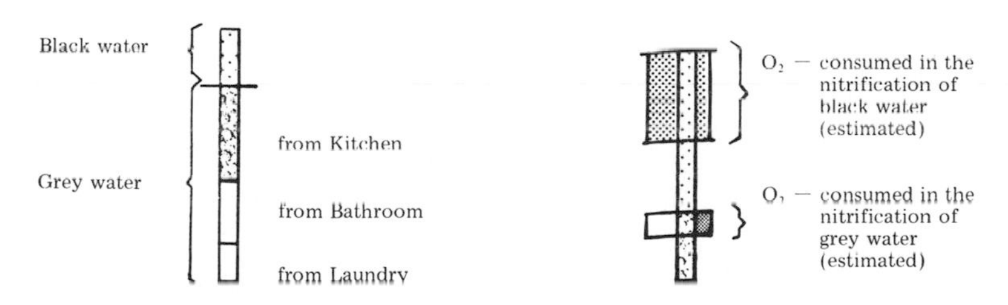

3. The amount of pollution in the wastewaters from kitchens, bathroom, vacuum toilets, and laundry expressed in terms of bio-oxidizable substances per person per day (of 24 hours).

bined system is still under development with funds from the National Swedish Board for Technical Development. At this time, the water purification equipment still requires considerable maintenance to keep it operational. In most cases, especially in rural areas, a simple infiltration technique does the same work.

Table 4
Comparison between the Quantities of Pollution per Person per Day in Grey Water and Black Water

					% division	
Analysis	*Unit*	*Grey Water*	*Black Water*	*Total Grey + Black Water*	*Grey Water*	*Black Water*
BOD_5	g/c.d	25	20	45	56	44
$KMnO_4$-O_2		48	72	120	40	60
Total P		2.2	1.6	3.8	58	42
Kjeldahl N		1.1	11.0	12.1	9	91
Total residue		77	53	130	59	41
Fixed total residue		33	14	47	70	30
Volatile total residue		44	39	83	53	47
Nonfilterable residue		18	30	48	38	62
Fixed nonfilterable residue		3	5	8	38	62
Volatile nonfilterable residue		15	25	40	38	62
Plate c 35°	Number of bacteria/c.d	$83.0 \cdot 10^9$ [a]	$62.2 \cdot 10^9$	$145.2 \cdot 10^9$	57	43
Coli 35°		$8.5 \cdot 10^9$ [a]	$4.8 \cdot 10^9$	$13.3 \cdot 10^9$	64	36
Coli 44°		$1.7 \cdot 10^9$ [a]	$3.8 \cdot 10^9$	$5.5 \cdot 10^9$	31	69
Flow	l/c.d	121.5	8.5	130	93	7

[a]The average of the weighted arithmetic average values for the grey water from kitchens, bathrooms, and the combination group kitchens and bathrooms. The laundry is not included.

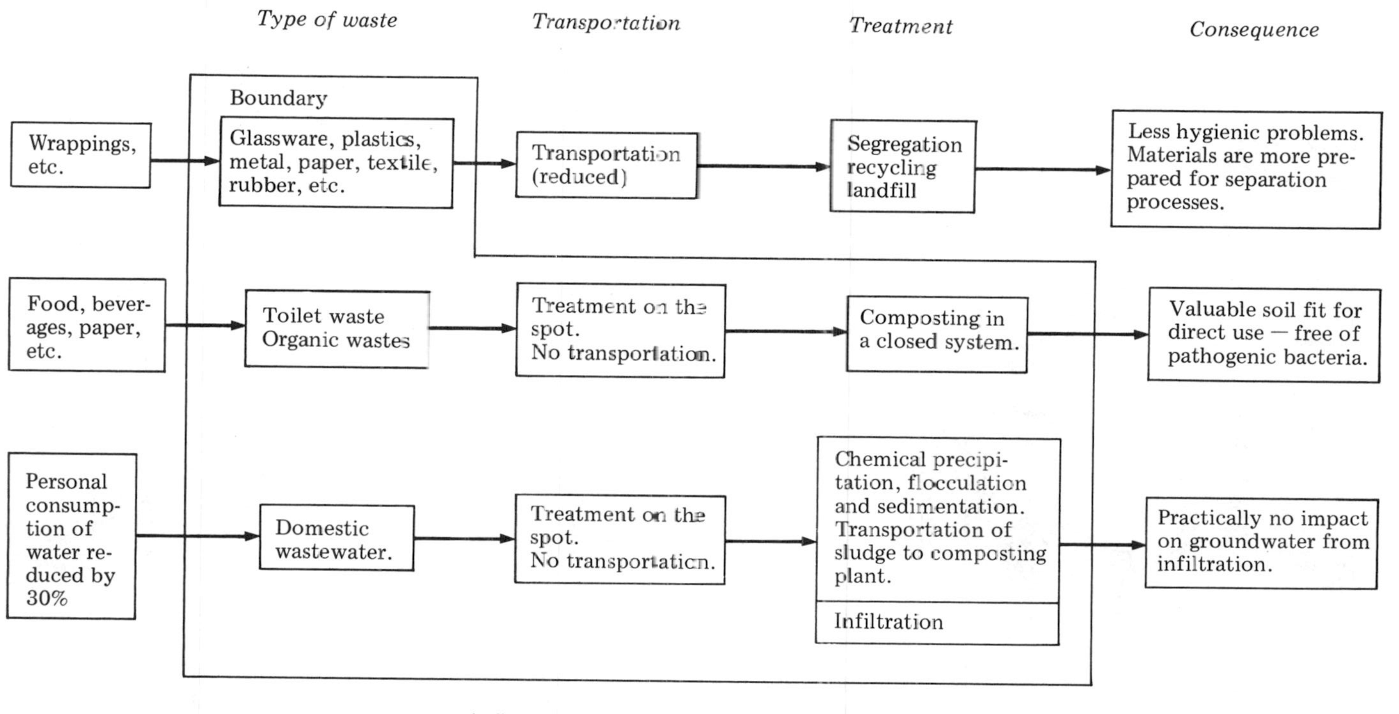

4. Selective decentralized waste treatment.

5. The Multrum system.

Compost Equipment

The compost equipment is the main new piece in the system. The Clivus compost reactor works on the principles of gravity, capillary action, and a special ventilation system. The composting material is enclosed in an impervious container (Figure 6). This container is connected to the toilet and kitchen refuse openings by means of two chutes. A vent allows the gases produced in the process to escape. A layer of topsoil placed on the sloping bottom contains the bacteria

Table 5

	Toilet Waste	*Grey water Sludge*	*Organic Kitchen Waste*	*Total*
Average Energy content	1070	850	3450	5370
KWs/person/day				
Energy demand to evaporate water-content	3000	5000 (2–3% dry substance)	100	8100
KWs/person/day		1250 (15% dry substance)		4350

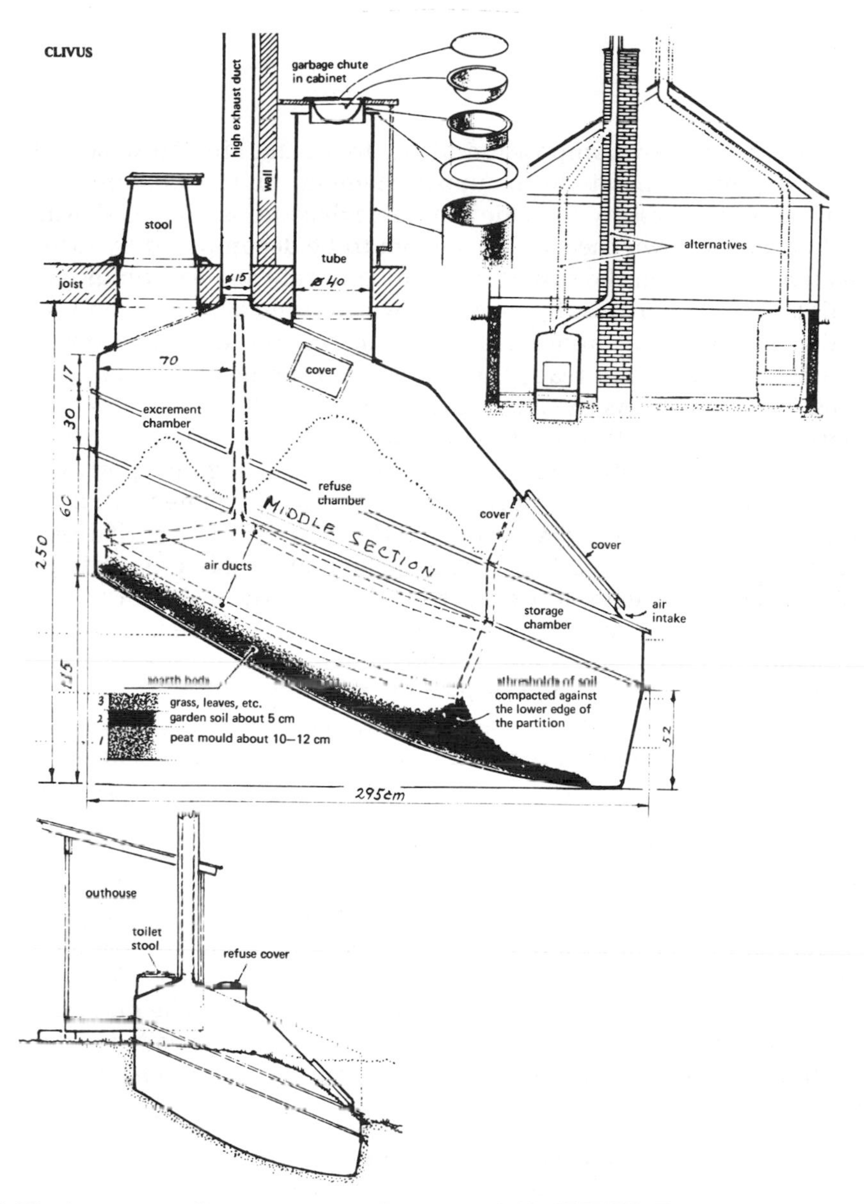

6. The impervious Clivus container. Courtesy of AB CLIVUS, Tonstigen 6, S-135 00 Tyresö, Sweden.

Container consists of a top and bottom section and, where a greater capacity is required, also a middle section.

Container in functioning position: height = 220 cm or 250 cm or 280 cm, length = 295 cm max, width = 120 cm.

For adequate draft, exhaust outlet high and insulated against cold. Sideward exhaust duct (tube with smooth inner surface) gives maximum rising and no sharp bends.

Decomposition generates a certain amount of heat. The container and the parts of the exhaust ducts which are exposed to cold air should be insulated so that a temperature favourable to ventilation and to the process can be maintained.

necessary for the decomposition of the toilet and kitchen wastes. The feces which accumulate in the upper chamber (on the left) also contain bacteria necessary to achieve aerobic decomposition. The urine is drained off down the sloping bottom to be decomposed by nitrobacteria existing in the soil, thus forming nitrate and carbon dioxide. This is the same process by means of which urine deposited on the soil in nature is decomposed. As the urine drains out of the excrement chamber, air is able to reach the oxygen-consuming organisms that break down feces. As the material decomposes, CO_2 and water vapor leave the container through the vent and the volume decreases simultaneously. As the kitchen refuse decomposes, the high proportion of cellulose present in this material produces heat which helps to evaporate the liquids, thus further reducing the volume. The ventilation duct is either connected to a vent in a chimney or has a separate path of its own and extends above the roof. Because the inside of the converter is kept at a lower pressure than the outside, there is no odor either in the bathroom or in the kitchen, which are both actively ventilated when the covers are opened. The first stage of the process in which the larger particles are broken down and during which the greatest heat is produced is rapid, whereas the secondary process, the mineralizing of the material, takes months and years. The total amount of finished compost produced per person per year is about ten gallons of soil. By contrast, a conventional toilet uses and pollutes 10,000 gallons of clean water per person per year for transportation alone.

According to tests made at the National Swedish Bacteriological Laboratory on eight different samples, there are no E. coli bacteria in the final product, most likely because of the environment close to the bottom, where the composted soil slowly comes out. The material is almost completely mineralized and can supply only so-called earth bacteria, which also consume most pathogens. The composter is practically failsafe because:

1. The unit has no mechanical turning or stirring.
2. The "turnover" speed is very low. It takes approximately two to three years for the solid particles to run through the reactor.
3. The construction makes it almost impossible to take out material which is not mineralized.

Experiences

The National Swedish Board of Health and Welfare say the following in their *Advice and Instructions about Composting*, 5061: 862:

Mouldering is a slow process of biological combustion, implying that organic material is converted into a stable humus-like substance through the action of various microorganisms under aerobic conditions. Unlike anaerobic putrefaction, this process does not emit offensive odors. In addition to a considerable decrease in weight and volume, mouldering results in the conversion (mineralization) of organic-bound mineral substances into inorganic nutrients that are available to plants through direct uptake. Through the temperature rise taking place at biological degradation and through the antibiotic effect of certain microorganisms, the material is of a very high standard from the viewpoint of hygiene. Complete mouldering under natural conditions takes from one to several years. The length of time depends on the kind of material. With easily degradable substance, a usable humus-like product could be obtained after only about half a year under optimum conditions.

The possibility of applying the mouldering method in the treatment of night soil is primarily dependent on the following factors:

(a) Oxygen supply. This must be good. The material cannot be compacted or compressed.
(b) Moisture. The process demands a moisture content of 40 to 60 percent with the optimum range of 50 to 55 percent.
(c) Nutrient balance. This effects primarily the rate of turnover. Most important is the balance between carbon and nitrogen (C/N ratio), which should be around 30.

In addition to what has been said above, it ought to be noted that variations in load may disturb the process, especially in small latrines where the amount of waste matter is too small to have a balancing effect on occasional overloads.

Latrines do not have any of the above-mentioned conditions for obtaining a good mouldering effect. They have a fairly dense consistency of feces and a high moisture content. Aeration is obstructed by the supply of urine. The ability of latrines to fix excess liquid is small, if anything. Owing to its high nitrogen content (low C/N ratio), there is a risk of considerable nitrogen losses through evaporation of ammonia.

In order to achieve mouldering, therefore, liquid-absorbing substances must be added to the latrine. Carbohydrate-rich material, e.g. garbage, paper, leaves, and grass, will facilitate mouldering in all the above-mentioned respects. The waste product can be permeated by air; the moisture content is stabilized and less sensitive to variations in load. Moreover, the rate of turnover is favorably affected through a better C/N balance.

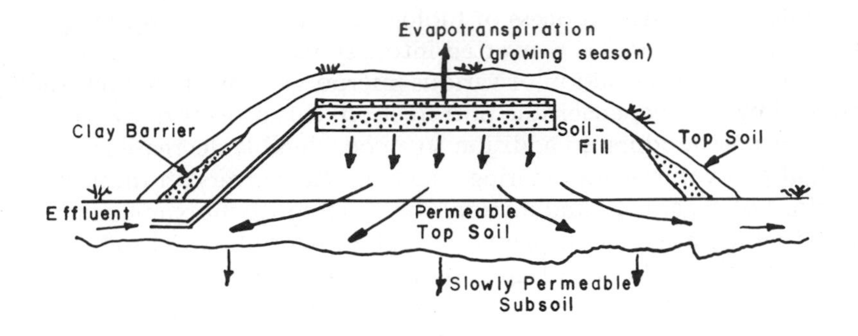

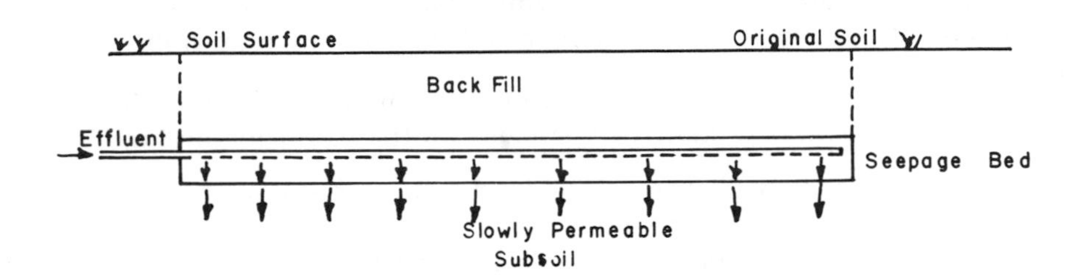

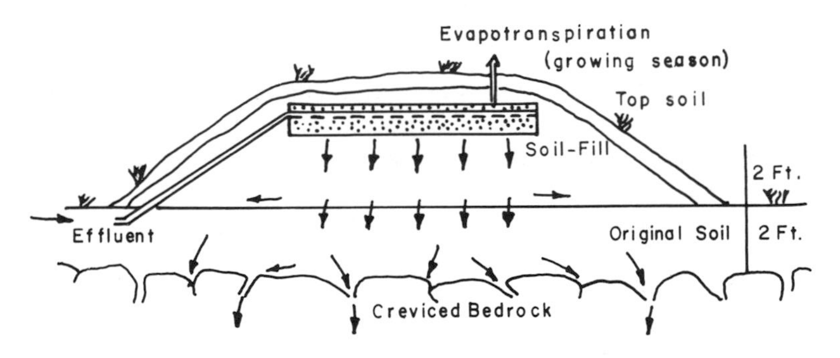

7. A ground infiltration system:
   I. Slowly permeable soil
      A. Mound
      B. Subsurface seepage bed
   II. Creviced bedrock mound

The method of composting latrines with carbohydrate-rich wastes is the basis of the most tested and oldest toilet for waste-matter mouldering in Sweden, the Multrum. Through its large volume and the long residence time of the waste matter, this system offers thorough mouldering and a high hygienic standard. The process is completely self-regulating, but the activity may decrease when the outer temperature is too low (approximately 0°C). The risk of serious disturbance of the process is very small. The final product resembles soil.

History

The prototype of the Multrum was built in 1939 for private use. Needless to say, World War II was not the ideal period to sell new ideas about waste disposal to the public. Consequently, the idea was not developed until 1964, when an environmental consciousness was awakened in Sweden, and the process was first described in a public paper later that year. After that, some new prototypes were built and studied for improvements. Typical early problems were the continuity of the process, inhibition of salt concentrations, ventilation, and, of course, general lack of long-range experiences. The present design has been installed in about 1,300 Scandinavian homes.

Present Development Work

Present developments, partially funded by the National Board for Technical Development, have the following aims:

1. to design and test an integrated waste treatment system for multifamily dwellings;
2. to develop techniques for internal transportation of waste without using water;
3. to improve small-scale water purification techniques;
4. to improve disposal techniques.

The multi-apartment project is being carried out in Finland in a four-apartment, two-story building.

Summary

Properly constructed a ground infiltration system (see Figure 7) for wastewater from the bath, sink, and laundry has little or no impact on groundwater. Such systems are described elsewhere in this text by different writers, e.g. J. Bouma, "Use of soil for disposal and treatment of septic tank effluent." Toilet and organic kitchen wastes can be processed in a compost reactor into a small volume of humus soil. Other household waste products, such as glass, plastics, paper, and metal, do not attract insects and rodents and can be saved longer, thus reducing transportation costs considerably.

28. Treatability of Septic Tank Sludge

William J. Jewell,
James B. Howley, and
Douglas R. Perrin

Introduction

Although the most common domestic sewage treatment process in the United States is the septic tank, the collection, treatment, and disposal of septic-tank sludge has received only passing attention from environmental quality control engineers and scientists (USDA, 1968; Patterson et al., 1971; Escritt, 1950 and 1954; Flood, 1944). The satisfactory operation of septic tanks depends, in part, on the periodic removal of the accumulated particulate materials from the tank. The frequency of sludge removal as recommended by various regulatory agencies varies from one to five years, with most agencies suggesting three years as a desirable accumulation period (USPHS, 1967). The magnitude of this problem can be illustrated by estimating the quantity of sludge which must be disposed of, assuming that cleaning occurs every three years. In the country as a whole, the eighty million people who depend on the septic tank for waste disposal potentially generate twice as much sludge as that produced in all domestic secondary treatment facilities.

The fate of this material in the environment is essentially unknown, and there are few proven methods of wastewater treatment which can be used to neutralize the polluting potential of this form of sewage sludge (Patterson et al., 1971). The major burden of responsibility for safe and desirable disposal of the contents of septic tanks has been left to the private septic-tank cleaner. This responsibility is difficult to meet because: (1) little detailed guidance is provided by regulatory authorities, (2) most existing wastewater-treatment facility operators do not allow addition of septage on account of fear that it may upset unit operations and/or processes, and (3) many regional authorities prohibit land disposal. Many sewage-treatment facility operators have observed upsets and deterioration of effluent quality when septage was added to their plants under varying field conditions. But there is no readily apparent explanation for these problems. Septage should be composed of materials which, when applied to various unit processes at "controlled rates," could be treated with little impact on the unit. It was the general goal of this study to provide basic engineering data on the treatability of septic tank sludge in order to provide a rational basis for handling and disposal of this material.

Characteristics of Sludge Pumped from Septic Tanks

The particulate materials that are retained by septic tanks are highly variable in composition because of the user history on the system. The characteristics indicate that this sludge (or septage as it is generally called) is very dilute (97 percent water) and has high dissolved organic concentration with COD values of 3000 mg/l or more (Kolega, 1971). Thus, because of the dilute nature of the sludge and the degree of soluble contamination, treatment of this material should be considered prior to final disposal. The strong odor of septage is also a problem with disposal in populated areas. Alternative handling and treatment methods are illustrated in Figure 1.

Biological, chemical, and physical methods of treatment were examined in this study with the object of providing information on alternatives for final disposal of sewage. The simple unit processes that might already exist as municipal treatment facilities and those which could be easily constructed were considered so that the data would be applicable to small communities as well as individual rural dwellings. The various unit processes examined in this study are summarized in Figure 2. The detailed objectives of this study were to determine:

1. the feasibility of biological treatment of domestic septage using some conventional wastewater treatment processes;
2. the fraction of soluble and suspended organics that are susceptible to aerobic and anaerobic decomposition, with emphasis on aerobic processes;
3. the kinetics of biological stabilization and septage in terms of the rate of decomposition of the biodegradable organics;
4. the degree of treatment required for odor control;
5. septage dewatering characteristics of raw sludge and sludge treated with biological and chemical conditioning;
6. a simple parameter to relate dewatering characteristics to the ease of handling using sand drying beds;
7. and, formulated from the above information, recommendations for the handling and disposal of septage to minimize its impact on environmental quality.

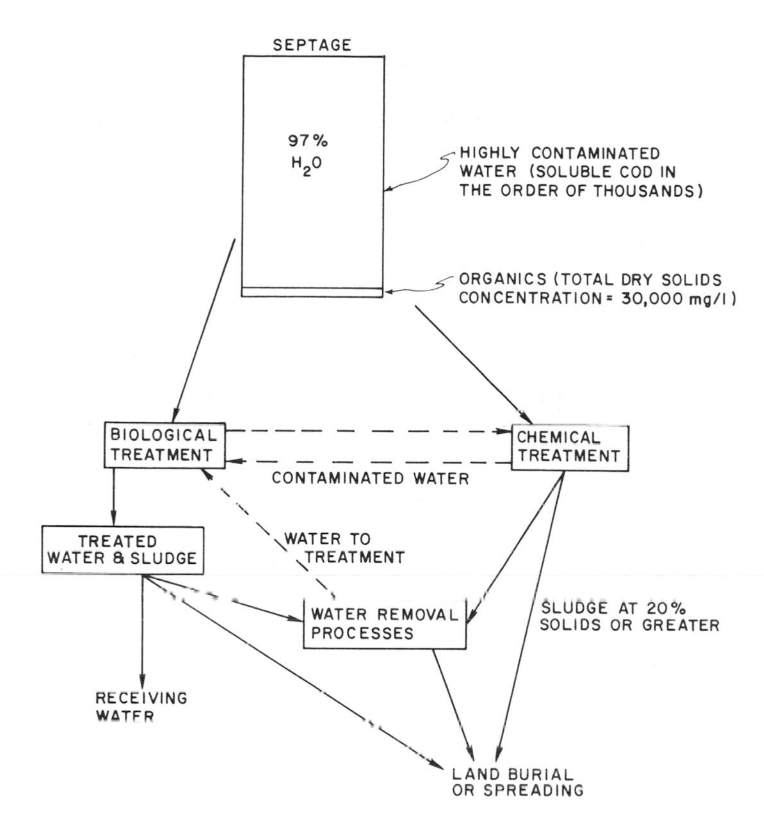

1. Alternatives for treatment and disposal of septage.

Materials and Methods

Treatability Studies

The capabilities of common sewage treatment processes to accept septage (see Figure 2) were studied. Sedimentation characteristics were tested, but this treatment alternative was not considered in depth, since most septage did not settle even with the addition of large doses of coagulant aids. The second major unit process reviewed was treatment in an aerobic reactor. The majority of data was developed on this process. However, because of the high concentration of solids and oxygen utilizing material in septage, processes involving recycling of sludge, such as activated sludge, were not analyzed. Instead, data were

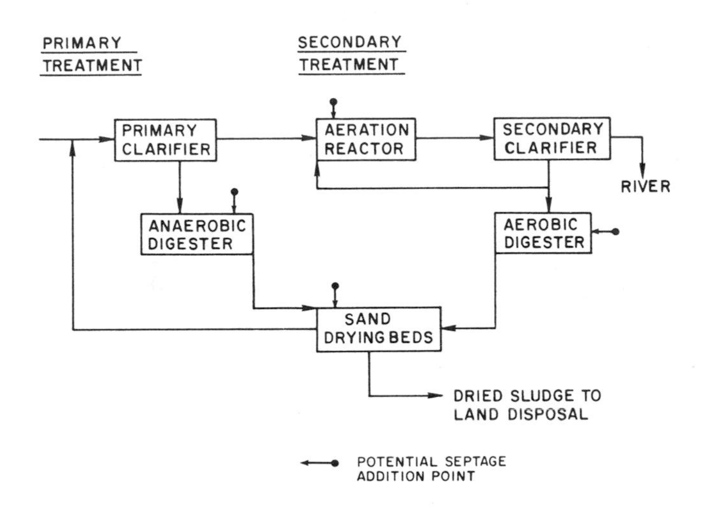

2. Alternatives for septage treatment in conventional primary and secondary treatment facilities.

developed on the biological reactivity of septage using processes where the sludge retention time equaled the hydraulic detention period, as is the case with aerobic digestion and aerated lagoons. Both batch units and semicontinuous feed units, which were fed once per day, were examined. Soluble and particulate organic decomposition rates and oxygen uptake rates were measured in order to define the decomposition kinetics. Anaerobic digestion kinetics were determined with only one septage sample. Finally, the potential use of sand drying beds to concentrate and dry the sludge was examined for a wide variety of septage samples. All biologically treated septage samples were characterized as to their dewaterability and then compared with the untreated dewatering characteristics. All studies were conducted using actual septage from domestic residences. A more detailed description of experiments has been given by Howley (1973) and Perrin (1974).

Methods of Measuring Septage Treatability

The biodegradability of septage should be considered in terms of the soluble and particulate organic material, since the soluble fraction is often in the order of several thousand mg/l COD. Both organic portions are composed of two fractions, a biodegradable portion and an organic fraction that is resistant to microbial decay—the refractory portion. The quantities of material in septage which are biodegradable or refractory were estimated, and the kinetics of their decay reactions were related using a simple model (Jewell & McCarty, 1971; Stein et al., 1972). This model can be written as follows:

$$\ln \frac{(M - fM_o)}{(M - fM_o)} = -kt \tag{1}$$

where f is the fraction of the initial mass which is refractory, M is the total mass which is measured at any time, t, and k represents the rate of decomposition of the biodegradable fraction only (day^{-1}). The particulate organic decay rates are expressed on a VSS basis and the soluble organics on a total COD basis. All reaction rate coefficients (k) in this paper are expressed to the natural logarithmic base e.

In order to determine whether sand drying beds could be utilized to dewater and concentrate septage, it was first necessary to develop a simple test that would relate dewaterability to the rate at which septage would lose water on a typical sand bed. A simple and rapid dewatering test referred to as the Capillary Suction Time (CST) was used to measure the dewaterability, or rate at which the sludge will give up its water.

Septage Supply and Preparation

A total of 24 different samples of septage were obtained from a local septic tank contractor. A brief history of each sample was obtained from the user on detergent use, time between cleanings, and other details. This was necessary so that sludge which was obviously not characteristic of domestic origins would not be included. For reasons such as oil contamination, more than twenty separate truckloads were also rejected for use in this study. Special efforts were also taken to be sure that the samples represented one source and that the material used in this study was a homogeneous representative sample of the material taken from the septic tank. In order to simplify sampling and testing problems, each sample was strained through a No. 5 soil sieve (0.157 inch mesh) to remove twigs, leaves, and any inordinately large household debris such as large masses of hair, cigarette butts, gum, and plastics. The straining action was assisted by forcing most of the material through the sieve with a rubber spatula. Although the quantity removed from each sample varied, it was not felt that the major quality characteristics of the sludge were changed significantly.

Laboratory Model Descriptions and Operation

Aerated lagoons and batch units. Five semicontinuously fed aerated lagoons were operated, each with different detention periods: 1, 5, 10, 20, and 30 days. The 1 and 10 day lagoons were operated with two different septage samples, while all others received daily feed from the same septage source. This feeding method was accomplished by freezing large quantities of the septage at the beginning of each test. The volume of the units was varied between 1.75 and 10 liters.

Fourteen aerobic decomposition batch units of septage were tested using an initial volume varying between 3 and 6 liters. The septage used in the batch tests was not frozen. Aerobic conditions were maintained in all units by the use of stone diffusers to maintain a minimum dissolved oxygen concentration of 1 mg/l. Evaporation was compensated for by the addition of deionized water.

Anaerobic digestors. The anaerobic digestion portion of this study involved only one septage sample, which was diluted with de-ionized water to obtain a variety of digester loadings at the same hydraulic detention period. One sample of raw septage and four dilutions (septage: water) (1:1, 1:3, 1:6, and 1:12) were used to feed two-liter anaerobic digesters on a semicontinuous basis to achieve hydraulic detention periods of 15 days. The septage used in the anaerobic digestion experiments was also used in a batch aerobic test (batch unit #7).

A summary of the operating conditions for the biological treatment portions of this study is shown in Table 1.

Sand drying beds were simulated using two-inch-diameter tubing 24 inches long. Two inches of four-inch stone were placed in the column followed by six inches of sand. Although the column bed depth was less than the 12 to 36 inch depth of full-scale sand beds, it was sufficient to measure the dewatering rates. The sand had an effective size of 0.62 mm and a uniformity coefficient of 4.0.

The coagulant aids were limited to two polelectrolytes (Dow Purifloc C–31 and C–42), ferric chloride and alum. All aids were added in stock solution concentrations of 50 mg/ml.

Testing and Analysis

One of the most difficult measurement problems occurred in attempting to obtain an accurate monitoring of the solid and dissolved fractions in the septage. The most rapid and accurate method, which was final-

Table 1
Range of Operating Variables for Treatability Studies of Septage

		Aerobic Treatment	
	Anaerobic Digestion	*Aerated Lagoons*	*Batch Units*
Number of septage samples used	1	7	14
Preservation daily feed portions	Freezing	Freezing	None
Hydraulic detention period, days	15	1, 5, 10, 20, 30	22 to 63
Temperature	30 ± 1	22 ± 2	22 ± 2
pH control	$NaHCo_3$	None	None
Waste loading, (lb VSS/ft^3-day)	0.004 to 0.08	.03 to 0.6	—

ly chosen, was to dilute samples 10:1 with de-ionized water and to filter various quantities through glass fiber filters (Whatman GF/C grade, 4.25 cm) in a Millipore Filter apparatus. Procedural accuracy was usually 5 percent or better at total suspended solids concentrations of 40,000 mg/l.

Dissolved oxygen concentrations and uptake rates were measured with a dissolved oxygen probe (Yellow Springs Instrument Company).

The biodegradable organics such as BOD_5 were measured on the filtrate obtained from the SS analysis. A nitrification inhibitor (allylthiourea) was used to limit the reactions to the organic carbon demand. Other parameters measured for control and analysis were pH, total solids (TSS), volatile suspended solids (VSS), total COD of the sludge, dissolved COD (COD_s) obtained from the filtrate of the SS analyses, and settling rate. After free drainage from the sand drying beds had ceased, the moisture content of the sludge cake was also determined. Generally, all analyses were performed according to *Standard Methods* unless otherwise noted.

Results

Raw Septage Characteristics

Analysis of raw septage samples confirmed the complex nature of this material. A summary of some of the characteristics and their variability is given in Tables 2 and 3. It is surprising that the septage was as dilute as 1.5 percent solids with a maximum concentration of nearly 8 percent. These solid concentrations, combined with soluble

Table 2
Variability of the 24 Different Samples of Septage Utilized in This Study. CST in Seconds, All Other Values Expressed as mg/1

Value	*TS*	*TSS*	*VSS*	COD_s	CST^a
Minimum	16,700	15,200	12,750	1,300	112
Maximum	70,800	78,800	51,500	7,650	449
Average	41,900	39,100	30,100	3,360	223
Median	44,700	39,400	33,000	2,700	208

[a]Does not include three samples from marina toilet-waste holding tanks.
(Note that values do not correlate between different parameters in the same sample.)

organic concentrations (COD) varying from 1,300 to 6,200 mg COD/l, indicate that this material would be difficult to treat. In general, no settling of raw septage occurred after a settling period of 60 minutes in a 1000 ml graduated cylinder. The dewaterability measurements are indicative of sludges which resist water removal and cannot be treated with sand drying beds. Excepting the dilute marina holding tank wastes, the CST of most septage samples was greater than 200 seconds. Typical CST values of sewage treatment sludge that can be dewatered by sand drying beds vary up to a maximum of about 70 seconds.

Anaerobic and Aerobic Treatment

Most of the processes, except anaerobic digestion, were effective in handling septage. A summary of the characteristics of the septage sample fed to the anaerobic digesters was as follows (all values in mg/l):

$$COD_T = 41{,}000,\ COD_s = 3100,\ TSS = 24, 50,\ VSS = 19{,}600$$

The digester with the highest loading rate (0.1 lb/ft^3 - day) was not able to function, a fact suggesting the presence of toxic materials (Figure 3). All other dilutions produced gas at comparable rates of about 4 ft^3 per lb. of total COD added. In order to determine whether an acclimated microbial population might be more effective at higher organic loadings, the reactor which was producing gas at a high rate (Unit 2) was loaded at the same rate as Unit #1 had been loaded when it failed. As is shown in Figure 3, the reactor failed under this higher loading. Since it was suspected that detergents caused the inhibition

Table 3
Some Characteristics of Septage Used in Biological Treatment and Dewatering Experiments

Unit	*COD$_s$ (mg/1)*	*TSS (mg/1)*	*CST (sec)*
Batch Units			
1	2800	43,400	165.5
2	5600	19,950	192.6
3	3400	44,600	186.9
4	2400	46,200	142.1
5	4160	45,500	275.4
6	1450	15,200	111.5
7	3100	24,150	286.2
8	3450	38,700	287.4
9	2700	73,500	283.6
10	1950	39,400	376.2
11	1640	49,900	330.4
12	2200	78,750	208.8
13[a]	1986	30,290	57.9
13A[a]			51.6
13B[a]			60.9
13C[a]			71.1
14	6190	47,430	339.8
Aerated Lagoons			
10-Day	7200	50,100	180.7
	1300	47,000	265.7
20-Day	2400	46,200	142.1
30-Day	1370	18,000	124.0
5-Day	7650	35,200	448.7
1-Day[a]	1340	26,900	60.1
	6190	47,400	339.8
Mean	3360	39,100	223.2 (199.6)[b]
Median	2700	39,400	208.8 (186.9)[b]

[a]Septage taken from marina holding tank.
[b]Value including marina holding tanks.

of microbial activity, amine was added to neutralize this effect (Bruce et al., 1966; *Notes on water pollution*, 1968; Swanwick and Shurben, 1969; Truesdale and Mann, 1968). No neutralization was observed, but the quantity of amine added was probably less than the concentration required to neutralize detergents if this had been the inhibiting substance. A summary of the results obtained with the digestion experiments is shown in Table 4. The conversion of organic material to gas averaged about 3.9 ft^3 of gas produced per lb of total

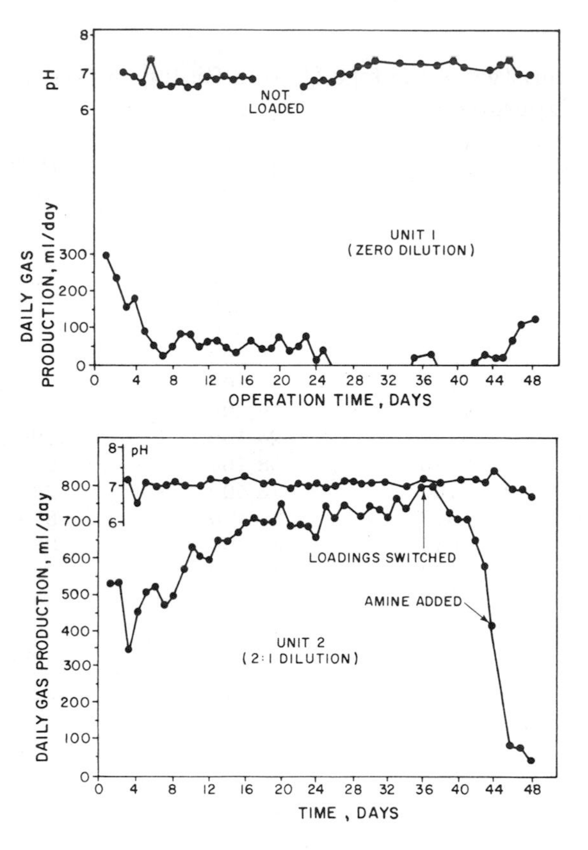

3. Gas production in digesters with organic loading of 0.10 lbs. TSS per ft^3 per day (Unit 2). These loadings were switched on day 30.

Table 4
Summary of the Results of Anaerobic Digestion of Septage

	Waste Loading (lb/ft^3-day)		*Gas Production (ft^3/lb COD added)*	*Waste Reduction Efficiency*		
Unit	*COD_T*	*TSS*		*COD_T*	*COD_s*	*VSS*
1	0.17	0.10	0.15	1.5	—	—
2	0.08	0.06	4.27	44	54	45
3	0.04	0.03	3.82	39	64	—
4	0.02	0.01	3.52	36	47	—
5	0.01	0.006	2.94	30	29	—
Average			3.87			

COD added to the digester. If it is assumed that the gas was composed of 65 percent methane, this gas-production rate would be theoretically equivalent to a COD conversion efficiency of about 40 percent. As is noted in Table 4, the average measured COD conversion was 37 percent. The batch aerobic unit (no. 7) experienced approximately the same removal efficiencies as the anaerobic unit.

Typical data on the reduction of VSS and COD_s and variation of pH, DO, and oxygen uptake rates for a batch unit and two continuous aerated lagoons are shown in Figures 4, 5, and 6. The soluble organic reduction observed in batch 2 (Figure 4) was higher than normal, but the initial soluble organic concentration was also high. The VSS reduction was quite low, with a significant fraction remaining after long periods of aeration. The pH of the units varied little in the same unit but varied between units because of differences in microbial activity. In general, the pH ranged between 6.5 and 8.5 for both batch and continuous feed experiments. The oxygen uptake rates were initial 2 to 3 mg/l-min but declined to less than 1 mg/l-min after ten or fifteen days in the batch units. The oxygen uptake rates in the aerobic lagoons varied in proportion with the detention period. Average DO uptake rates increased from 0.3 mg/l-min to 1.5 mg/l-min as the hydraulic detention period was decreased from thirty days to one day. In most cases the continuously fed lagoons were monitored after feeding had terminated to determine whether a significant amount of material remained to be stabilized. As can be seen in Figure 5, fifteen days of aeration after feeding ceased resulted in little change in the VSS and COD_s concentrations, thus indicating a high degree of stabilization in the continuously fed units. In two cases, the aerated lagoons were tested with two different septage samples. The data in Figure 6 represents two separate runs with different septage aerated for a one-day hydraulic period. In several cases, the VSS reduction in the aerated lagoons with the shorter aeration periods was either zero or negative. These results can be explained in terms of the synthesis of new microbial mass on the soluble substrate. In cases where the soluble COD reduction was 5,000 mg/l, the newly synthesized microbes could have been as high as 2,000 mg/l.

The data summarized in Table 5 indicates the biodegradability of the septage in the aerobic biological treatment processes. More than 75 percent of the soluble organic material was removed in most batch and continuous units. However, only a little more than 40 percent of the particulate organic material was oxidized in a period longer than thirty days under batch conditions. The reduction of volatile solids in the aerated lagoon was much less than that observed in the batch units, with an average stabilization less than 25 percent.

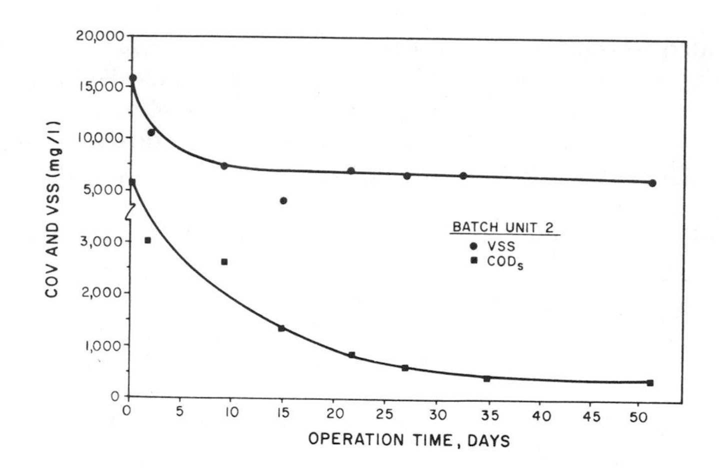

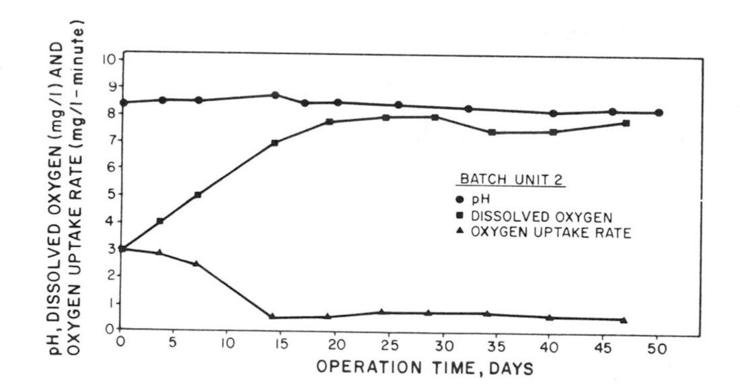

4. Typical waste reduction and variation of operating conditions for batch composition experiments.

Even in the case where the septage was aerated for periods of twenty to thirty days in the continuously fed units, the percentage of reduction remained less than 25.

The effluent COD_s from both the batch units was often greater than 600 mg/l, but the average BOD_5 was 29 mg/l in the batch units and 51 mg/l in the continuously fed units. The amount of COD_s or BOD_5 resulting from any particular treatment did not correlate with the period of detention or any other observed factor. The effluent BOD_5 from the 10 and 20 day aerated lagoons was less than 3.0 mg/l, whereas the effluent from the thirty-day lagoon contained 46 mg/l BOD_5. In six of the eleven batch units, the BOD_5 after more than thirty days of aeration was less than 20 mg/l.

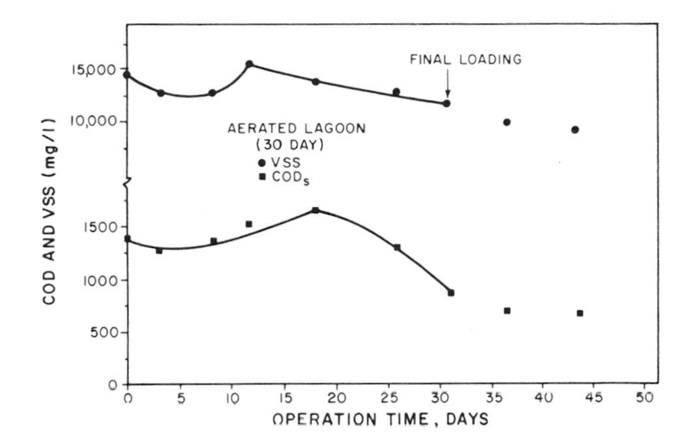

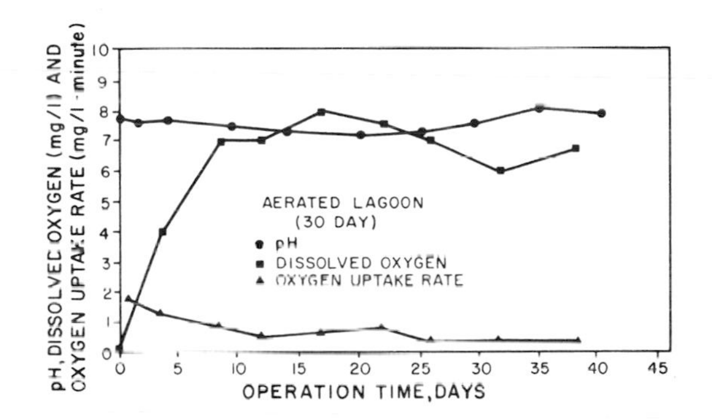

5. Typical waste reduction and variation of operating conditions for continuously fed aerated lagoons.

The rate of removal or decomposition of the soluble organic material and the organic particulate material was found to be a first-order decay relationship as described by Equation 1. A summary of the first-order decay rates is shown in Table 6, with the refractory fractions used to calculate the rate with Equation 1. The refractory fraction of the soluble organic matter (COD_s) in the batch reactors varied from 7 to 66 percent (average 25 percent), and the decomposition rate (k) varied from 0.048 to 0.535 day^{-1}, with an average of 0.086 day^{-1}. Refractory VSS in the batch reactors varied from 32 to 82 percent (average 57 percent), and decomposition rates varied from 0.019 to 0.248 day^{-1}, with an average of 0.085 day^{-1}. It is interesting to note that the solids and the soluble organic decay rates were equal in the batch units.

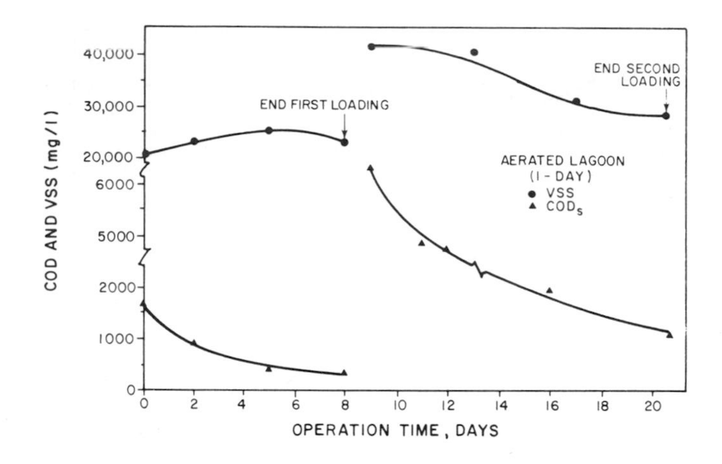

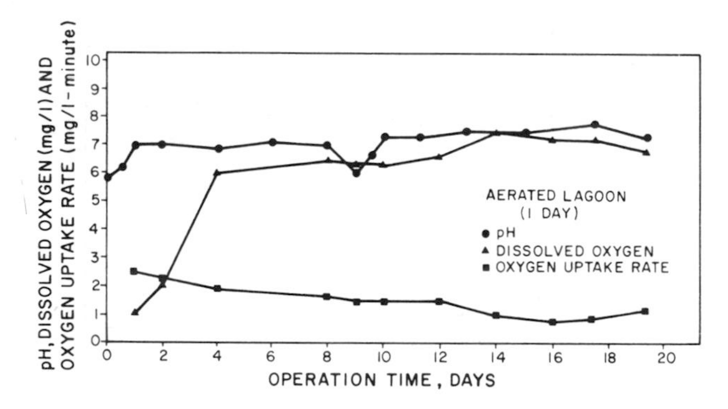

6. Waste reduction and variation of operating conditions in an aerated lagoon with very short hydraulic detention periods.

Foaming and Odor Control

A summary of the foaming and odor problem is shown in Table 7. Several techniques were tried to control foaming. Mixers were suspended in the air above the units to break the foam physically, and this was effective in most cases. However, the foam often occupied a volume equal to that of the liquid portion. Two commercial silicone defoamers (Dow-Corning DB-110 and DB-31) were applied to three samples in a dose of 50 ml of the defoamer (as received) to 3 liters of aerating septage. The foam was drastically reduced in several minutes. However, within twenty-four hours, the foam had developed to the volume it held before defoamer application.

Table 5
The Reduction or Biodegradability Measured in Aerobic Stabilization of Septage

Batch Unit	Period of Aeration (days)	Organic Loading (lb VSS) (ft^3-day)	Biodegradability, % Removed			
			COD_T	COD_s	TSS	VSS
1	35	—	54	69	45	49
2	35	—	56	93	54	62
3	35	—	47	71	54	59
4	40	—	26	72	30	38
5	37	—	30	84	35	44
6	29	—	22	34	10	20
7	63	—	76	92	55	68
8	28	—	25	77	17	18
9	29	—	42	78	39	45
10	25	—	35	79	30	29
11	22	—	35	73	42	43
Mean			39	75	40	43
Median			35	77	39	44
Aerated Lagoons						
1	30	0.03	21	48	14	24
2	20	0.10	11	69	22	21
3	10	0.20	—	94	6	20
4	10	0.30	23	71	15	8
5	5	0.4	42	91	5	26
6	1	0.6	43	74	negative	negative
7	1	1.3	—	83	31	26
Mean			28	74	16	20
Median			23	74	17	24

Foam fractionation of three of the samples was tested. After fifteen minutes of foam removal in an intensively aerating unit, no further foam was generated. Although this method was effective in controlling foam, it also removed more than 30 percent of the VSS and the COD_s.

In an attempt to identify the cause of the foaming, the household from which each septic tank sample was obtained was contacted about detergent use. Foaming occurred in all samples where laundry detergents were reported used, but in only one where they were reported not used. Foam persisted for varying lengths of time but on the average disappeared within eleven days in most batch units. The foaming problem did not cease in the continuously fed aerated units.

One of the most important parameters of septage disposal is that

Table 6
Summary of Refractory Organic Fractions and Decomposition Rates for Biological Treatment of Septage

		Soluble COD		*VSS*	
Unit Batch	*Period of Decay, Days*	*Refractory Fraction, f, %*	*Decay Rate k, Base "e" Day^{-1}*	*Refractory Fraction, f, %*	*Decay Rate k, Base "e" Day^{-1}*
1	53	—	—	50	0.019
	35	31	0.057	—	—
2	53	—	—	38	0.075
	35	7	0.109	—	—
3	52	—	—	41	0.023
	35	29	0.057	—	—
4	48	28	0.074	62	0.072
5	37	16	0.092	56	0.034
6	29	66	0.048	80	0.030
7	85	—	—	32	0.051
	63	8	0.059	—	—
8	14	—	—	82	0.248
	28	23	0.535	—	—
9	14	—	—	55	0.123
	29	22	0.053	—	—
10	25	21	0.134	71	0.039
11	22	27	0.182	57	0.217
Mean		25	0.086[a]	57	0.085
Median		23		56	

[a]Obtained by deleting unit 8.

of odor, since septage is one of the most repulsive smelling wastes. Fortunately, this appeared to be one of the more easily controlled parameters. As is shown in Table 7, odor changed from a septage odor to an earthy odor similar to that of activated sludge five days after aeration was initiated. In the continuously fed units, no odor was observed after an average acclimation period of three or four days, except at the time of feeding, in all units including the one-day detention-period lagoon. Thus odor appears to be easily controlled by aerobic treatment.

Effect of Aerobic Treatment on Dewaterability

All biological treatment units were monitored by the CST analysis to determine the effect of aeration on the rate at which septage would give up water. The change of CST with aeration time is illustrated in

Table 7
Period of Aerobic Treatment Required to Control Odor and Foam in Treatment of Septage

Unit Identification	*Aerobic Period to Eliminate Odor (days)*	*Aerobic Period to Eliminate Foam (days)*	*Survey of Household Washer Detergent Use*
Batch Units			
1	4	9	Yes
2	9	10	Yes
3	4	13	Yes
4	3	19[c]	Yes
5	10	12	Yes
6	5	No foam	No
7	5	6	Yes
8	1	FF[c]	Yes
9	No septic odor	14[c]	Yes
10	3	7	Yes
11	3	11	Yes
12	5	—	—
13	4	—	—
14	8	—	—
Mean	4.9	10.9	
Median	4.0	10.5	
Aerated Lagoons			
1 day	3	No foam	No
	7	No foam	No
5 day	4	No foam	Yes
10 day	2	Foam present[b]	Yes
	—	Foam present[b]	Yes
20 day[a]	1	Foam present[b]	Yes
30 day	2	No foam	No
Mean	3.2		
Median	2.0		

[a]Same sample as Batch 4.
[b]10 and 20-day lagoons foamed throughout their loading periods.
[c]FF Batch 4 was foam fractionated after 19 days, batch 8 at start-up, batch 9 after 14 days.

Figure 7 for both batch and continuous feed units. In most cases, the CST increased initially, but after five to fifteen days decreased rapidly and reached a value between seven and twenty seconds after thirty to forty days of aeration. Because of the unique increase of CST observed during the first loading condition with the 10 day aerated lagoon (Figure 7), another septage sample was fed to this unit.

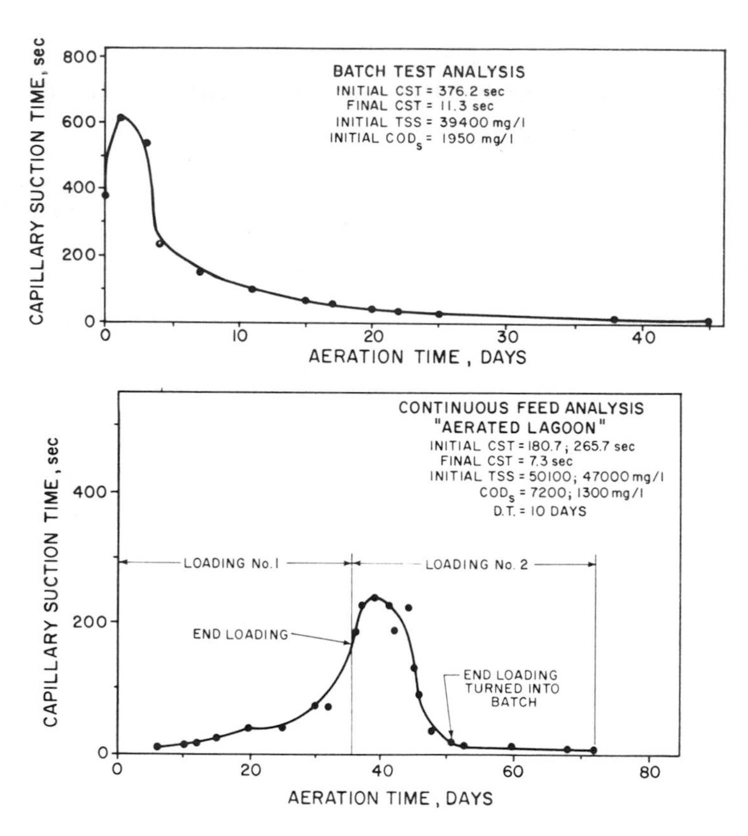

7. Typical effect of aeration upon filterability of septage in a batch unit (10) and in an aerated lagoon.

In all cases with the aerated lagoons, after an initial period of acclimation the CST decreased below fifty seconds even in the short detention lagoons. Thus the data indicates a marked improvement of dewatering rates with aeration, with minimum capillary suction time obtained after thirty to forty days of aeration.

Effect of Biological Treatment on Settleability

The rate of sedimentation of the raw and treated septage was measured using a 1000 ml graduated cylinder. The majority of raw septage samples did not exhibit any settling properties. In general, long periods of aeration (thirty to seventy days) improved the settleability such that the average improvement in zone settling rates was 2 ft per

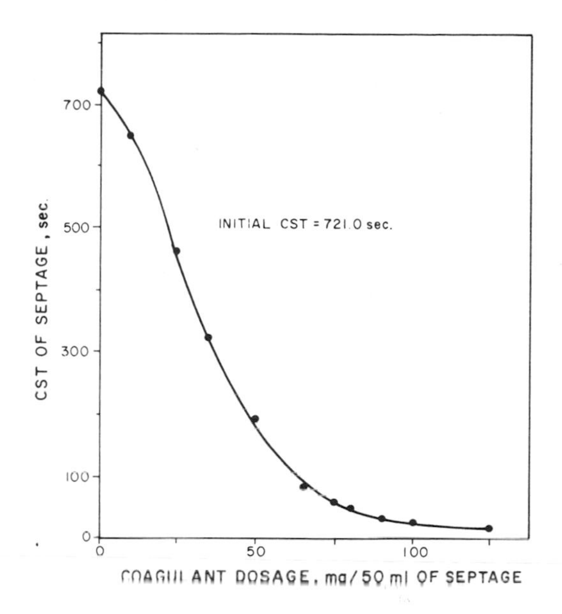

8. Typical relationship between chemical dosage and capillary suction time.

hour. After sedimentation for one hour, the maximum solids concentration approached 100,000 mg/l.

Chemical Coagulation of Septage

The purposes of this portion of the study were to determine whether chemical coagulation would increase the dewaterability of septage so that untreated sludge could be treated with sand drying beds, and to develop a relationship between a simple dewaterability measurement parameter (CST) and chemical dosage. Initial CST values of the septage ranged from 120 seconds to 825 seconds. Figure 8 illustrates a typical relationship between chemical coagulant dosage and CST for a very slow dewatering sample.

In order to determine whether coagulants could be used with various types of septage, seven different samples with widely varying initial CST values were examined. As will be explained, a CST of 50 seconds was estimated to be the maximum value for effective dewatering of the sludge by sand beds. Figures 9 through 12 illustrate the effects of various dosages of four different coagulants with a wide variety of sludges. Note that in all cases improvement of the dewaterability to the 50 second CST value was achieved. The difference be-

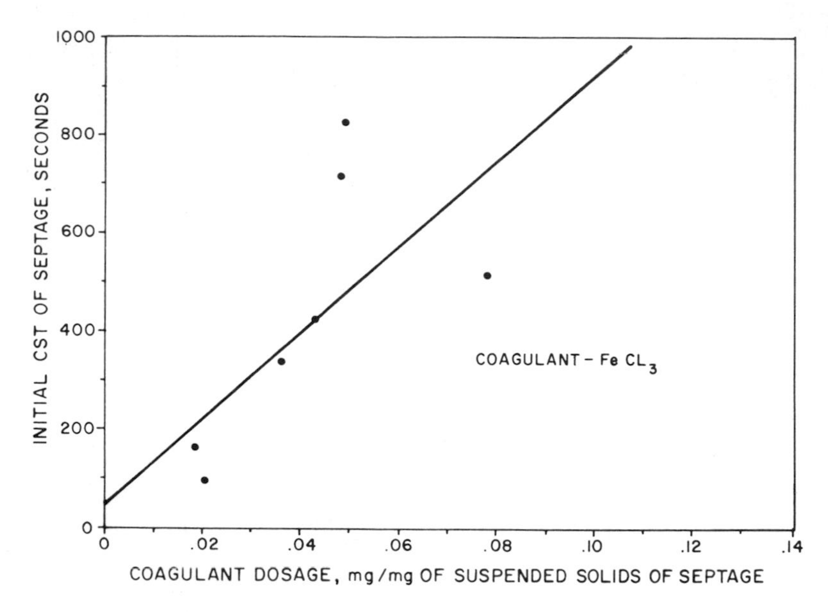

9. Relationship between CST and $FeCl_3$ dosage to reduce CST to 50 seconds for seven different septage samples.

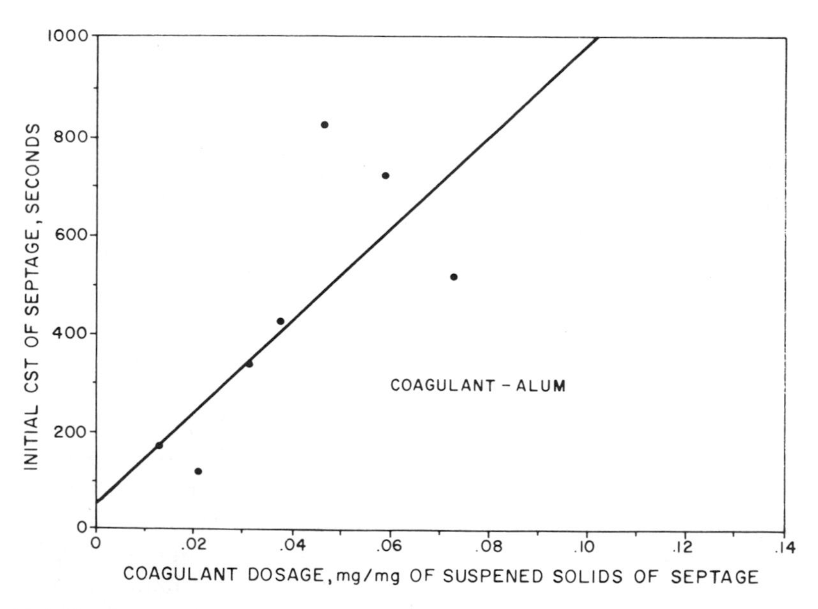

10. Relationship between CST and alum dosage to reduce CST to 50 seconds for seven different septage samples.

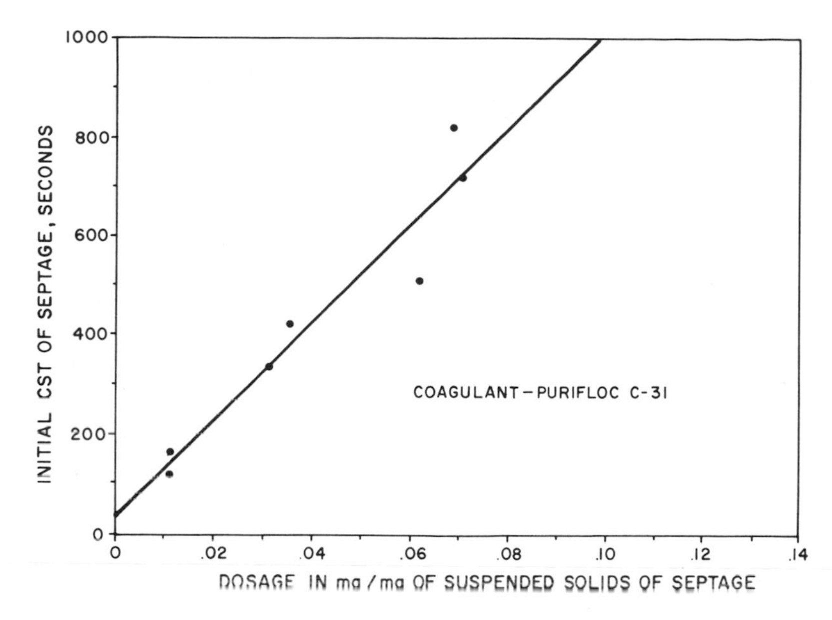

11. Relationship between CST and Purifloc C-31 dosage to reduce CST to 50 seconds for seven different septage samples.

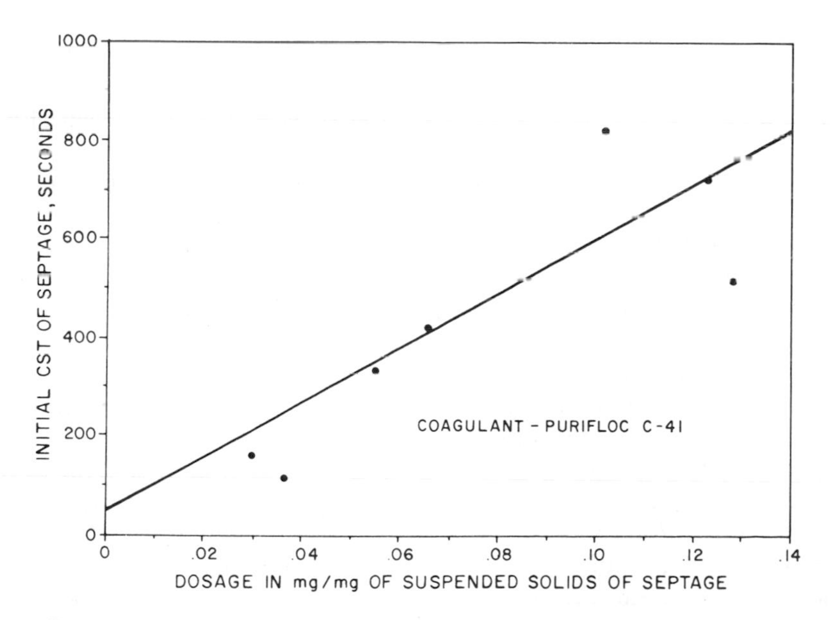

12. Relationship between CST and Purifloc C-41 dosage to reduce CST to 50 seconds for seven different septage samples.

Table 8
Range of Variables Included in Experiments on Dewatering of Septage on Sand Drying Beds

Value	*Period of Aeration (days)*	*CST (sec)*	*TSS (mg/1)*	*COD_s (mg/1)*	*Initial Moisture Content (%)*
Low	0	11	9,590	290	92.15
High	89	537	78,750	2200	99.04
Mean	24	134	31,350	1060	96.88
Median	14	66	31,780	940	96.32

tween the doses required by all four coagulants to decrease the CST to 50 seconds is relatively small. In most cases, septage can be chemically treated to improve dewatering characteristics by adding 2 to 4 percent (by solids dry weight) of any of the coagulant aids tested.

Dewatering Septage by Sand Drying

The sand drying column portion of the study was conducted to determine if a relationship existed between dewatering rates as measured by the CST and drying time and rates on sand drying beds. Half-liter samples of raw and treated septage were taken from units described in the previous paragraphs and placed on the laboratory sand columns, thus making a ten-inch deep sludge application. The range of variables included in this dewatering study are summarized in Table 8. The choice of the thirty samples used in this portion of the study was based on a desire to cover a wide range of CST values. In order to develop information on the rate of water removed from the septage by the sand drying beds, numerous tests were conducted to measure the rate of water draining from a sand column with various types of septage. Figure 13 contains a typical sand column drainage curve, using a 500 ml sample of septage. In most cases a small volume of liquid drained instantaneously as the sludge was added to the column. The linear drainage rate following this initial release was assumed to be the dewatering rate and was measured by the slope of this portion of the curve. The point in the curve where the linear portion starts to become horizontal represents the elapsed time at which free drainage of liquid from the column ceased. Data developed by this method was used to relate initial CST to the dewatering rate and the drainage time to reach cessation of free drainage (Figures 14 and 15). Since it has been reported that the maximum period of free drainage should

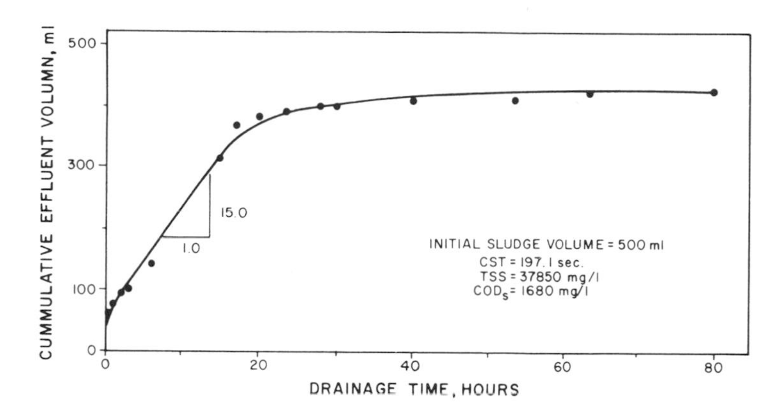

13. Relationship between elapsed time and cumulative effluent volume for sand drying columns (curve for column 5).

be about 48 hours, the data in Figure 15 can be used to determine a limiting CST value (Tang et al., 1969; Nebiker, 1967; Nebiker et al., 1969). A free drainage period of two days corresponds to a CST of about 50 seconds. As indicated by Figure 14, this CST also represents a dewatering rate of about 20 ml/hr. The average concentration of the sludge cake was 20 percent solids after a drainage period of about three days.

Discussion

General

The magnitude of the septage disposal problem is very large in proportion to the information that is available upon which to base rational decisions. There are several possible explanations for this apparent lack of information. First, most septic tank users may not clean their tanks until obvious operational problems develop. If septic tanks are not cleaned every three years, then the quantity of sludge removed is not as great as this study has estimated. Second, the present disposal methods may not adversely affect the quality of the environment. Third, the difficulties of studying this highly odoriferous material are extreme. Although all three factors may be partially responsible, the problems of conducting studies on septage are likely the main reason for the lack of good problem definition. Two studies conducted simultaneously with actual domestic septage are combined in

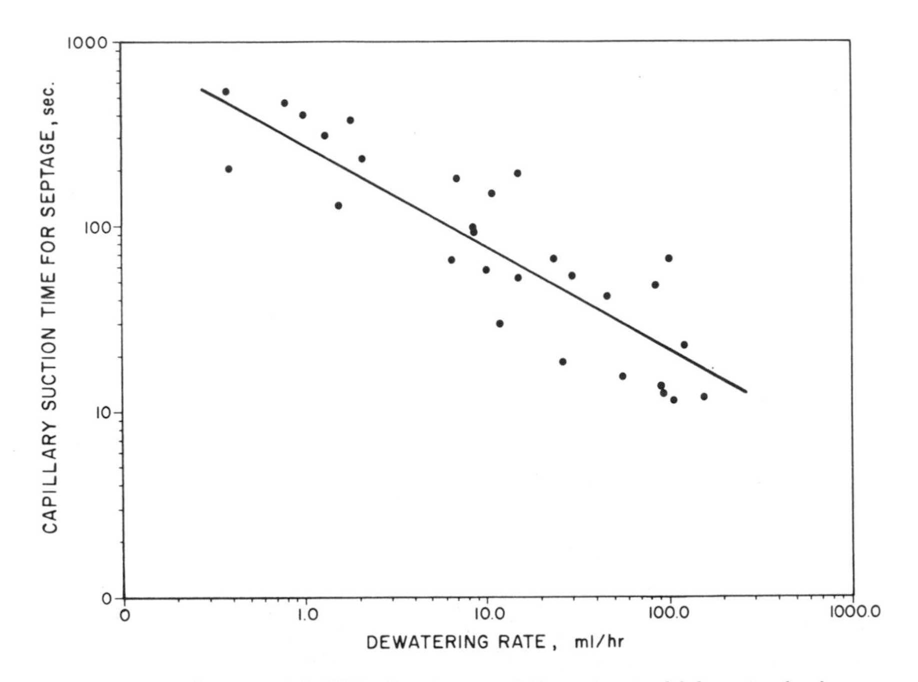

14. Relation of the initial CST of septage and the rate at which water is given up when placed on a 2-inch diameter sand column.

this paper in order to provide information on the engineering aspects of alternative solutions to the problem of treatment and disposal of septage (Howley, 1973; Perrin, 1974). It is recognized that the solution to each septage disposal problem is limited by varying conditions and that alternative treatment and disposal methods will be applicable under different conditions. It was the objective of these two studies to provide engineering data to evaluate a number of treatment and disposal alternatives. It should also be noted that septage is, by nature, highly variable in composition and that a limited number of samples were analyzed in this study. Thus sound engineering judgment should be used in the application of the design values derived from these studies.

Biological Treatment

Significant reduction of most gross pollution parameters was achieved by biological, chemical, and/or physical treatment of septage. Anaerobic digestion of septage was found to give similar reductions of COD_s and VSS as aerobic treatment except at high loadings (0.10 lb total

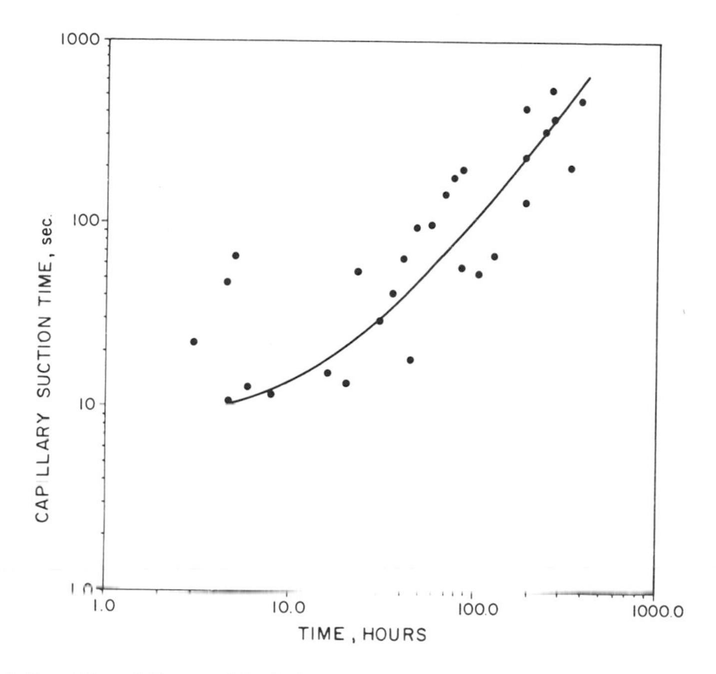

15. Relationship of the sand bed dewatering period to achieve cessation of free drainage to the initial CST.

solids per ft^3 per day). Results at high loadings indicated that the septage contained a material which prevented gas formation. The literature has suggested that this material is detergent in high concentration (Swanwick and Shurben, 1969; Truesdale and Mann, 1968). Since detergent concentration may be as high as 3 percent of the solids (dry weight) and since most detergents are not decomposed by anaerobic microorganisms, anaerobic digestion appears to be an unfavorable unit-process for septage treatment.

Aerobic treatment was found to be very effective in controlling most of the pollution parameters of septage. Batch aeration tests were used to determine the maximum quantity of the organic material that could be decomposed, and aerated lagoons were used to simulate expected results with continuously fed biological treatment units. The volatile solids reductions were relatively low in the batch units, with the maximum biodegradable fractions between 40 to 45 percent of the initial VSS. Because of the long retention period of the sludge in the septic tank, some decomposition has already taken place before its removal. This reduction was estimated to be 35 to 40 percent by Truesdale and Mann (1968) in a carefully controlled study of septic tank operation. Since microorganisms are often composed of material considered to be 20 percent nonbiodegradable or refractory, the same should be true of septage. Thus if 35 to 40 percent of the VSS is decomposed in the septic tank, the maximum further reduction that should be possible under ideal decomposition conditions would be an additional 40 to 45 percent VSS reduction. That value agrees well with the data obtained in this study.

The VSS reduction in the continuously fed units was always much less than that obtained in the batch units, and in several cases was negative. This was not entirely unexpected, since the high soluble organic concentration would result in synthesis of particulate organic material. However, this data does emphasize the problems of estimating continuous unit treatment efficiency with batch unit data.

The soluble organic matter decomposition was found to be similar in both batch and continuous units, with a reduction of about 75 percent of that originally present. This reduction often produced an effluent COD_s of 600 to 700 mg/l. Most of this, however, was nonbiodegradable, since the BOD_5 was usually less than 20 mg/l.

Kinetic Description of Biological Treatment

The rates of decomposition of the biodegradable soluble and particulate material were equal ($k = 0.085^{-1}$) in the batch units. It was not possible to obtain a corresponding rate coefficient with the aerated lagoons because the refractory fraction was not determined. The decay rates measured in the aerated lagoons were very different, however, from those obtained in the batch unit. The soluble organic material in the one-day aerated lagoon decomposed at more than ten times the average batch decay rate, while the rate in the thirty-day lagoon was only one quarter the batch rate. In both lagoons the quantity degraded was similar and the percentage of reduction in most lagoons was similar. Thus it can be expected that the decay rate would decrease with increased detention periods. The average reduction of VSS in aerated lagoon units with long detention periods was less than half that observed in batch units with comparable detention periods. In the one-day detention lagoon, which received a volumetric loading of 0.6 lb VSS per ft^3 per day, a slight increase in VSS was observed. Thus it can be expected that significant reduction of soluble organics can be easily achieved in continuously operating units, although significant VSS decomposition would be very difficult to achieve.

The oxygen uptake rates in both the batch units and the continuously fed units varied from 30 mg/l-hr to 200 mg/l-hr. In the short detention period lagoons, the oxygen uptake rates were significantly greater than those usually measured in conventional activated sludge. If an activated sludge facility has a design sludge retention time of twenty days, the addition of each load of septage (assumed to be 1,000 gallons) to the aerated reactor adds an oxygen load equivalent to 1,000 people. These oxygen usage rates indicate the necessity of an adequate oxygenation capacity when septage is applied to an activated sludge reactor.

Odor and Foam Control

Two of the most obvious and persistent problems of septage and its treatment are the odor of raw septage and the development of large quantities of foam during aerobic treatment. Of all pollution parameters measured, odor was found to be the easiest to control. Very short hydraulic detention periods (one day) in continuously fed units were capable of converting the strong septage odor to an earthy odor similar to that of activated sludge. The development of large quantities of foam in aerating units was another indirect indication that high detergent concentrations were present. The volume of foam often equaled the volume of the reactor in the case of the ten-liter aerated lagoons. Although the foam was degraded in the batch reactors within about twelve days, it was a continuous problem when it appeared in the aerated lagoons. Several attempts to remove the foam by fractionation indicated that this might be an effective method of control.

Dewatering Characteristics of Septage

An alternative method of decreasing the magnitude of the septage-handling problem would be to dewater septage using sand beds. It is not feasible to use this method without prior treatment, however, because of the poor dewaterability of raw septage. Both chemical coagulation and aerobic biological treatment improve the dewatering rates of septage enough so that it can be applied to sand drying beds.

Although there are numerous complex variables that affect the dewatering characteristics of sludges, the relations developed between the rate of sand-bed removal of water and the capillary movement of moisture from septage are quite uniform. Data from Figures 13 to 15 were used to determine that septage would require a CST of fifty seconds or less, assuming that a sludge should cease free or rapid drainage within forty-eight hours after placement on a sand bed.

These relationships could also serve as simple control tests to determine the degree of treatment required to condition septage. The simple measurement of the CST of a sludge (which should require no more than five minutes) would make possible estimates of the rate at which water could be removed from the sludge, the time required to achieve cessation of free drainage on a sand drying bed, and the value of a coagulent aid. With the use of Figures 8–12, the dose necessary to achieve a desired dewatering rate can be estimated. For example, a septage with an initial CST of 400 seconds would have a dewatering rate on a two-inch-diameter column of less than one ml per hour (Figure 14) and a time to free drainage greater than ten days (Figure 15), and it would, therefore, require chemical coagulants or biological treatment before dewatering could be effective. Addition of 0.035 gms alum per gm of suspended solids would reduce the CST of fifty seconds (Figure 10) which would be required in order for free drainage on a sand drying bed to cease in two days (Figure 15).

Although chemical conditioning of sludge is capable of improving the dewatering characteristics of all types of septage, the odor is unaffected and prevents the use of sand beds with raw septage. Aerobic biological treatment handles the odor problem effectively and also increases the dewaterability of the sludge. After an initial period of increasing CST values in batch reactors, the CST would drop rapidly to values less than 50 after twenty days of aeration. In all cases the CST was reduced below 50 seconds in the continuously fed aerated lagoons. In the unit with a one-day retention, the initial CST of 340 seconds increased to 800 seconds after five days of operation and then rapidly decreased to a constant value of fifteen seconds.

Recommended Guidelines for Septage Treatment

Recommended treatment to achieve a high degree of control of the five parameters examined in this study is shown in Table 9. Anaerobic treatment is not included because it appears that this process cannot function when any significant concentration of detergents is present. Some of the recommended controls listed in Table 9 are not compatible with others. For example, if odor control is the main concern, twenty days of aeration to minimize the soluble BOD_5 should not be required. However, short periods of aeration will probably result in

Table 9
Recommended Treatment of Septage to Minimize the Problem Indicated

Parameter	*Degree of Control*	*Recommended Treatment*
Odor	Complete Elimination	1 to 3 days of aeration
Foam	Complete Elimination	Foam fractionation or more than 12 days of aeration
Soluble BOD_5	Less than 20 mg/1	20 days aeration
VSS	40% Reduction	More than 30 days of aeration
Solids Concentration	20% cake solids	Addition 2 to 4 percent (by dry solid weight) FeC1 of alum, and 3 days of sand bed drying or 1 to 5 days aeration and sand bed drying for 3 days

serious foaming problems unless some other control such as foam fractionation is also utilized.

Treatment of Septage in Existing Municipal Wastewater Treatment Facilities

One of the most important questions in the treatment of septage is whether existing treatment facilities have the capacity to treat this material. In considering this question, it should be noted that in many small communities the population using septic tanks equals that served by central collecting systems. Examination of the oxygenation capacity of activated sludge facilities indicates that unless a facility is tremendously overdesigned, it will not have the aeration capacity to handle the entire population using septic tanks. And even if the aeration capacity were sufficient to maintain the desired aerobic conditions, the sludge-handling facilities would often not have sufficient capacity. Addition of sludge with 30,000 to 40,000 mg/l solids to the aeration tank of an activated sludge facility would shorten the designed sludge retention period. This might decrease the dewaterability of the sludge and hence decrease the effectiveness of the solids concentration processes. Finally, untreated septage cannot be dewatered by sand drying beds. Chemical coagulants can increase the dewatering characteristics enough so that raw septage could be concentrated on sludge drying beds. This approach should be considered, however, only where the strong odor of raw septage would not create secondary problems. Also, the filtrate from the sand beds would require treatment.

Several other factors are also apparent in considering existing municipal sewage treatment facilities as receptors of septage. The governing tax structure often cannot provide equitable support between village and county residences. Furthermore, the rural or recreational areas where septic tanks are especially abundant are also the areas in which eutrophication of receiving waters is a great concern. Septage contains large quantities of nitrogen and phosphorus that would enter receiving waters.

Centralized Septage Disposal Area

The alternative to the use of existing facilities for septage disposal would be to construct central septage treatment facilities. Based on the information obtained here, it is recommended that regional treatment facilities be considered to take advantage of economics of size. An effective facility would consist of the following unit processes:

1. a covered receiving-storage tank;
2. an aerobic digester with a five-day hydraulic detention period, a capacity for foam removal or storage, and a capacity for the addition of chemical coagulants;
3. a system for land disposal of treated sludge.

Future Research

Several significant questions were not examined in this study and should be considered for future research. The nutrient characteristics of septage are important parameters in considering either surface water or land discharge of this material. Pathogen movement and survival in land disposal systems and the presence of heavy metals are topics of concern in determining final treatment disposal recommendations. High concentrations of detergents in septage indicate that boron may be present. Since boron severely inhibits some types of plant growth, its effect should be examined where land disposal is considered. Finally, subsurface injection of raw septage into the soil may be an alternative to treatment and surface-land disposal.

Conclusions

1. Sludge drained from septic tanks is a highly complex waste but is amenable to treatment by common biological, chemical, and physical processes.

2. Significant soluble COD (COD_s) reductions were observed with aerobic and anaerobic treatment. Reduction in the aerated lagoons averaged 74 percent, while that in anaerobic digesters averaged 64 percent for similar hydraulic detention periods. These reductions resulted in effluent BOD_5 values less than 20 mg/l in most units, even though the COD_s concentration was 580 to 1,000 mg/l.

3. The average decay rates (k, day^{-1}, base *e*) of the biodegradable COD_s and VSS determined for the aerobic batch units were the same (0.085 day^{-1}).

4. Foaming and odor were major problems in the aerobic treatment of septage. Fractionation of the foam or a minimum of twelve days aeration were required to eliminate the foaming problem. One-day hydraulic detention in aerobic semicontinuously fed units was found effective in eliminating odor.

5. The composition of septage may inhibit bacterial activity in anaerobic digesters.

6. Raw septage should not be applied to sand drying beds because of its slow dewatering rate. In all cases, aerobic treatment significantly increased the dewatering rates. Detention periods in aerated lagoons of one and five days resulted in sludge that could be dewatered to 20 percent solids concentration within three days on a sand drying bed.

7 A linear relationship exists between dewaterability of septage as measured by the capillary suction time (CST) method and the rate of dewatering on a sand drying bed. This relationship indicated that a septage sample with a CST of 50 seconds or less could be treated on a sand drying bed without excessively long drainage periods.

8. Chemical coagulant aids were found to be effective at concentrations between 2 and 4 percent (by weight of the dry solids) in increasing the dewaterability of all septage samples to the CST of 50 seconds required for sand bed utilization. A linear relationship between chemical dosages required to obtain a CST of 50 seconds for a wide variety of septage samples was developed for four coagulants.

9. In reviewing the treatability characteristics, it appeared that the use of existing municipal treatment facilities to treat septage should not be recommended. Instead, a centrally located treatment facility combining the following unit processes would be a feasible alternative:

covered receiving tank, aerobic lagoon or aerobic digestion, and sand beds or direct land disposal.

References

American Public Health Association. 1971. *Standard methods for the examination of water and wastewater*. New York: American Public Health Association.

Baskerville, R. C., and Gale, R. S. 1968. A simple automatic instrument for determining the filterability of sewage sludges. *Journal of the Institute of Water Pollution Control* no. 2: 233-244.

Bruce, A. M., Swanwick, J. D., Ownsworth, R. A. 1966. Synthetic detergents and sludge digestion: some recent observations. *Journal and proceedings of the Institute of Sewage Purification* part 5: 427-447.

Escritt, L. B. 1950. Small private septic tanks. *Surveyor* 109: 491-492.

Escritt, L. B. 1954. Further thoughts and a recapitulation on small private sewage treatment works. *Water and sanitary engineer* 5: 101-104.

Flood, F. L. 1944. Problems in the design, construction, and operation of very small sewage disposal plants. *Sewage works journal* 16: 90-103.

Howley, J. B. 1973. Biological treatment of septic tank sludge. Master of science thesis, University of Vermont.

Jewell, W. J., and McCarty, P. L. 1971. Aerobic decomposition of algae. *Environmental science and technology* 5: 1023-31.

Kolega, J. J. 1971. Design curves for septage. *Water and sewage works* 118: 132-135.

Kolega, J. J., Cosenza, B. J., Chuang, F. S., Dhodi, J. 1973. Anaerobic-aerobic treatment of septage (septic tank) pumpings. Proceedings of the 28th annual Purdue University Industrial Waste Conference. In press.

Nebiker, J. H. 1967. Drying of wastewater sludge in the open air. *Journal of the Water Pollution Control Federation* 39: 608-626.

Nebiker, J. H., Sanders, T. G., Adrian, D. D. 1969. An investigation of sludge dewatering rates. *Journal of the Water Pollution Control Federation* 41: R255-R266.

Patterson, J. W., Minear, R. A., Neyed, T. K. 1971. *Septic tanks and the environment*. Washington, D.C.: National Technical Information Service publication no. PB-204-519.

Perrin, D. R. 1974. Physical and chemical treatment of septic tank sludge. Master of science thesis, University of Vermont.

Stein, R. M., Jewell, W. J., Eckenfelder, W. W., Adams, C. E. 1972. A study of aerobic sludge digestion comparing pure oxygen and air. *Proceedings of the 27th annual Purdue University Industrial Waste Conference*, ed. John M. Bell, pp. 492-500. Lafayette: Purdue University Press.

Swanwick, J. D., and Shurben, D. G. 1969. Effective chemical treatment for inhibition of anaerobic sewage sludge digestion due to anionic detergents. *Journal of the Institute of Water Pollution Control* no. 2, pp. 3-15.

Tang, N. H., Schnelle, K. B., Jr., and Parker, F. L. 1969. *Moisture transport in sludge dewatering and drying on sand beds*. Technical report no. 18. Nashville: Vanderbilt University.

Truesdale, G. A., and Mann, H. T. 1968. Synthetic detergents and septic tanks. *Surveyor and municipal engineer* 131: 28–30, 33.

United States Department of Agriculture. 1968. *Status of water and sewage facilities in communities without public systems.* Agricultural economic report no. 143. Washington, D.C.: USDA, Economic Research Service.

United States Public Health Service. 1967. *Manual of septic tank practice.* Washington, D.C.: USPHS publication no. 526.

Index